ANNUAL PLANT REVIEWS
VOLUME 35

ANNUAL PLANT REVIEWS VOLUME 35

Plant Systems Biology

Edited by

Gloria M. Coruzzi

Department of Biology and Center for Genomics and Systems Biology
New York University, New York, NY, USA

Rodrigo A. Gutiérrez

Department of Biology and Center for Genomics and Systems Biology
New York University, New York, NY, USA
Departamento de Genética Molecular y Microbiología
Pontificia Universidad Católica de Chile, Santiago, Chile

WILEY-BLACKWELL

A John Wiley & Sons, Ltd., Publication

This edition first published 2009
© 2009 Blackwell Publishing Ltd

Blackwell Publishing was acquired by John Wiley & Sons in February 2007.
Blackwell's publishing programme has been merged with Wiley's global Scientific,
Technical and Medical business to form Willey-Blackwell.

Registered office
John Wiley & Sons Ltd, The Atrium, Southern Gate, Chichester, West Sussex, PO19 8SQ,
United Kingdom

Editorial office
9600 Garsington Road, Oxford, OX4 2DQ, United Kingdom
2121 State Avenue, Ames, Iowa 50014-8300, USA

For details of our global editorial offices, for customer services and for information about
how to apply for permission to reuse the copyright material in this book please see our
website at www.wiley.com/wiley-blackwell.

The right of the author to be identified as the author of this work has been asserted in
accordance with the Copyright, Designs and Patents Act 1988.

Library of Congress Cataloging-in-Publication Data is available

A catalogue record for this book is available from the British Library.

Annual plant reviews (Print)   ISSN 1460-1494
Annual plant reviews (Online)   ISSN 1756-9710

Set in 10/12 pt Palatino by Aptara® Inc., New Delhi, India
Printed and bound in Singapore by Markono Print Media Pte Ltd

1   2009

Annual Plant Reviews

A series for researchers and postgraduates in the plant sciences. Each volume in this series focuses on a theme of topical importance and emphasis is placed on rapid publication.

Titles in the series:

1. **Arabidopsis**
   Edited by M. Anderson and J.A. Roberts
2. **Biochemistry of Plant Secondary Metabolism**
   Edited by M. Wink
3. **Functions of Plant Secondary Metabolites and their Exploitation in Biotechnology**
   Edited by M. Wink
4. **Molecular Plant Pathology**
   Edited by M. Dickinson and J. Beynon
5. **Vacuolar Compartments**
   Edited by D.G. Robinson and J.C. Rogers
6. **Plant Reproduction**
   Edited by S.D. O'Neill and J.A. Roberts
7. **Protein–Protein Interactions in Plant Biology**
   Edited by M.T. McManus, W.A. Laing, and A.C. Allan
8. **The Plant Cell Wall**
   Edited by J.K.C. Rose
9. **The Golgi Apparatus and the Plant Secretory Pathway**
   Edited by D.G. Robinson
10. **The Plant Cytoskeleton in Cell Differentiation and Development**
    Edited by P.J. Hussey
11. **Plant–Pathogen Interactions**
    Edited by N.J. Talbot
12. **Polarity in Plants**
    Edited by K. Lindsey

CONTENTS

CONTRIBUTORS

Réka Albert
Department of Physics, Pennsylvania State University, University Park, PA, USA
Huck Institutes of the Life Sciences, Pennsylvania State University, University Park, PA, USA

Sarah M. Assmann
Department of Biology, Pennsylvania State University, University Park, PA, USA

Ivan Baxter
Bindley Bioscience Center, Purdue University, West Lafayette, IN, USA

Philip N. Benfey
Department of Biology and the Institute for Genome Science and Policy, Duke University, Durham, NC, USA

Richard Bonneau
Department of Biology and Center for Genomics and Systems Biology, New York University, New York, NY, USA
Computer Science Department, Courant Institute for Mathematical Sciences, New York University, New York, NY, USA

Siobhan M. Brady
Department of Biology and the Institute for Genome Science and Policy, Duke University, Durham, NC, USA

Alejandro Burga
Departamento de Genética Molecular y Microbiología, Pontificia Universidad Católica de Chile, Santiago, Chile

Gloria M. Coruzzi
Department of Biology and Center for Genomics and Systems Biology, New York University, New York, NY, USA

Ian M. Ehrenreich
Department of Biology and Center for Genomics and Systems Biology, New York University, New York, NY, USA
Department of Genetics, North Carolina State University, Raleigh, NC, USA

Kathleen E. Engelmann
University of Bridgeport, Bridgeport, CT, USA

Anita Fernandez
Department of Biology and Center for Genomics and Systems Biology, New York University, New York, NY, USA
Department of Biology, Fairfield University, Fairfield, CT, USA

Pamela J. Green
Department of Plant and Soil Sciences and Delaware Biotechnology Institute, University of Delaware, Newark, DE, USA

Erich Grotewold
Department of Plant Cellular and Molecular Biology and Plant Biotechnology Center, The Ohio State University, Columbus, OH, USA

Mary Lou Guerinot
Department of Biological Sciences, Dartmouth College, Hanover, NH, USA

Kristin C. Gunsalus
Department of Biology and Center for Genomics and Systems Biology, New York University, New York, NY, USA

Rodrigo A. Gutiérrez
Department of Biology and Center for Genomics and Systems Biology, New York University, New York, NY, USA
Departamento de Genética Molecular y Microbiología, Pontificia Universidad Católica de Chile, Santiago, Chile

Yoshie Hanzawa
Department of Biology and Center for Genomics and Systems Biology, New York University, New York, NY, USA

Thadeous Kacmarczyk
Department of Biology and Center for Genomics and Systems Biology, New York University, New York, NY, USA

Manpreet S. Katari
Department of Biology and Center for Genomics and Systems Biology, New York University, New York, NY, USA

Blake C. Meyers
Department of Plant and Soil Sciences and Delaware Biotechnology Institute, University of Delaware, Newark, DE, USA

Scott C. Peck
Division of Biochemistry, University of Missouri-Columbia, Columbia, MO, USA

Fabio Piano
Department of Biology and Center for Genomics and Systems Biology, New York University, New York, NY, USA

Chris Poultney
Courant Institute for Mathematical Sciences, Computer Science Department, New York University, New York, NY, USA

Michael D. Purugganan
Department of Biology and Center for Genomics and Systems Biology, New York University, New York, NY, USA

Christina L. Richards
Department of Biology and Center for Genomics and Systems Biology, New York University, New York, NY, USA

David E. Salt
Center for Plant Environmental Stress Physiology, Purdue University, West Lafayette, IN, USA
Bindley Bioscience Center, Purdue University, West Lafayette, IN, USA

Dennis Shasha
Courant Institute for Mathematical Sciences, Computer Science Department, New York University, New York, NY, USA

Nathan Springer
Department of Plant Biology, University of Minnesota, St. Paul, MN, USA

Peter Waltman
Computer Science Department, Courant Institute for Mathematical Sciences, New York University, New York, NY, USA

Wolfram Weckwerth
Department of Molecular Plant Physiology and Systems Biology, University of Vienna, Vienna, Austria
GoFORSY, Potsdam, Germany
Max Planck Institute of Molecular Plant Physiology, Potsdam, Germany

PREFACE

'Just as we cannot think of spatial objects at all apart from space, or temporal objects apart from time, so we cannot think of any object apart from the possibility of its connections with other things'

Ludwig Wittgenstein – Tractatus Logico-Philosophicus (2.012)

This volume captures the avant-garde of biological systemic research and aims to be an introductory material for undergraduate and graduate students, as well as researchers who wish to immerse themselves in the relevant questions and problems that system biology is facing today. But, what is systems biology? Herein, we provide the opinion of experts in fields impacting systems biology ranging from statistics to ecology, with emphasis given to case studies where the concepts of systems biology are applied to particular problems such as the study of development, environmental response, metabolism in plants and diverse model organisms.

In the first part of this volume, an overview of the systems biology field is presented with a focus on plant systems biology. A fundamental conceptual framework such as Network Theory is covered as well as the progress achieved for diverse model organisms: prokaryotes, due to their 'simplicity', and *C. elegans*, one of the most tractable animal models. The second part of this volume deals with the diverse sources of information necessary for a systemic understanding of plants. Insights are given into the software tools developed for systems biology and how they can be applied for plants and a comprehensive analysis of the data that can be integrated, that is, genome, transcriptome, proteome, metabolome and ionome. Finally, an interesting case study regarding root development is presented as well as important ecological and evolutionary considerations regarding living systems.

Despite huge advances in technology in the genomic era, we are far from having a complete description of the molecular components of biological systems and the ways they interact. In the particular case of Arabidopsis, advances in plant systems biology studies lag behind other model organism in terms of the data sets available (e.g. interactomes). However, this lack of complete information has not precluded creative researchers from taking advantage of systems biology approaches to conduct research to develop testable hypotheses, which in turn fuel a new cycle of systems level research.

After reading this volume, we hope you have more questions about systems biology than when you started reading it. We will consider our book a success if, after reading it, you discover novel ideas and ways to apply systems biology approaches in your own area of research. When used as an educational platform, we hope this volume will inspire the next generation of young scientists to enter the field of systems biology in the post-genomic era.

ACKNOWLEDGEMENTS

We would like to thank all the authors who contributed their time and expertise to this volume on systems biology. We believe that the individual chapters, like the field of systems biology, have great synergies and interconnections. We appreciate the enormous amount of insight and information encapsulated by each of the authors. We would like to especially acknowledge Alexis Cruikshank for her superb organizational and technical editing skills on this volume, as well as for her unending patience on completing this project. We also acknowledge the superb artwork of Suzan Runko who created the cover art for this volume, which conveys the science of systems biology in an exceptionally artistic fashion.

Part I

Systems Biology:
An Overview

Annual Plant Reviews (2009) **35**, 3–40
doi: 10.1111/b.9781405175326.2009.00001.x

www.interscience.wiley.com

Chapter 1

SYSTEMS BIOLOGY: PRINCIPLES AND APPLICATIONS IN PLANT RESEARCH

Gloria M. Coruzzi,[2] Alejandro R. Burga,[1]
Manpreet S. Katari[2] and Rodrigo A. Gutiérrez[1,2]

[1] *Departamento de Genética Molecular y Microbiología, Pontificia Universidad Católica
de Chile, Santiago, Chile*
[2] *Department of Biology and Center for Genomics and Systems Biology, New York
University, New York, NY, USA*

'We cannot solve our problems with the same thinking we used when we created them'

Albert Einstein

Abstract: Plants have played a major role in the geochemical and climatic evolution of our planet. Today, in addition to their fundamental ecological importance plants are essential for humans as the main source of food, provide raw materials for many types of industry and chemicals for medical applications. It is thus daunting to realize how little we understand about plant systems. To date, only approximately 15% of the genes of *Arabidopsis thaliana*, the most explored model system for plant biologists, have been characterized experimentally. Systems biology offers the opportunity to increase our understanding of plants as living organisms, by generating a holistic view of the organism grounded at the molecular level. In this chapter, we discuss the basics of systems biology, the data and tools we need for systems research and how it can be used to produce an integrated view of plant biology. We finish with a discussion of case studies, published examples of plant systems biology research and their impact on our knowledge of plants as integrated systems.

Keywords: network; Arabidopsis; genomics; bioinformatics; modelling; data integration

1.1 Introduction

1.1.1 Systems thinking

What is systems biology? We advocate a definition anchored in the general systems theory: 'The exercise of integrating the existing knowledge about biological components, building a model of the system as a whole and extracting the unifying organizational principles that explain the form and function of living organisms.' Systems thinking is not a new trend, but dates back to the end of the 1800s and the beginning of the 1900s. One of the pioneers of systems thinking was the Russian philosopher Alexandr Bogdanov (1873–1928). His interests and writings ranged from social, to biological and physical sciences. His work anticipated in many important ways Norbert Weiner's *'Cybernetics'* and Ludwig von Bertalanffy's *'General Systems Theory'*. Bogdanov (1980) proposed that all physical, biological and human sciences could be unified by treating them as systems of relationships and by seeking the organizational principles that underlie all systems.

We would like to illustrate systems thinking with the following example. Imagine you are standing in front of 'La Grande Jatte', the painting by the famous pointillist artist George Seurat. If we are close to the painting, we can easily distinguish the small coloured spots, but we lose their pattern, in fact we may not even realize they are part of a composition. Only when we are standing back far enough, we can appreciate the subject and beauty of the painting, as people standing by the lake. Most scientists today are standing very close to their subjects. They know their area of research extremely well, but cannot easily place their knowledge in the global context. The aim of systems biology is to move away from the detail and instead produce a global view of the system under scrutiny. This is achieved not solely by system-wide data collection, integration and analysis. It is also about a change in the research focus from the elements to the interactions and to the discovery of higher levels of organization and the emerging properties at these higher levels of organization.

1.1.2 Complexity and robustness in biological systems

Everyone agrees that biological systems are complex. But what does complexity mean? Is complexity in biological systems just a consequence of the inherent difficulty in understanding them? In this chapter, we would like to use a more precise definition of complexity. We refer to a system as a complex when it exhibits the following characteristics: (1) It is composed of many different elements. (2) The constituent elements interact. (3) The elements and their interactions are dynamic, and are often governed by non-deterministic rules. (4) The elements and/or interactions can be influenced by external factors. As a consequence of these properties, complex systems exhibit unexpected or emergent behaviour that cannot be understood from studying the parts in isolation. In addition, mathematical models that describe them usually involve nonlinear behaviour. Even the simplest life forms (e.g. mycoplasma) are complex by this definition. Living systems are composed of a large number of elements with diverse chemical properties (nucleic acids, proteins, carbohydrates, lipids, ions and many small molecules or metabolites) that are intricately interconnected. We all know that these elements and connections are dynamic and respond to internal or external factors. For instance, the messenger RNA (mRNA) levels of hundreds of genes that code for hundreds of different proteins can vary over time due to the action of signal transduction cascades activated by developmental or environmental cues.

In addition to complexity, and perhaps because of it, biological systems are robust. This key and ubiquitous feature of biological systems refers to their ability to maintain proper functions despite internal and/or external perturbations and uncertainty (Stelling et al., 2004; Kitano, 2007). Robustness is an example of a property that emerges at the system level and that cannot be understood at the individual part level (Kitano, 2004). The types of perturbations encountered by living systems are varied and include: stochastic noise (e.g. due to low copy number of cellular components), physiological and developmental signals, environmental change and genetic variation. It is important to distinguish robustness from homeostasis. As indicated above, robustness is related to preserving function. In contrast, homeostasis refers to maintaining the state of the system. A system is robust as long as it maintains functionality, even if this is achieved by moving to a new steady state. For instance, during the diauxic shift yeasts drastically changes from anaerobic to aerobic metabolism. This transition involves genome-wide changes in gene expression and the reprogramming of metabolic pathways (DeRisi et al., 1997). In this example, the state of the system is dramatically altered. However, the system is robust as it continues to produce adenosine triphosphate and grow. Robustness allows for changes in the structure and components of the system in response to perturbations, but maintains specific functions (Kitano, 2004).

Robustness can be achieved by multiple mechanisms (for a review see Kitano, 2004). First, negative or positive feedbacks allow for robust dynamic responses in regulatory contexts as diverse as the cell cycle, circadian clock and chemotaxis. Second, robustness can also be achieved by providing multiple means to achieve a specific function. These alternative or fail-safe mechanisms encompass the typically observed phenomena in living systems of redundancy, overlapping function and diversity. Third, the modularity of living systems is an effective mechanism of containing perturbations and damage locally to minimize the effect on the whole system. Fourth, decoupling isolates low-level variation from high-level functionalities. All four different mechanisms are typically found in a living organism.

1.1.3 Robust evolving systems

How can complex biological systems be both robust to mutational perturbations and at the same time accumulate heritable changes that produce new adaptive traits during evolutionary time? Intuitively, robustness appears to act against evolutionary change. The extent of phenotypic change caused by natural selection depends on phenotypically expressed genetic variation that is buffered in a robust system. This is true in the short term, but in the long term silent mutations accumulate that can express under certain conditions (e.g. a specific environmental stress). This phenomenon was originally postulated and studied by Waddington (1959). He showed that once the phenotypic buffering capacity of an organism is exhausted by severe perturbations, altered phenotypes can emerge, that is, the phenotype is no longer 'canalized'. Molecular evidence for Waddington's original observations was presented first in *Drosophila melanogaster* (Rutherford and Lindquist, 1998) where Hsp90 (a heat shock-induced chaperone) was shown to be an evolutionary conserved protein affecting phenotypic variation. Inhibiting Hsp90 by mutation or pharmacological means caused an increase in a wide range of altered phenotypes in Drosophila. Those phenotypes could be fixed through breeding and interestingly became robust and independent of environmental perturbations (temperature) or Hsp90 inhibition. Later studies in *Arabidopsis thaliana* (Queitsch *et al.*, 2002) showed that in plants, reduction of Hsp90 function resulted in an increase in phenotypic variation in the absence of genetic variation. Probably, epigenetic mechanisms are involved in this phenomenon (Sollars *et al.*, 2003); however, the exact molecular mechanisms are still a mystery (Salathia and Queitsch, 2007). Robustness and phenotypic variability are not antagonistic forces, but are intimately related concepts. Isalan *et al.* (2008) recently addressed this relationship in *Escherichia coli*. They systematically explored the effect of adding new edges in the *E. coli* gene network by expressing hundreds of promoter-open reading frame combinations. The open reading frames coded for transcription factors working at different levels of the gene network hierarchy. Remarkably, nearly 95% of the new edges were tolerated (i.e. *E. coli* shows a robust gene network) and some of the new strains grew better than the wild type under diverse selective pressures (Isalan *et al.*, 2008).

1.2 Network biology

How do we approach the complexity of biological systems? In 1736, the famous mathematician Leonhard Euler found a solution to an old problem: 'The seven bridges of Königsberg.' In those days, Königsberg was a city of the Kingdom of Prussia, set on the Pregel River. The city included two large islands that were connected to each other and the mainland by seven bridges. The problem was to decide whether it was possible to walk by a route that crossed each bridge exactly once. Euler did not solve the problem empirically, but instead made a fundamental abstraction of the problem. First, he eliminated all the features except the landmasses and the bridges. Then, he replaced each landmass with a dot (vertex or node) and each bridge with a line (edge or link). The resulting structure was a graph. Euler realized that the degrees of the nodes were an important property of the graph and he could later demonstrate in terms of degrees that the hypothetic walk was impossible. The degree tells us how many links a particular node has to other nodes. The key to Euler's success was that he could abstract the fundamental elements of the system and their relationships and represent them in a way that favoured their analysis. He knew, for instance, that parameters such as the length of the bridges or whether they were made of stone or wood were irrelevant for his purpose. Similar principles and approaches can apply to complex systems such as living organisms. By abstracting and focusing on the important features for system form and function, a detailed and comprehensive yet tractable view of the system can be obtained. A highly successful and now widespread abstraction to represent complex biological systems uses graphs. The following sections will discuss the basics of how graphs are applied in biological research. (Note: For more details regarding the subjects covered in this section please see Chapter 2 of this volume.)

1.2.1 General principles

When networks, or graphs in a more formal mathematical language, are applied to molecular biology nodes (or vertices) represent the molecules present inside a cell (e.g. proteins, RNAs and/or metabolites) and links (or edges) between nodes represent their biological relationships (e.g. physical interaction, regulatory connections, metabolic reactions). For example, in protein interaction networks, protein complexes can be represented as networks where nodes represent proteins and their edges represent the physical interactions between them. In metabolic networks, nodes can represent metabolites and the edges the metabolic reactions that transform one metabolite into another (Wagner and Fell, 2001). In genetic networks, nodes can represent genes and the link between them a genetic interaction, such as synthetic lethality. But other abstractions are possible. For instance, nodes can also be used to represent biological processes and the edges connecting them indicate the functional relationship between the processes. This network simplification has uncovered basic principles of the structure and organization of different types of molecular networks in many different organisms (Barabasi and Oltvai, 2004) and has been successful for biological research. Examples of the use of networks in plant biology are discussed later in this chapter.

1.2.2 Network properties

Networks have diverse structures and topologies. Prior to the work by Barabasi and Albert (Barabasi and Albert, 1999; Barabasi and Oltvai, 2004), networks were generally considered as having either regular (e.g. with a square lattice) or random topologies. To construct a random network according to the classical Erdös–Rényi model, two nodes are chosen randomly from a pool of N nodes and a link is established between them (Erdös and Rényi, 1960). This procedure results in a network where the number of edges for each node (degree) follows a Poisson distribution. In random networks, there are a few nodes that are lowly or highly connected and most nodes have roughly the same number of links. In contrast to random networks, one of the common architectural features of naturally occurring networks including molecular networks is to present a scale-free topology (Barabasi and Albert, 1999). Networks with scale-free topology are characterized by a power law degree distribution: $P(k) \sim k^{-\gamma}$, where $P(k)$ is the probability that a selected node has k links and γ is the degree exponent. The γ value for most naturally occurring networks varies between 2 and 3 (Albert *et al.*, 2000). Whereas in random networks most nodes have approximately the same degree or number of connections, in scale-free networks there are many nodes that are poorly connected and few nodes that are very highly connected (Albert *et al.*, 2000). The highest degree nodes (highest number of edges) are typically referred to as 'hubs' and are important for the architecture and function of the network.

One important consequence of the scale-free topology of naturally occurring networks is robustness. Scale-free networks show an extraordinary tolerance to perturbations as compared to random networks of equivalent size. For example, 5% of the nodes in a scale network can fail without affecting the mean path length (Albert *et al.*, 2000). In contrast, an informed targeting of 5% of the most connected nodes results in a doubling of the mean path length (Albert *et al.*, 2000). This phenomenon can be understood based on the low frequency of highly connected nodes. And therefore the probability of randomly targeting a hub in a scale-free network is very low. This observation is correlated with findings in biological networks. In yeast, approximately 20% of proteins with less than 5 connections are essential, in contrast to the 62% when considering proteins with more than 15 connections (Jeong *et al.*, 2001). Removing a hub protein has a high probability of resulting in a lethal phenotype, supporting their fundamental role. Similarly, a malicious hacker trying to affect Internet function may have a higher chance of succeeding by damaging servers that are hubs. It was recently shown that human proteins targeted by the Epstein–Barr virus were enriched for hub proteins (Calderwood *et al.*, 2007). Viruses could have evolved to attack key hub proteins in their host's proteomes.

Another interesting feature of naturally occurring networks is their 'small world effect'. In order to understand it, we must first introduce the concept of 'path'. Distance in networks is measured by the number of links that we need to pass when travelling from node A to node B (path from A to B). The path with the smallest number of edges between nodes A and B is the 'shortest path'. If we take all possible pair of nodes in a network and calculate the average length of all the shortest paths, we obtain a measure called the 'mean path length'. The diameter of the network is the maximal distance (from the shortest path set) between any pair of nodes. Studies regarding social networks performed by Stanley Milgram in 1967 showed that the mean number of individuals required to connect one arbitrary person to another arbitrary person anywhere in the USA is only six; thus the concept of six degrees of separation. Another example of small mean path lengths within a large network is the World Wide Web, with over 800 million nodes, it has a mean path length of only 19 (Albert *et al.*, 1999). The property that every node in these large networks is separated by a few links from all the other nodes in the network is known as the 'small world effect'. The architecture of *Caenorhabditis elegans* nervous system, the power grid of the western United States and the collaboration graph of film actors also show a 'small world effect' (Watts and Strogatz, 1997).

An additional interesting property of networks is clustering. Clustering in networks can be intuitively explained using social networks as an example. In social networks, two of your friends will have a greater chance of knowing one another than two people chosen at random from the entire population. This is due to their common acquaintance with you. This property is quantified with the clustering coefficient. The clustering coefficient for a node k measures how connected are its neighbours. Mathematically, it is calculated as the ratio of the number of edges that exist between the neighbours of k and the total number of edges possible between the neighbours. The clustering coefficient is 1 in a fully connected network. In real-world networks, the clustering coefficient typically has values of 0.1–0.5. These values are much higher than what obtained from random networks.

The combined presence of scale-free topologies and the high degree of clustering in real-world networks is a consequence of their hierarchical structure. Many real-world networks are fundamentally modular, where groups of highly connected nodes can be identified that are poorly connected to nodes outside of the group (for a review see Ravasz *et al.*, 2002). Hartwell *et al.* (1999) defined a biological module as a discrete entity of the cell whose function is separable from those of other modules. Biological network modules have been correlated with biological function in various organisms from yeast to animals to plants (Han *et al.*, 2004; Gunsalus *et al.*, 2005; Gutierrez *et al.*, 2007). An interesting question to be addressed is whether a central control module exists or the control is shared between all the modules in a cell. In addition, an evolutionary implication of modularity is that shifting the connections between modules could potentially change the phenotype of individuals. Therefore, emphasis must be given in the following years to the study of hubs proteins (most highly connected nodes) and their regulation from a systemic perspective. These proteins can occupy key places connecting diverse modules and their study could provide important insights about cellular regulatory mechanisms.

1.3 Experimental approaches for plant systems biology

A systems approach to biology is possible today because of the breakthrough in technologies that allow us to measure and connect the entire complement of defined molecules inside an organism (e.g. transcriptome, proteome). The new influx of high-throughput data is shifting biology from a reductionist to a systems-level view. Practically speaking, systems biology generates and integrates different types of genome-wide data in living organisms to produce models of the system as a whole (Kitano, 2002; Shannon *et al.*, 2003). Ideker, Galitski and Hood (Shannon *et al.*, 2003) described a fundamental framework to carry out research in systems biology: (1) define all the components of the system, (2) systematically perturb and monitor components of the system, (3) reconcile the experimentally observed responses with those predicted by the model and (4) design and perform new perturbation experiments to distinguish between multiple or competing hypotheses. Below, we describe the different data types currently available for plant systems biology.

1.3.1 Enumerating the parts

1.3.1.1 The genome
The first step to understand a living organism from a systems point of view is to enumerate its basic components. The genome of an organism is the complete hereditary information encoded in the DNA (or RNA in some viruses). The first genome of a complex free-living organism to be completed was that of *Haemophilus influenza* (Fleischmann *et al.*, 1995). The tremendous advances in sequencing technologies since then, has catapulted us into the genomic era. Today, the scientific community benefits from the complete genome sequences of many organisms including several plants. In 2000, the complete genome sequence of the plant *A. thaliana* was published (Initiative, 2000). In 2005, the first monocotyledoneous plant was sequenced, *Oryza sativa* (Locke *et al.*, 2005). To date, in addition to Arabidopsis and rice, the complete sequenced genomes of *Chlamydomonas reinhardtii* (Merchant *et al.*, 2007), *Populus trichocarpa* (Tuskan *et al.*, 2006) and *Vitis vinifera* (Jaillon *et al.*, 2007) are available. In addition, at the beginning of 2008, the first genome sequence of a genetically modified plant was reported, for the SunUp *Carica papaya* (Ming *et al.*, 2008). Genomes provide the parts lists as far as genes and their products (proteins and RNAs) are concerned.

Due to lack of financial support or simply due to the large size, the genomes of many species are not and may not be completely sequenced. In addition, much of the DNA in large genomes is thought to correspond to repetitive DNA and transposable elements (Bennet and Ilia, 1995). There are two main approaches for 'sampling' the gene space of genomes without sequencing them completely: (1) sequencing expressed sequence tags (ESTs) (Adams *et al.*, 1991) and (2) sequencing genomic sequences that have been filtered based on either hypomethylation (Bennetzen *et al.*, 1994; Rabinowicz *et al.*, 1999, 2003a) or High-Cot (Yuan *et al.*, 2003). EST sequencing involves preparing cDNA of the transcribed messages and then randomly sequencing them from either the 3' or the 5'. The advantage of EST sequencing is that the sequences obtained represent the transcribed regions of the genome. The disadvantages are that ESTs typically do not represent the entire cDNA (only about 500 bp) and they tend to represent only roughly 50% of the total genes in the genome. Large stretches of repetitive DNA and transposable elements tend to be hypermethylated compared to gene coding regions which are unmethylated, for example 95% of the maize exons are thought to be unmethylated (Rabinowicz *et al.*, 2003b). Genomic libraries can be prepared enriched for unmethylated DNA by cloning in bacterial hosts that destroy methylated DNA. Alternatively, the High-Cot method removes repetitive DNA by depending on the rapid association of repetitive DNA. The advantage of sequencing filtered genomic sequences over EST sequencing is that it is possible to obtain the introns and regulatory regions of the gene rather than just the transcribed portion of the gene. Palmer *et al.* (2003) annotated genes from methyl-filtrated sequences and compared them to random genome sequences from maize. They observed a twofold reduction in the number of repetitive reads and fivefold increase in the number of exonic regions in the methyl-filtrated sequences (Palmer *et al.*, 2003). The drawback to this approach is the significant amount of repetitive DNA obtained. Despite the limitations, the different genome sampling methods are still the best alternative to sequencing the entire genome for very large genome sequences. With the advent of highly parallel sequencing technologies, such as the platforms developed by 454 (Margulies *et al.*, 2005) or Solexa Illumina (Bennett, 2004), these limitations may be less relevant in the near future. In addition, in the years to come, the focus may shift from sequencing depth within individual genomes to sampling as many genomes as possible to determine the gene complement of the biosphere.

The first step after obtaining the genome sequence of an organism is the annotation of its genes (identification of their basic elements as exons, introns, regulatory sequences). Several strategies are available for gene discovery: *in silico* gene prediction, information obtained from ESTs, full-length cDNA, tilling and expression arrays, MPSS (massive parallel signature sequencing), SAGE (serial analysis of gene expression) (Alonso and Ecker, 2006). Today, many specialized software are available in the market as well as freeware. Even though *in silico* predictions are an excellent start, they have to be used critically because they are not always accurate. Indeed, gene predictions and genome annotations are typically dynamic improving over time. For example, at least 40% of the original predictions made for the Arabidopsis genome were subsequently found to be wrong and corrected in later releases of the annotation. The next step after gene discovery is gene functional annotation. This functional analysis can be done in several ways: computational predictions, gene expression, mutant analysis and protein–protein interactions, just to name some of the most common approaches (for an excellent review of this topic please read Alonso and Ecker, 2006). However, high-throughput screens of mutants are fundamental for accessing plant gene functions on a genomic scale (Alonso and Ecker, 2006).

1.3.1.2 Epigenome

Epigenetics is the study of heritable traits that do not involve changes in the DNA sequence. These changes alter the chromatin structure which affect regulation of gene expression. Two well-studied forms of epigenetic regulation in plants are methylation of the DNA molecule at the cytosine base and post-translational modifications of histones (Henderson and Jacobsen, 2007). Heterochromatic regions of plants tend to be highly methylated (Rabinowicz *et al.*, 2003b) which is why sequencing methods such as methyl filtration produce sequences that are enriched in gene coding regions. To understand the importance of DNA or histone modifications for gene regulation, we need to look at DNA or histone modification patterns across the genome and correlate them with gene expression data. Several research groups have carried out such studies in plants. The experimental approach involves isolating nuclear DNA and enriching for the methylated DNA fraction. The methylated DNA is then hybridized to an Arabidopsis whole genome tiling microarray to determine global DNA methylation patterns (Zhang *et al.*, 2006; Zilberman *et al.*, 2007). These studies have uncovered several interesting aspects of the epigenome in Arabidopsis. For example, the heterochromatic regions are heavily methylated, approximately a third of Arabidopsis genes are methylated within the transcribed region, methylation is biased away from the 5′ and 3′ regions in Arabidopsis genes, genes that are methylated in the transcribed regions tend to be expressed at high levels and genes that are methylated in the promoter region tend to be tissue specific (Zhang *et al.*, 2006; Zilberman *et al.*, 2007). Additional studies have been carried out to look at histone modifications (Zhang *et al.*, 2007). Epigenetics adds another level to the regulatory complexity of gene expression. Elucidating the importance and function of epigenetic modifications in the years to come will be crucial to understanding how genomes respond to internal and external perturbations.

1.3.1.3 Transcriptome (including RNAs and small RNAs)

The transcriptome is the set of all the parts of the genome that are expressed as RNA transcripts in one or several populations of cells in a given time and a given environmental condition. Not so many years ago, it was unthinkable to propose measuring the expression of thousands of genes in one experiment. Northern blots were the only choice for molecular biologists and biochemists. Microarray technology has been around since the early 1990s, but the high cost of this technology precluded widespread utilization by the scientific community. Patrick Brown and colleagues made this technology cheaper and easier to do (Schena *et al.*, 1995) and triggered a revolution in transcriptome studies. Interestingly, Arabidopsis was chosen as a case study to demonstrate microarray technology in 1995. Arabidopsis was utilized because of its small genome and rich EST collection. At the present time, one of the most widely used systems for gene expression studies in Arabidopsis is the ATH1 Genome Array, designed by Affymetrix in collaboration with The Institute for Genome Research. It contains more than 22 000 probes sets representing approximately 24 000 gene sequences on a single array. This technology allows the simultaneous quantitative determination of the expression level of thousands of genes.

Today, the Arabidopsis community benefits from the success of microarray technology. The bottleneck of research is not longer the acquisition of data, but often its analysis. There are several public databases containing microarray data: Genevestigator (https://www.genevestigator.ethz.ch/), NASC-Arrays (http://affymetrix.arabidopsis.info/narrays/experimentbrowse.pl), ArrayExpress (http://www.ebi.ac.uk/arrayexpress), The Gene Expression Omnibus (http://www.ncbi.nlm.nih.gov/geo/) and Stanford Microarray Database (http://genome-www5.stanford.edu) perhaps among the most popular. Hundreds of microarray experiments (available through these resources) have been performed comparing diverse developmental stages, soil conditions and compositions, pathogen infections, oxidative stress and response to diverse chemicals. The information obtained from these experiments has being useful in the detection of new genes or new gene functions involved in particular processes.

It is important to emphasize that the transcriptome does not refer just to messenger RNAs. In addition to mRNAs, ribosomal RNAs (rRNAs) and tRNAs (transfer RNAs), there is great interest in studying the expression of the large and heterogeneous population of small RNAs in plants (Finnegan and Matzke, 2003). These small RNAs can have important roles for regulation of gene expression as well as other roles (Finnegan and Matzke, 2003). Unfortunately, EST libraries and microarray gene chips were not designed to detect small RNAs such as microRNAs (miRNAs). However, with the advancement of sequencing technologies, we are now able to measure miRNAs and other small RNAs (sRNAs), quantify their expression in different cell types and treatments and begin to understand their functional roles in plants (Lu *et al.*, 2005). (Note: For more on the role of sRNA see Chapter 7 of this volume.)

1.3.1.4 Proteome

The proteome is the set of all the proteins that are expressed in a given system in a given time and under a defined environmental condition. This term was coined by Mark Wilkins and colleagues in 1995 when studying the smallest known self-replicating organism *Mycoplasma genitalium* (Wasinger and Corthals, 2002). There are many questions one would like to address regarding proteomes: How abundant are the proteins? Where are the proteins located? What are the post-translational modifications in these proteins? Answering these questions is not an easy task. Low-abundance proteins can be extremely difficult to detect and there is no polymerase chain reaction

(PCR)-like method for protein amplification. In addition, the great variability of structures and chemical properties of proteins compared with nucleic acids makes their separation and identification a challenge. The most common approach in proteomics is the use of two-dimensional (2D) gel electrophoresis for separating the proteins (which typically resolves proteins based on charge and molecular weight). The separated proteins are then excised from the gel, fragmented and the fragments analyzed by mass spectrometry. The mass fingerprint of the peptides is then compared with databases for identification (Yates, 2000).

In Arabidopsis, large-scale proteomic efforts have been carried out, for example to determine the proteome of organelles such as the chloroplasts (Friso et al., 2004), mitochondria (Heazlewood et al., 2004) and vacuoles (Mitreva et al., 2004). These studies have shown that many in silico predictions of protein sub-cellular localization are wrong and they must be verified experimentally. These studies also suggest that cellular trafficking is more complex than previously thought.

Another problem in proteomic studies has been the coverage of the available techniques. Giavalisco et al. (2005) carried out a large-scale study in the hope of achieving a complete coverage of Arabidopsis cells proteome using 2D gel electrophoresis and matrix-assisted laser desorption/ionization-time of flight (MALDI-TOF) mass fingerprinting. Despite sampling different tissues in order to increase the probability of finding novel peptides, they could only find 663 different proteins from 2943 spots. Recently, a proteomic study of A. thaliana was carried out (Baerenfaller et al., 2008), giving valuable insights into the proteomic map of diverse organs, developmental stages and undifferentiated cultured cells. The authors identified around 13 000 proteins, which accounts for almost 50% of all Arabidopsis predicted gene models. This study not only allowed the corroboration of gene annotations, but also enriched the genome annotation. For example, the authors provided expression evidence for 57 genes models that were not present in the TAIR7 database. Another interesting result was the identification of 571 organ-specific proteins, which could be useful for the mapping of gene regulatory networks that drive the differentiation programmes in plants. (Note: For more on proteomics in plants see Chapter 8 of this volume.)

1.3.1.5 Metabolome

The metabolome refers to the complete set of small molecule metabolites (such as metabolic intermediates, hormones and other signalling molecules, and secondary metabolites) to be found within a biological system (Oliver et al., 1998). One of the main problems when studying the metabolome is the extreme heterogeneous chemical nature of the compounds considered as constituents of the metabolome. This is especially true in plants. Plants are known to have a tremendous enzymatic capacity, allowing for the production of an estimated ∼200 000 different small molecules (Fiehn, 2002). Finding the equilibrium between coverage and accuracy of measurement is therefore key in the metabolomics field.

With respect to the technology used to explore the metabolome, classical approaches use gas chromatography–mass spectrometry (GC–MS). This technology detects >300 metabolites in plant tissues (Hirai *et al.*, 2004). One of the first studies of this kind was carried out to determine the mechanism of action of herbicides like acetyl CoA carboxylase and acetolactate synthase inhibitors, using GC–MS (Schauer and Fernie, 2006). They could obtain the metabolic profiles of the treated and control seedlings. These metabolic profiles have also been achieved recently using H-1 NMR (proton nuclear magnetic resonance) (Ott *et al.*, 2003). A new technology used in metabolomics is the Fourier transform-ion cyclotron MS that separates metabolites based on differences in their isotopic masses (Hirai *et al.*, 2004). Crude plant extracts without prior separation of metabolites by chromatography are injected directly in this system. The mass resolution (>100 000) and accuracy (<1 ppm) is superior to other technologies.

Metabolite profiling has evolved from a diagnostic tool used in agriculture to a valuable source of information for gene function prediction in plants. Classically, gene function prediction was made based on studies at the mRNA or protein levels. In the year 2000, Fiehn and co-workers developed a novel tool for plant functional genomics (Fiehn, 2002). Using GC–MS, they automatically quantified 326 distinct compounds from *A. thaliana* leaf extracts, half of those whose chemical structure could be assigned. Four Arabidopsis genotypes were compared (two homozygous ecotypes and a mutant of each ecotype) and distinctive metabolite profiles of each genotype were reported. Data mining tools enabled the assignment of 'metabolic phenotypes'. Metabolite profiling studies have been performed in a diverse array of plant species (Schauer and Fernie, 2006) including: Arabidopsis (Fiehn, 2002), tomato (Schauer *et al.*, 2005), potato (Roessner *et al.*, 2001), rice (Young *et al.*, 2005), strawberry (Aharoni *et al.*, 2002) and eucalyptus (Merchant *et al.*, 2007). Metabolic profiling is also being used in studies of environmental perturbations in order to understand the complex shifts under nutrient limitation and biotic stress. The use of forward and reverse genetics in conjunction with metabolite profiling plays an important role in gene annotation, characterization of biochemical pathways and the identification of new genes for their use in biotechnological applications. (Note: For more details on metabolomics in plants, see Chapter 9 of this volume.)

1.3.1.6 Ionome

The ionome was first described to include all the metals, metalloids and non-metals present in an organism (Lahner *et al.*, 2003), which are involved in a broad range of important biological phenomena including: electrophysiology, signalling, enzymology, osmoregulation and transport (Hesse and Hoefgen, 2006). There is a small and sometimes diffuse boundary between ions and metabolites. Take for example macronutrients (essential elements used by plants in relatively large amounts for plant growth) like nitrogen, sulfur and phosphorus and metals like manganese, iron, zinc and copper essential for metalloprotein's activity. High-throughput ion-profiling strategies for genomic scale profiling of nutrient and trace element have been utilized for ionome studies (Lahner *et al.*, 2003). Using inductively coupled plasma mass spectrometry (ICP-MS) technology, it was possible to profile shoots from more than 40 000 plant samples (at a rate of 1000/per month) including diverse transferred DNA (T-DNA) insertional mutants and also mutants generated by fast neutron treatment. They have also characterized the ionome of Arabidopsis shoot and seed under standard soil growth conditions. (Note: For more information on ionomics in plants, see Chapter 10 of this volume.)

1.3.2 Systematic characterization of interactions between components

In the previous section, we briefly outlined some of the components that constitute a biological system and for which we have a substantial amount of global experimental data. However, even if we had a complete set of accurate measurements for every molecular component of the cell, we would not be able to reconstruct the system. We cannot think of these components as apart in space and time. The essence of biological systems is to understand not only what the parts are, but also how these parts interact. As discussed in other sections and chapters in this book, it is from the interactions of the parts that the properties typically associated with biological systems emerge. In this section, we shall briefly review the interactions that are better understood at a genome-wide level. It is important to emphasize that Arabidopsis and plants, in general, lag behind other systems in the systematic characterization of the interactions between molecular components. As a community, we should attack this limitation in order to advance plant systems research.

1.3.2.1 Metabolic pathways

Metabolic pathways were classically studied by isotopic radiolabelling of metabolites and biochemical isolation and characterization of the enzymes involved in those pathways. All these efforts resulted in the establishment of linear or cyclic pathways, where metabolites are continuously transformed by enzyme-catalyzed reactions with a defined stoichiometry. One way to represent all those reactions is constructing a network where each node represents a metabolite and a link between two nodes represents the enzymatic reaction that converts one into the other. These metabolic networks show a scale-free topology and small world which probably reflects the need of the system to have a wide and plastic response to environmental perturbations (Wagner and Fell, 2001; Ravasz *et al.*, 2002). Metabolic pathways for plants are stored in public databases such as AraCyc (Mueller *et al.*, 2003) and KEGG (Aoki and Kanehisa, 2005). These databases are built from knowledge in Arabidopsis but also from the enzymes and pathways studied in other species. Another important source of information is the Arabidopsis Reactome (www.arabidopsisreactome.org), inspired by the human Reactome project. It covers biological pathways ranging from the basic processes of metabolism to high-level processes such as hormonal signalling, and it includes also metabolic pathways for other plant species.

1.3.2.2 Co-expression networks

At the present time, one of the major sources of genome-wide information for Arabidopsis is microarray technology. By comparing gene expression patterns obtained in experiments across a wide range of treatments, networks of co-expression relationships can be built. Basically, the similarity of gene expression between two genes is calculated using a statistic such as the Pearson correlation coefficient. The nodes in the network represent the transcripts and the link between two nodes corresponds to the correlation coefficient between these two genes. Typically, a similarity cut-off is used to decide whether two expression patterns are similar or not and to draw or not the edges between nodes. Alternatively, weighted edges can be used to record the value of the statistic. Gene expression networks are useful to hypothesize gene function. But how many microarray experiments do we need in order to have a robust co-expression network? This issue was recently addressed for Arabidopsis (Aoki *et al.*, 2007). Aoki *et al.* (2007) found that the density of networks derived from microarray experiments from Arabidopsis reached equilibrium when more than 100 arrays were used. This result suggests that 100 experiments are sufficient to build a robust co-expression network. It remains to be determined whether this is insensitive to the experimental factors tested in the microarray experiments utilized. Is the Arabidopsis co-expression network different from the network of another model organism? Bergmann *et al.* (2004) compared the global features of co-expression networks from *Saccharomyces cerevisiae, C. elegans, E. coli, A. thaliana, D. melanogaster* and *Homo sapiens*. The co-expression networks built for all these organisms showed a power law degree distribution indicative of scale-free networks. In addition, all these networks exhibited high modularity. The modules were defined as sets of co-expressed genes that share a common function. A number of these core modules were conserved throughout evolution. Examples of core modules identified were: rRNA processing machinery, heat shock response and the proteosome. More recently, it was reported that hub genes from co-expression networks in *A. thaliana* tend to be single-copy genes (Wei *et al.*, 2006). For the positive co-expression network studied, 65% of the hub genes (having 20 or more connections) were found to be single-copy. On the other hand, only 37% of genes having fewer than 20 connections were single-copy genes. Co-expression networks are powerful as they provide a mechanism to relate genes without known function to known genes and help build testable hypothesis.

1.3.2.3 The interactome

The interactome is defined as the complete physical interaction map between the proteins in an organism. Molecular machines (e.g. ribosome, proteasome,

DNA or RNA polymerases), structural protein complexes, signalling cascades are just a few of the many examples of situations in which these interactions are relevant. Thus, defining protein complexes is critical to understanding virtually all aspects of cell function. Interactomes can be generated by various experimental approaches. The most widely used approaches are high-throughput yeast 2-hybrid binding assays and co-affinity purification followed by MS. In Arabidopsis, Van Leene *et al.* (2007) recently reported the development and application of a high-throughput tandem affinity purification (TAP)/MS platform for cell suspension cultures to analyze cell cycle-related protein complexes. Key for their methodology was the fast generation of transgenic cultures overproducing tagged fusion proteins, TAP adapted for plant cells. Using this strategy, they were able to validate 14 interactions that were previously reported and identified 28 new molecular associations. The building of interactomes (protein–protein interaction network) has been useful in the study of many model organisms (Schwikowski *et al.*, 2000; Walhout *et al.*, 2000; Giot *et al.*, 2003) as well as humans (Lehner and Fraser, 2004). Due to the lack of an experimentally determined interactome network for Arabidopsis, several strategies have been tried to predict interactomes. For example, Yu *et al.* proposed transferring annotations between genomes (Yuan *et al.*, 2003). Based on sequence similarity, they could transfer protein–protein and protein–DNA interactions from model organisms (whose interactome was partially available) to Arabidopsis. A similar approach was utilized to predict interactomes for Arabidopsis based on orthologs in *S. cerevisiae, C. elegans, D. melanogaster* and *H. sapiens* (Geisler-Lee *et al.*, 2007; Gutierrez *et al.*, 2007). Predictions are useful for systems research but far from satisfactory. It is known that experimental approaches for determining protein–protein interactions are not perfect and high-throughput data sets contain a high rate of false positives. Even if they were completely error free, transferring this information to another organism will certainly introduce errors. In addition, it has been reported that at least 14% of Arabidopsis proteins are likely to be found just in the plant lineage (Gutierrez *et al.*, 2004) and homology-based approaches will not predict interactions for these proteins. Small-scale interactome efforts have been reported. For example, the protein interaction map of the MADS Box family of transcription factors (de Folter *et al.*, 2005). It is not necessary to explain this acronym as it is well known in plant biology. Explaining its origin in this context confuses the discussion about protein interactions. A promising new strategy is the use of protein arrays (Popescu *et al.*, 2007). A high confidence Arabidopsis protein interactome defined experimentally is a must in order to advance plant systems biology.

1.3.2.4 Regulatory interactions

The transcriptional circuits underlying genetics programmes are characterized by interactions between transcription factors (TFs) and *cis*-regulatory regions in the promoter regions of target genes. Large-scale efforts to characterize regulatory interactions in Arabidopsis have not been performed. Regulatory interaction predictions have been attempted and at least in a few cases the predictions have been experimentally validated (e.g. Gutierrez *et al.*, 2008). To build regulatory networks, a very useful tool for the plant community is the Arabidopsis Gene Regulatory Information Server (AGRIS; http://arabidopsis.med.ohio-state.edu/). AGRIS contains all Arabidopsis promoter sequences, TFs, and their target genes and functions (Palaniswamy *et al.*, 2006). AGRIS currently houses three linked databases: At*cis*DB (*Arabidopsis thaliana cis*-regulatory database), AtTFDB (*Arabidopsis thaliana* TF database) and AtRegNet (*Arabidopsis thaliana* regulatory network). At*cis*DB currently contains 25 516 promoter sequences of annotated Arabidopsis genes with a description of putative *cis*-regulatory elements (Molina and Grotewold, 2005). AtTFDB contains information on approximately 1770 TFs. These TFs are grouped into 50 families, based on the presence of conserved DNA-binding domains.

1.3.2.5 The phenome: large-scale phenotypic analysis of mutants

Classic genetic approaches aim to understanding gene function by the study of the phenotypes caused by mutations of the gene. Generally speaking, mutations in genes that result in similar phenotypes often imply that the genes function in the same process. This rationale has led to the development of large-scale initiatives to characterize phenotypes caused by the systematic inactivation of genes in the genome of model organisms. Integrating this phenome data with other genome-wide data has led to important advances, for example in the understanding of animal developmental processes (Gunsalus *et al.*, 2005). Currently, the Arabidopsis community benefits from large populations of T-DNA and transposon insertional mutants (Scholl *et al.*, 2000). The location of the T-DNA elements in the genome has been determined using PCR-based amplification of flanking sequences and sequencing (Bevan and Walsh, 2005). To date, there are ~320 000 sequenced mutant lines in the reference Arabidopsis Columbia genome. The *Arabidopsis* Biological Resource Centre and the Nottingham *Arabidopsis* Stock Centre (NASC) contain seed banks where the global community can request individual knockout lines for functional studies. Many protein-coding genes (around 1600) do not possess insertions in exonic or intronic sequences. According to a simulation (Bevan and Walsh, 2005), if the number of random insertions is doubled, a raise from 94% to 98% is expected in the number coding genes with insertions. RNA interference (RNAi) and targeting induced local lesions in genomes (TILLING) are alternatives to isolate mutants in individual genes. Recently, Shinozaki and co-workers (Kuromori *et al.*, 2006) carried out a pilot study to characterize the Arabidopsis phenome. They constructed 18 000 transposon-insertion lines of Arabidopsis. They selected approximately 4000 transposon-insertional lines which had a Ds transposon inserted in the coding region of genes and visually analyzed their phenotypes. About 140 lines were found with reproducible and distinguishable phenotypes. They established 8 categories and 43 secondary categories for the description of those mutant lines. The images of the morphological phenotypes observed have been entered into a searchable database (http://rarge.gsc.riken.jp/phenome/) (Kuromori *et al.*, 2006). Although this work is based mainly in observation and description of phenotypes, the information obtained is very important for the functional characterization of new genes. In addition, phenotypic information can be used together with transcriptome and interactome data sets for the construction of powerful functional networks (Gunsalus *et al.*, 2005). A promising approximation has been reported that uses 3D (three dimensional) laser scanners for the collection of morphological information (Kaminuma *et al.*, 2005). Phenome efforts like the ones described will undoubtedly play a key role for understanding gene function in Arabidopsis.

1.3.2.6 Genetic interactions

Some single mutations or gene deletions do not have an obvious altered phenotype, but the phenotype is uncovered when two of those are combined. These epistatic relationships are used to define interactions (edges) between genes (nodes) and build genetic interaction networks. Building genetic interaction networks is important because most heritable traits are affected by interactions between multiple genes. Most of the studies involving genetic interactions have focused on synthetic lethal interactions because they are much easier to score and detect as compared to other subtle changes in phenotype. A synthetic lethal interaction between two genes is defined when the survival of the combined mutations is less than the product of the survival of the two single mutations (Lehner, 2007). If these interactions are carried out in a systematic way, it is possible to construct genetic interaction networks. Nowadays, *S. cerevisiae* is the model organism in which most of the genetic interactions have been mapped (Ooi *et al.*, 2006; Ming *et al.*, 2008). There are two main experimental strategies in yeast: synthetic genetic arrays (SGA) and synthetic lethal analysis of microarrays (SLAM). In the first approach, the starting point is a yeast strain carrying a query mutation that is systematically mated to a library of viable deletion strains and subsequently evaluated phenotypically in parallel. On the other hand, in the second approach, double mutants are constructed (integrative transformation) and assayed as a single pool. Later, microarrays are used to detect the relative growth rates making use of a barcode that each double mutant has in its genome. In addition, genetic interaction screenings (Baugh *et al.*, 2005) and a genetic interaction network have also been generated for *C. elegans*, taking advantage of the bacterial RNAi feeding system (Lehner *et al.*, 2006). It is also possible, although less practical, to do this in mammalian and fly cell cultures (Wheeler *et al.*, 2004). Genetic interaction screenings have also been described in Arabidopsis (Parker *et al.*, 2000), although it is more complicated and laborious to develop a systematic test of all the possible genetic interactions in plants. Genetic interactions can be used to understand gene function. For instance, the higher the number of shared genetic inter-actions partners between two genes, the higher the probability that both genes interact physically and share a biological function (Ming *et al.*, 2008). However, a genetic interaction can have diverse explanations and a systemic interpretation of those is lacking. Although the transfer of genetic interac-tion between species using orthology relationships is a possibility in order to obtain a genetic interaction network in Arabidopsis, preliminary evidence suggests that this step may not be so direct. In contrast to gene essential-ity, genetic interactions seem not to be so widely conserved across species (Lehner, 2007). Another strategy consists in the prediction of genetic interac-tions. Zhong and Sternberg integrated information about anatomical expres-sion patterns, phenotypes, functional annotations, microarray co-expression and protein interactions to predict *C. elegans* genetic interactions (Zhong and Sternberg, 2006). With their model, they could identify 12 genetic interaction partners of the *let-60*/Ras gene and 2 of the *itr-1* gene. It would be a challenge worth pursuing to construct an analogous genetic interaction network for Arabidopsis.

1.4 Strategies for genomic data integration

Here is an interesting question: What do Belgium and systems biology have in common? Well, definitely not their 500 varieties of beer. Interestingly, they both share a motto: *'L'union fait la force'* ('Strength through unity'). Our uni-verse is a noisy place. By noise, we are not just talking about an unintended sound that reaches our ears, but it can also be something unpleasant, unex-pected and undesired. The word 'noise' can be traced back to the Latin word *nausea* (feeling of sickness), and there is no doubt that some scientists feel that way when facing it. Noise is a random and generally persistent disturbance that obscures or reduces the clarity of a signal or the result of an experi-ment. Nowadays, it is common to read in scientific literature that one of the biggest problems when constructing genome-scale models of biological sys-tems is that the underlying data is noisy. This is an inherent property of the high-throughput techniques used for the acquisition of massive data, interac-tions between proteins, etc. As a consequence of this noise, data sets contain false positives. This occurs, for example due to self-activators in the yeast 2-hybrid technique, and much effort has been invested in solving this problem (Margulies *et al.*, 2005). False-positive interactions can also be 'biological'. This means that the physical interaction between proteins A and B can be true, but maybe this interaction does not make sense in a cellular context, be-cause protein A and B are expressed in different tissues or different cell types or different organelles or even in different periods of times in the same cell during development. That is why special caution must be taken, for example when trying to predict the function of a gene based on a single global data set. When we say 'Strength through unity', we mean that it is possible to increase the confidence of the edges in a network by integrating different sources of information. The rationale behind this is that it is unlikely that false-positive interactions are reproducible in data sets acquired through different experimental or *in silico* approaches. Data integration is also important as we try to build comprehensive models of a system. The sheer amount of information published daily for any of the model organisms makes it almost impossible to manually store, classify and integrate the data for systems modelling. Bioinformatics tools are key for systems research. Ideally, a high quality integrated model of the organism of choice will speed up the discovery process by allowing experimental scientists to spend time addressing important biological questions in the laboratory instead of navigating through an ocean of genomic information on their computers.

1.4.1 Integration strategies

There are generally four methods that are used as a solution for data integration: (1) Hypertext Navigation, (2) Data Warehouse, (3) Unmediated MultiDB queries and (4) Federated Databases (Karp, 1996). Hypertext Navigation allows the user to query only one database but the results often contain hyperlinks to the equivalent entry in another database. This method is more common for websites that are based on information retrieval. A Data Warehouse retrieves data from multiple resources, translates the formats and puts them in one database. This allows for much faster and more complex queries taking advantage of all the data loaded in the one database; however, translating database formats from one to another is a challenge in itself. The major drawback of this method is that it would be very difficult to keep up with all the new resources becoming available and keeping it all updated. The Unmediated MultiDB Queries allow the databases to remain separate but the query itself extends across all of the databases. The Federated Database is a combination of Data Warehouse and Unmediated MultiDB Queries; it allows the databases to be separate but it contains a federated schema, which dynamically translates to queries in the individual schemas. For a detailed discussion about these methods see Peter Karp's (1996) review.

In addition to simply providing data, many websites also provide applications that analyze the data for the user. Integrating these services has the same obstacles as integrating data, but instead of only querying the data you are also launching an application. Two recent efforts, iSYS (Siepel *et al.*, 2001) and BioMOBY (Wilkinson *et al.*, 2003), provide infrastructure that allows integration of applications. Both are designed to be 'plug and play' and generic to allow for compatibility with practically any application or data.

1.4.2 Case studies on data integration

A good example of an integration strategy (Hwang *et al.*, 2005) is the data integration methodology called 'Pointillist'. This methodology handles multiple data types from technologies with different noise characteristics. Pointillist integrated 18 data sets containing information about the galactose utilization in yeast. The data included global changes in mRNA and protein abundance, genome-wide protein–DNA interaction data, database information and computational predictions. Even though, the galactose biochemical pathway has been studied for more than 40 years and is one of the best understood in eukaryotic systems, they managed to predict and corroborate experimentally a new relationship between the fructose and the galactose pathways.

Kelley et al. (2003) proposed an integration strategy to face the problem of yeast genetic interaction analysis. There are many types of genetics interactions. For example, synthetic lethal interactions in which mutations in two non-essential genes result in a lethal phenotype. Genetic interactions are useful to study pathway organization. Some high-throughput methods for discovering genetic interactions have been developed such as SGA or SLAM. Although these processes can be mostly automated, the bottleneck is the interpretation of the functional significance of each interaction found. In order to find a solution to this problem, Kelley et al. assembled a genetic interaction network and a physical interaction network. The former was generated by SGA large-scale screen data and interactions culled from Munich Information Center for Protein Sequences, and the latter was connected by interactions of three types: protein–protein interactions (A and B interact physically), protein–DNA (A binds to regulatory sequence of B) and shared-reaction metabolic relationships (enzymes A and B have a substrate in common). They defined three different interpretations of genetics interactions: (1) between-pathways in which the genetic interaction bridges genes operating in two pathways with redundant or complementary functions, (2) within-pathways in which genetic interactions occur between proteins subunits within a pathway and (3) indirect effects in which the lethal phenotype involves many diverse pathways. The researchers could uncover mechanisms behind many of the observed genetics interactions that were preferentially between-pathway and they could predict new functions for 343 yeast proteins based on the models generated (not validated experimentally).

Another practical example of an integration strategy is the recent work of Hirai et al. (2004) in plants in which they integrated transcriptome and metabolomic data. They focused on the study of glucosinolates (GSLs) which are produced by Brassicaceae family. GSLs apparently provide anticarcinogenic, antioxidative and antimicrobial activity. Despite the high biotechnological importance of these compounds, before this research, there was no previous report on the genes regulating methionine-derived aliphatic GSL biosynthesis. The authors used an integrated strategy based on transcriptome co-expression analysis for public data sets and their own data sets together with metabolic profiling. They first generated a data set of condition independent co-expression profiles by calculating the Pearson correlation between all paired combinations of 22 263 Arabidopsis genes. Then they visualized the co-expression network including only pairs of genes whose correlation coefficient was >0.65. They found that genes involved in aliphatic GSL biosynthesis were clustered in a discrete module together with two un-characterized genes that coded for TFs: Myb28 and Myb29. Further experiments revealed that Myb28 is a positive regulator for basal-level production of aliphatic GSL and Myb29 presumably plays an accessory function integrating JA signalling and GSL biosynthesis. Interestingly, they could induce the biosynthesis of large amounts of GSLs by overexpressing Myb28 in Arabidopsis-cultured suspension cells that do not normally synthesize them. In addition, they predicted many of the genes involved in the pathway by examining the obtained co-expression network. This clearly shows that it is possible to find genes involved in particular plant processes using integrative systems biology strategies.

1.4.3 Software tools for systems biology

The large-scale genomic data available to biologists today require the use of a variety of software tools to process and analyze the data. A systems approach to understanding biology involves an iterative process of data integration, building a model, designing experiments to support the model and generating new hypothesis based on new data (Gutierrez *et al.*, 2005). There are several tools that are specific for each stage, however, few of them span across several stages and even fewer allow for the iterative analysis. In Chapter 5, several systems biology tools are discussed in detail: Sungear (Poultney *et al.*, 2007), Genevestigator (Zimmermann *et al.*, 2004), MapMan (Thimm *et al.*, 2004) and Cytoscape (Shannon *et al.*, 2003). In this section, we will highlight some additional tools that help biologists perform systems-level analysis. Please note that in no way are we providing a comprehensive list of all software tools available. In addition, this is an active and dynamic field with new tools constantly being generated.

Several environments have been developed in the past years that permit data integration and modelling (Endy and Brent, 2001). Such software allows detailed mathematical representation of cellular processes (e.g. Gepasi (Mendes, 1997) and Virtual Cell (Loew and Schaff, 2001)), as well as qualitative representations of cellular components and their interactions (e.g. Cytoscape (Shannon *et al.*, 2003) and Osprey (Breitkreutz *et al.*, 2003)). The Arabidopsis eFP browser (Winter *et al.*, 2007) provides a graphical representation of gene expression plotted on drawings of the different tissue and developmental stages based on data from AtGenExpress (Schmid *et al.*, 2005). Generally, quantitative models are directed towards specific cellular processes of interest and are built by representing existing literature as a system of mathematical equations. Quantitative models are powerful because they describe a system in detail (Endy and Brent, 2001). Unfortunately, they require a very detailed understanding of the system under scrutiny. This information is not readily obtained for every component in a biological process, and less so in every model organism. In fact, there are still many gaps in our qualitative understanding of biological systems. For example, most of the genes in Arabidopsis have not yet been experimentally characterized. The crucial first stage in building quantitative models is constructing a map of the interaction network to be analyzed (Schwender et al., 2003).

Because genes encoding proteins that participate in the same pathway or are part of the same protein complex are often co-regulated in their expression patterns, clustering techniques (e.g. Eisen et al., 1998) are commonly applied to genome-wide expression measurements to hypothesize the role of uncharacterized genes. In addition, in experiments where the biologist is comparing a treated versus a control sample, statistical methods such as Rank Products (Breitling et al., 2004) can be used to determine the genes that are differentially expressed between the experimental conditions. To learn the biological significance of such lists of genes (e.g. co-regulated genes or differentially expressed genes), a common next step is to evaluate the frequency of occurrence of functional attributes using structured functional annotations such as the gene ontology (GO) (Ashburner et al., 2000). Several software packages to automate this type of analysis now exist (e.g. Onto-Express (Khatri et al., 2002), GoMiner (Zeeberg et al., 2003), GOSurfer (Zhong et al., 2004), FatiGO (Al-Shahrour et al., 2004)). While advanced data analysis tools for exploiting genomic data are rapidly emerging, one of the drawbacks of current software tools is that they are highly specialized to each platform. This translates in the need to travel through several different web services or to use various stand-alone applications to be able to analyze the large data sets characteristic of genomic research, a cumbersome and inefficient process not amenable to iterative in silico exploration and experimentation.

Due to the increased use of networks to represent interaction data, viewers and graph drawing software such as Pajek (Batagelj and Mrvar, 1998), daVinci (Wilkinson and Links, 2002), Graphlet (Wilkinson and Links, 2002) and Graphviz (Wilkinson and Links, 2002), developed for general purpose graph drawing and visualization, are now used to organize and display biological data as graphs. Other tools such as MintViewer (Zanzoni et al., 2002), GeneInfoViz (Zhou and Cui, 2004), PIMRider (Hybrigenics, 2004), GenMAPP (Dahlquist et al., 2002), MAPPFinder (Doniger et al., 2003), TopNet (Yu et al., 2004), MAPMAN (Thimm et al., 2004) and PaVESy (Ludemann et al., 2004), each developed in the context of specific applications, allow certain data types to be displayed and provide some data analysis tools. Other advanced software environments such as Cytoscape (Shannon et al., 2003), Osprey (Breitkreutz et al., 2003), VisANT (Hu et al., 2004), PathwayAssist (Wilkinson and Links, 2002) and PathBlazer (Wilkinson and Links, 2002) include various methods to layout the network graphs, support connection to external databases and include more sophisticated data analysis tools (Shannon et al., 2003).

In many cases, biologists need to use many if not all of the tools mentioned and data management becomes an issue. Several institutions, for example MetNet (Wurtele et al., 2007) and The Bio-Array Resource for Arabidopsis

Functional Genomics who created Arabidopsis efP browser (Winter *et al.*, 2007), have created entire suites of wonderful applications to analyze genomic data but there is absolutely no interaction between them. VirtualPlant is an attempt to provide a platform for systems biology where biologists can maintain their genomic data and execute different data visualization and analysis tools. VirtualPlant's GeneCart supports the iterative nature of systems biology analysis and allows users to save results from one analysis and feed it into another. Applications available on VirtualPlant range from simple list functions, microarray data analysis to gene network visualization and analysis.

1.5 Systems biology in plant research

The development of systems biology approaches and tools for Arabidopsis has enabled the generation of testable biological hypotheses derived from the integration and analysis of genomic data within a network context. In this section, we briefly discuss proof of principle studies where a systems approach has uncovered testable hypotheses for regulatory networks in Arabidopsis.

1.5.1 Qualitative network models and genome-wide expression data define carbon/nitrogen molecular machines in Arabidopsis

In plants, there is ample evidence that C-signals interact with nitrogen (N) status (Gutierrez *et al.*, 2007) to control genes in metabolic pathways, including N-assimilation and amino acid synthesis (e.g. Gutierrez *et al.*, 2007; Palenchar *et al.*, 2004) as well as broad kinds of developmental mechanisms such as germination (Alboresi *et al.*, 2005), shoot growth (Walch-Liu *et al.*, 2000, 2005; Rahayu *et al.*, 2005; Krouk *et al.*, 2006), root growth (Forde, 2002) or flowering (Raper *et al.*, 1988; Rideout *et al.*, 1992). While these studies confirm the existence of a complex CN-responsive gene regulatory network in plants, the possible mechanisms for CN sensing and signalling remain unknown. While transcriptome studies conducted in the pre-systems biology era, confirmed the existence of genome-wide CN-responses (Palenchar *et al.*, 2004; Price *et al.*, 2004), those studies were not able to model responses or reveal underlying mechanisms. For example, at least four general mechanisms for CN sensing can be proposed: (1) N-response independent of C, (2) C-response independent of N, (3) C- and N-interaction or (4) a unified CN-response (Gutierrez *et al.*, 2007). To support or reject these models for C/N sensing, plants subjected to a systematic matrix of C and N treatments were analyzed using a systems approach (Gutierrez *et al.*, 2007). Hierarchical cluster analysis (≥ 0.5 correlation) of transcriptome data generated in this study uncovered many gene clusters whose average expression patterns showed statistically significant CN interactions (analysis of variance, $p < 0.01$), suggesting that

C and N interaction (Model 3) is a prominent mode of regulation by N in Arabidopsis roots. To gain a global but detailed view of how transcriptional responses to carbon and N nutrient treatments affects plants as a system, an Arabidopsis multi-network was created and used to query gene clusters whose members respond to C/N according to Models 1–3, for highly connected sub-networks defined using a graph clustering algorithm developed by (Ferro *et al.*, 2003). This analysis revealed a set of 'molecular machines' comprised of highly connected genes involved in metabolic, cellular or signalling pathways, whose expression is regulated by C and N metabolites. One such CN-responsive sub-network is involved in responses to auxin, as it contains 13 genes in the auxin response pathway (including the auxin receptor), 5 auxin efflux carriers and 2 auxin transport proteins. Validation using time-course studies suggest that the phytohormone auxin acts as a regulator of plant growth in response to C and/or N availability. In addition, the CN-responsive gene network was shown to contain a significant proportion of regulatory proteins including: 299 TFs and 27 genes that are known targets of miRNAs. This latter result implicated a new role for miRNAs in post-transcriptional regulation of gene expression by CN metabolite signals in plants.

1.5.2 A systems approach identifies an organic nitrogen-responsive gene network regulated by the master clock gene CCA1

A second systems biology study, addressed the mechanisms by which distinct forms of N are sensed as signals for N status (Gutierrez *et al.*, 2008). There is growing evidence that the N 'input' (nitrate) and 'output' (Glu/Gln) of the N-assimilatory pathway serve not only as N-metabolites, but also as N-signals that are sensed and transduced, to control genes regulating plant metabolism and development (for NO_3^- see Forde, 2002; Sollars *et al.*, 2003; Scheible *et al.*, 2004; for Glu/Gln see Oliveira and Coruzzi, 1999; Rawat *et al.*, 1999). Studies with nitrate reductase nulls, unable to reduce nitrate, have shown that nitrate can serve as a 'metabolic N-signal', to regulate genes in N-uptake/assimilatory pathway and to control root development in response to nitrate availability in the soil (Forde, 2002). There is also ample, though less direct evidence, that the assimilated forms of organic-N such as glutamate (Glu) or glutamine (Gln) serve as signals for levels of organic-N or C:N status, to repress further N-assimilation, or lateral root growth, when organic-N stores are replete (Oliveira and Coruzzi, 1999; Rawat *et al.*, 1999). To identify the regulatory networks regulated by these N-signals, transcriptome analysis identified a set of 834 genes regulated in response to inorganic-N and/or organic-N treatments. To identify the global processes regulated by these N-signals, the Arabidopsis multi-network (Gutierrez *et al.*, 2007) was queried for network connectivity between these N-regulated genes. Of the 834 N-regulated genes, 368 were connected by a variety of edges including metabolic and regulatory connections. To identify potential 'master' regulators of this N-responsive gene sub-network, N-regulated TFs were ranked based on their number of predicted regulatory connections. At the top of the list, with 47 connections to targets in the N-regulated gene network, is a Myb family TF which is the central clock control gene *CCA1* (Daniel *et al.*, 2004). Exploration of the network 'neighbourhood' surrounding the *CCA1* hub predicts that *CCA1* is an important regulator of an N-assimilation gene network. Specifically, the model predicts that overexpression of *CCA1* would induce the expression of the *GLN1.3* gene and repress the expression of *ASN1* and *GDH1*, a result which was validated using 35S::*CCA1* lines (*CCA1-ox*) (Sollars *et al.*, 2003). Chromatin-IP assays using CCA1 antibodies confirmed binding of CCA1 to the promoter regions of *GLN1.3*, *GDH1* and *bZIP1* and ASN1. These results indicate that the circadian clock regulates N-assimilation (Farre *et al.*, 2005) in response to N availability by transcriptional control by *CCA1* which in turn targets genes central to the N-assimilatory pathway in plants. Specifically, in this model Glu-repression of CCA1 leads to downregulation of GLN1.3, and upregulation of ASN1. This regulation leads to the conversion of metabolically reactive Gln to inert Asn, used for N-storage, when levels of Glu/Gln are abundant.

CCA1 is a key component involved in a negative feedback loop at the centre of the circadian clock (Locke *et al.*, 2005; McClung, 2006). The finding that *CCA1* mRNA levels are regulated by organic N-sources, suggest that N-signals act as an input to the circadian clock. To test this new hypothesis derived from the systems approach, Arabidopsis seedlings were subject to pulses of inorganic-N or organic-N at intervals spanning a circadian cycle and determined the effects on the phase of the oscillation in *CCA1::LUC* expression. Each N-treatment resulted in subtle (2 h) but stable phase shifts in CCA1::LUC expression, indicating that N-status serves as an input to the circadian clock. These alterations in phase response curves are consistent with N-signals mediating a weak (type 1) clock resetting observed for other metabolic signals (Bunning and Moser, 1973; Bollig *et al.*, 1978; Kondo, 1983). The emerging view of the circadian clock as a key integrator of metabolic and physiologic processes (Farre *et al.*, 2005; McClung, 2006) is that it receives input not only from environmental stimuli but also from metabolic pathways, many of which are subject to circadian regulation. Thus, the clock is known to regulate genes in N-assimilation (Pilgrim *et al.*, 1993; Farre *et al.*, 2005), this systems study suggests a new hypothesis that N in turn feeds back to the circadian clock, at least in part through the N-regulation of *CCA1* expression.

1.5.3 Cell-specific nitrogen responses mediate developmental plasticity

In this application of systems biology to Arabidopsis development, cell-specific transcriptome analysis was analyzed using a systems approach to ask how N-nutrient signals coordinate expression of gene networks in specific cell types of roots (Gifford *et al.*, 2008). *Green fluorescent protein* (GFP)-marked transgenic lines spanning the cell types of the root were transiently treated (2 h) with nitrate (5 mM) or control KCl treatments, and GFP expressing cells were sorted and isolated using fluorescence-activated cell sorting (Gifford *et al.*, 2008). Five thousand three hundred ninety-six N-regulated transcripts were identified in which 87% were significantly N-regulated in at least one, but not *all* cell-types profiled. Only 771 transcripts responded to N-treatment across *all* cell-types examined. Thus, cell sorting greatly increased the sensitivity to detect transcriptional regulation in specific cell types by N treatment that were largely hidden in studies of whole roots from nitrate-treated plants (Sollars *et al.*, 2003; Scheible *et al.*, 2004; Gutierrez *et al.*, 2007). This cell-specific N-response transcriptome data also revealed new mechanisms by which N regulates developmental processes in roots. One example is the cell-specific regulation of a transcriptional circuit derived from a systems analysis of the data and validated to mediate lateral root outgrowth in response to N. Importantly, the N-regulated response of the miR167 targets (Wu *et al.*, 2006) was only detectable with cell-specific resolution. Validation studies showed that miR167 is specifically expressed and N-regulated in the pericycle (PmiR:GUS line and mature RNA) and that N-regulation of its target ARF8 is abrogated in a mutant in which the miR167 binding site of ARF8 is mutated (Wu *et al.*, 2006). These findings fill a gap in our knowledge of how multi-cellular organisms cope with N-responses at the cellular level.

1.5.4 Quantitative models of defined molecular processes

One of the best examples of systems biology in action in plants was demonstrated by James Locke and colleagues (reviewed in Ueda, 2006) where they studied the circadian clock in Arabidopsis. The first molecular component of the Arabidopsis central oscillator identified was the feedback loop between the genes *LHY* (*late elongated hypocotyl*), *CCA1* (*circadian clock associated*) and *TOC1* (*timing of cab expression*) (for a historical perspective on the clock see (McClung, 2006). *LHY* and *CCA1* are partially redundant genes that are activated by light and their products repress *TOC1*. Locke and colleagues analyzed experimental data from various circadian studies (Matsushika *et al.*, 2000; Mizoguchi *et al.*, 2002; Kim *et al.*, 2003) including *cca1/lhy* double mutant (Locke *et al.* 2005), and determined that the feedback loop was not sufficient to explain many of the circadian rhythms observed. For example, the feedback loop does not explain why there is a 12-h delay in *LHY/CCA1* after activation of *TOC1*. The feedback model also does not explain why TOC1 levels drop much earlier (dusk) than *LHY/CCA1* activation. Locke and colleagues proposed a new model with two loops and two new factors, X and Y, to account for the experimental observations. In the new model, factor Y is light induced and activates *TOC1* and factor X activates *LHY/CCA1* and itself is activated by TOC1. They next looked at experimental data to identify candidate genes to carry out the X or Y functions. The data analyzed suggested that *GI*

(*GIGANTEA*) could play the role of Y. X remains to be identified. Circadian study of *gi/lhy/cca1* showed a decreased level of expression of TOC1 (Locke *et al.*, 2006). Similar studies allowed Locke and colleagues to create another model with three loops to incorporate other genes *Pseudo Response Regulatory* (*PRR7*) and *PRR9* that have also circadian rhythmic expression patterns (Farre *et al.*, 2005). The new loop is a feedback loop between *PRR7/PRR9* and *LHY/CCA1*. The work by Locke and colleagues contains an iterative nature of analyzing data, building a model, validating a model and then developing new hypotheses, which is a trademark of systems biology.

1.5.5 Modelling the behaviour and tissue and organs

One of the beautiful aspects of development is the creation of shapes and patterns. How do single cells develop into multi-cellular organisms? This is a key question in plant biology that can benefit from systems approaches. Several examples of tissue, organ or plant modelling have been reported. For example, the growth dynamics underlying the development of *Antirrhinum* petals were modelled (Prusinkiewicz and Rolland-Lagan, 2006). They determined three types of parameters: the rate of increase in size (growth rate), the degree to which growth occurs preferentially in any direction (anisotropy) and the orientation angle of the main direction of growth (direction). The model allowed them to conclude that the key parameter determining the petal asymmetry of the *Antirrhinum* is the direction of growth rather than the regional differences in growth rate. In a different study, Mundermann *et al.* (2005) systematically took thousands of measurements, such as length, width and other parameters that describe shapes, in frequent time intervals and created a model representing plant development. In their model the plant is broken into four basic parts: apices, internodes, leaves and flowers, and each flower is further broken into pedicel, carpel, sepals, petals and stamens. The measurements are taken for each of the individual parts and the simulation of the model assembles all the components into a 3D growing plant. The model is written in L-system-based modelling language L +C (Karwowski and Prusinkiewicz, 2003). The validation of the model was the generation of a simulated image that looked very similar to a wild-type Arabidopsis. A similar model is also available for rice (Watanabe *et al.*, 2005). These plant models can serve as references for future studies in plant development and morphology.

There are several different computational techniques that have been developed for processing of data, such as images and experimental measurements, and construction of simulation models (reviewed in Prusinkiewicz and Rolland-Lagan, 2006). One such model is reaction-diffuse where the pattern forming substance is assumed to diffuse across cells (Turing, 1952; Gierer and Meinhardt, 1972). Meyerowitz and colleagues used this model to study the shape and size of the shoot apical meristem (SAM) and expression domains of the *WUSCHEL* (*WUS*) gene (Jonsson *et al.*, 2005). *WUS* is known to be involved in SAM development and is known to induce the ligand *CLAVATA3* (*CLV3*) which binds to the *CLAVATA1* (*CLV1*) receptor kinase. *CLV1* then activates a signalling cascade that represses *WUS* (Sharma *et al.*, 2003). In this study, the authors used in vivo confocal microscopy and image-processing algorithms to obtain quantitative measurements of gene expression. The measurements were subsequently used as input for the reaction-diffuse model, which allowed them to predict the outcome of laser cell ablation experiments (Reinhardt *et al.*, 2003) and removal of *CLV3* signal (Fletcher *et al.*, 1999). The model correctly simulated creation of two new *WUS* domains after the ablation of the central cells from the SAM and the expansion of *WUS* expression with the removal of *CLV3*. (Note: For more on systems biology applications to plant development and evolution, see Chapters 11 and 12 of this volume.)

1.6 Conclusion

Hodgkin and Huxley (1952) described a model that explained in mathematical terms the ionic mechanisms underlying the initiation and propagation of action potentials in the squid giant axon. They could even predict with some success how a neuron works. Both researchers integrated their chemical, biological, physical and mathematical knowledge in order to understand this phenomenon, and they can be fairly called pioneers of systems biology. One of the main problems today when integrating diverse disciplines like biology, physics or computer science is actually the integration of the people who work in those fields. These scientists have, for example diverse backgrounds and cultures for publishing, communicating their results and doing research. In addition, their 'languages' can differ substantially, making communication a difficult task. This is another challenge systems biology must overcome in the coming years, and new educational strategies are definitely needed in universities. Wilson (2000) wrote: 'The love of complexity without reductionism makes art; the love of complexity with reductionism makes science.' This is clearly a challenge for a new generation of scientists and students who must coherently and elegantly integrate the vast amount of biological information available in order not only to make and validate predictions, but to understand the underlying principles of living matter. Perhaps art and science are different worlds, but they have something in common: the creative process. You may call it inspiration or scientific induction, but those will only favour the prepared minds. Are you prepared for systems biology?

Acknowledgements

We regret that many interesting articles could not be cited due to space limitations. Research in RAG's laboratory in this area is supported by grants from FONDECYT (1060457), ICGEB (CRPCHI0501) and ICM-MIDEPLAN (MN-PFG P06–009-F). Research in GC's laboratory in this area is supported by grants from NIH (GM032877), NSF (IOB 0519985), DOE (DEFG02–92ER20071). Collaborative work by RAG's and GC's labs on systems biology is supported by NSF (DBI-0445666).

References

Adams, M.D., Kelley, J.M., Gocayne, J.D., Dubnick, M., Polymeropoulos, M.H., Xiao, H., *et al.* (1991) Complementary, D.N.A sequencing: expressed sequence tags and human genome project. *Science* **252**, 1651–1656.

Aharoni, A., Ric de Vos, C.H., Verhoeven, H.A., Maliepaard, C.A., Kruppa, G., Bino, R., *et al.* (2002) Nontargeted metabolome analysis by use of Fourier Transform Ion Cyclotron Mass Spectrometry. *Omics* **6**, 217–234.

Albert, R., Jeong, H. and Barabasi, A.L. (1999) Diameter of the World-Wide Web. *Nature* **401**, 130–131.

Albert, R., Jeong, H. and Barabasi A.L. (2000) Error and attack tolerance of complex networks. *Nature* **406**, 378–382.

Alboresi, A., Gestin, C., Leydecker, M.T., Bedu, M., Meyer, C. and Truong, H.N. (2005) Nitrate, a signal relieving seed dormancy in Arabidopsis. *Plant Cell Environ* **28**, 500–512.

Alonso, J.M. and Ecker, J.R. (2006) Moving forward in reverse: genetic technologies to enable genome-wide phenomic screens in Arabidopsis. *Nat Rev Genet* **7**, 524–536.

Al-Shahrour, F., Diaz-Uriarte, R. and Dopazo, J. (2004) FatiGO: a web tool for finding significant associations of gene ontology terms with groups of genes. *Bioinformatics* **20**, 578–580.

Aoki, K., Ogata, Y. and Shibata, D. (2007) Approaches for extracting practical information from gene co-expression networks in plant biology. *Plant Cell Physiol* **48**, 381–390.

Aoki, K.F. and Kanehisa, M. (2005) Using the KEGG database resource. *Curr Protoc Bioinformatics* Chapter 1, Unit 1 12.

Ashburner, M., Ball, C., Blake, J., Botstein, D., Butler, H., Cherry, J., *et al.* (2000) Gene ontology: tool for the unification of biology. The Gene Ontology Consortium. *Nat Genet* **25**, 25–29.

Baerenfaller, K., Grossmann, J., Grobei, M.A., Hull, R., Hirsch-Hoffmann, M., Yalovsky, S., *et al.* (2008) Genome-scale proteomics reveals Arabidopsis thaliana gene models and proteome dynamics. *Science* **320**, 938–941.

Barabasi, A.L. and Albert, R. (1999) Emergence of scaling in random networks. *Science* **286**, 509–512.

Barabasi, A.L. and Oltvai, Z.N. (2004) Network biology: understanding the cell's functional organization. *Nat Rev Genet* **5**, 101–113.

Batagelj, V. and Mrvar, A. (1998) Pajek – program for large network analysis. *Connections* **21**, 47–57.

Baugh, L.R., Wen, J.C., Hill, A.A., Slonim, D.K., Brown, E.L. and Hunter, C.P. (2005) Synthetic lethal analysis of Caenorhabditis elegans posterior embryonic patterning genes identifies conserved genetic interactions. *Genome Biol* **6**, R45.

Bennet, M.D. and Leitch, I.J. (1995) Nuclear DNA amounts in angiosperms. *Ann Bot* **76**, 113–176.

Bennett, S. (2004) Solexa Ltd. *Pharmacogenomics* **5**, 433–438.

Bennetzen, J.L., Schrick, K., Springer, P.S., Brown, W.E. and SanMiguel, P. (1994) Active maize genes are unmodified and flanked by diverse classes of modified, highly repetitive DNA. *Genome* **37**, 565–576.

Bergmann, S., Ihmels, J. and Barkai, N. (2004) Similarities and differences in genome-wide expression data of six organisms. *Plos Biol* **2**, 85–93.

Bevan, M. and Walsh, S. (2005) The Arabidopsis genome: a foundation for plant research. *Genome Res* **15**, 1632–1642.

Bogdanov, A. (1980) *Essays in Tektology: The General Science of Organization* (Seaside, CA: Intersystems Publications).

Bollig, I., Mayer, K., Mayer, W.-E. and Engelmann, W. (1978) Effects of cAMP, theophylline, imidazole, and 4-(3,4-dimethoxybenzyl)-2-imidazolidone on the leaf movement rhythm of Trifolium repens – a test of the cAMP-hypothesis of circadian rhythms. *Planta* **141**, 225–230.

Breitkreutz, B.-J., Stark, C. and Tyers, M. (2003) Osprey: a network visualization system. *Genome Biol* **4**, r22.1–r22.4.

Breitling, R., Armengaud, P., Amtmann, A. and Herzyk, P. (2004) Rank products: a simple, yet powerful, new method to detect differentially regulated genes in replicated microarray experiments. *FEBS Lett* **573**, 83–92.

Bunning, E. and Moser, I. (1973) Light-induced phase shifts of circadian leaf movements of phaseolus: comparison with the effects of potassium and of ethyl alcohol. *Proc Natl Acad Sci USA* **70**, 3387–3389.

Calderwood, M.A., Venkatesan, K., Xing, L., Chase, M.R., Vazquez, A., Holthaus, A.M., *et al.* (2007) Epstein-Barr virus and virus human protein interaction maps. *Proc Natl Acad Sci USA* **104**, 7606–7611.

Dahlquist, K., Salomonis, N., Vranizan, K., Lawlor, S. and Conklin, B. (2002) GenMAPP, a new tool for viewing and analyzing microarray data on biological pathways. *Nat Genet* **31**, 19–20.

Daniel, X., Sugano, S. and Tobin, E.M. (2004) CK2 phosphorylation of CCA1 is necessary for its circadian oscillator function in Arabidopsis. *PNAS* **101**, 3292–3297.

de Folter, S., Immink, R.G., Kieffer, M., Parenicova, L., Henz, S.R., Weigel, D., *et al.* (2005) Comprehensive interaction map of the Arabidopsis MADS Box transcription factors. *Plant Cell* **17**, 1424–1433.

DeRisi, J.L., Iyer, V.R. and Brown, P.O. (1997) Exploring the metabolic and genetic control of gene expression on a genomic scale. *Science* **278**, 680–686.

Doniger, S., Salomonis, N., Dahlquist, K., Vranizan, K., Lawlor, S. and Conklin, B. (2003) MAPPFinder: using Gene Ontology and GenMAPP to create a global gene-expression profile from microarray data. *Genome Biol* **4**, R7.

Eisen, M.B., Spellman, P.T., Brown, P.O. and Botstein, D. (1998) Cluster analysis and display of genome-wide expression patterns. *Proc Natl Acad Sci USA* **95**, 14863.

Endy, D. and Brent, R. (2001) Modelling cellular behaviour. *Nature* **409**, 391–395.

Erdös, P. and Rényi, A. (1960) On the evolution of random graphs. *Publ Math Inst Hung Acad Sci* **5**, 17–61.

Farre, E.M., Harmer, S.L., Harmon, F.G., Yanovsky, M.J. and Kay, S.A. (2005) Overlapping and distinct roles of PRR7 and PRR9 in the Arabidopsis circadian clock. *Curr Biol* **15**, 47–54.

Ferro, A., Pigola, G., Pulvirenti, A. and Shasha, D. (2003) Fast clustering and minimum weight matching algorithms for very large mobile backbone wireless networks. *Int J Found Comput Sci* **14**, 223–236.

Fiehn, O. (2002) Metabolomics – the link between genotypes and phenotypes. *Plant Mol Biol* **48**, 155–171.

Finnegan, E.J. and Matzke, S.A. (2003) The small RNA world. *J Cell Sci* **116**, 4689–4693.

Fleischmann, R.D., Adams, M.D., White, O., Clayton, R.A., Kirkness, E.F., Kerlavage, A.R., *et al.* (1995) Whole-genome random sequencing and assembly of Haemophilus influenzae Rd. *Science* **269**, 496–512.

Fletcher, J.C., Brand, U., Running, M.P., Simon, R. and Meyerowitz, E.M. (1999) Signaling of cell fate decisions by CLAVATA3 in Arabidopsis shoot meristems. *Science* **283**, 1911–1914.

Forde, B.G. (2002) Local and long-range signaling pathways regulating plant responses to nitrate. *Annu Rev Plant Biol* **53**, 203–224.

Friso, G., Giacomelli, L., Ytterberg, A.J., Peltier, J.B., Rudella, A., Sun, Q., *et al.* (2004) In-depth analysis of the thylakoid membrane proteome of Arabidopsis thaliana chloroplasts: new proteins, new functions, and a plastid proteome database. *Plant Cell* **16**, 478–499.

Geisler-Lee, J., O'Toole, N., Ammar, R., Provart, N.J., Millar, A.H. and Geisler, M. (2007) A predicted interactome for Arabidopsis. *Plant Physiol* **145**, 317–329.

Giavalisco, P., Nordhoff, E., Kreitler, T., Kloppel, K.D., Lehrach, H., Klose, J., *et al.* (2005) Proteome analysis of Arabidopsis thaliana by two-dimensional gel electrophoresis and matrix-assisted laser desorption/ionisation-time of flight mass spectrometry. *Proteomics* **5**, 1902–1913.

Gierer, A. and Meinhardt, H. (1972) A theory of biological pattern formation. *Kybernetik* **12**, 30–39.

Gifford, M.L., Dean, A., Gutierrez, R.A., Coruzzi, G.M. and Birnbaum, K.D. (2008) Cell-specific nitrogen responses mediate developmental plasticity. *Proc Natl Acad Sci USA* **105**, 803–808.

Giot, L., Bader, J.S., Brouwer, C., Chaudhuri, A., Kuang, B., Li, Y., *et al.* (2003) A protein interaction map of drosophila melanogaster. *Science* **302** (5651), 1727–1736.

Gunsalus, K.C., Ge, H., Schetter, A.J., Goldberg, D.S., Han, J.D., Hao, T., *et al.* (2005) Predictive models of molecular machines involved in Caenorhabditis elegans early embryogenesis. *Nature* **436**, 861–865.

Gutierrez, R.A., Green, P.J., Keegstra, K. and Ohlrogge, J.B. (2004) Phylogenetic profiling of the Arabidopsis thaliana proteome: what proteins distinguish plants from other organisms? *Genome Biol* **5**, R53.

Gutierrez, R.A., Lejay, L.V., Dean, A., Chiaromonte, F., Shasha, D.E. and Coruzzi, G.M. (2007) Qualitative network models and genome-wide expression data define carbon/nitrogen-responsive molecular machines in Arabidopsis. *Genome Biol* **8**, R7.

Gutierrez, R.A., Shasha, D.E. and Coruzzi, G.M. (2005) Systems biology for the virtual plant. *Plant Physiol* **138**, 550–554.

Gutierrez, R.A., Stokes, T.L., Thum, K.E., Xu, X., Obertello, M., Katari, M.S., *et al.* (2008) Systems approach identifies an organic nitrogen-responsive gene network that is regulated by the master clock control gene *CCA1*. *Proc Natl Acad Sci USA* **105** (12), 4939–4944.

Han, J.D., Bertin, N., Hao, T., Goldberg, D.S., Berriz, G.F., Zhang, L.V., *et al.* (2004) Evidence for dynamically organized modularity in the yeast protein-protein interaction network. *Nature* **430**, 88–93.

Hartwell, L.H., Hopfield, J.J., Leibler, S. and Murray, A.W. (1999) From molecular to modular cell biology. *Nature* **402**, C47–C52.

Heazlewood, J.L., Tonti-Filippini, J.S., Gout, A.M., Day, D.A., Whelan, J. and Millar, A.H. (2004) Experimental analysis of the Arabidopsis mitochondrial proteome highlights signaling and regulatory components, provides assessment of targeting prediction programs, and indicates plant-specific mitochondrial proteins. *Plant Cell* **16**, 241–256.

Henderson, I.R. and Jacobsen, S.E. (2007) Epigenetic inheritance in plants. *Nature* **447**, 418–424.

Hesse, H. and Hoefgen, R. (2006) On the way to understand biological complexity in plants: S-nutrition as a case study for systems biology. *Cell Mol Biol Lett* **11**, 37–56.

Hirai, M.Y., Yano, M., Goodenowe, D.B., Kanaya, S., Kimura, T., Awazuhara, M., *et al.* (2004) Integration of transcriptomics and metabolomics for understanding of global responses to nutritional stresses in Arabidopsis thaliana. *Proc Natl Acad Sci USA* **101**, 10205–10210.

Hodgkin, A.L. and Huxley, A.F. (1952) A quantitative description of membrane current and its application to conduction and excitation in nerve. *J Physiol* **117**, 500–544.

Hu, Z., Mellor, J., Wu, J. and DeLisi, C. (2004) VisANT: an online visualization and analysis tool for biological interaction data. *BMC Bioinformatics* **5**, 17.

Hwang, D., Smith, J.J., Leslie, D.M., Weston, A.D., Rust, A.G., Ramsey, S., *et al.* (2005) A data integration methodology for systems biology: experimental verification. *Proc Natl Acad Sci USA* **102**, 17302–17307.

Hybrigenics (2004) http://pim.hybrigenics./.

Initiative. T.A.G. (2000) Analysis of the genome sequence of the flowering plant Arabidopsis thaliana. *Nature* **408**, 796–815.

Isalan, M., Lemerle, C., Michalodimitrakis, K., Horn, C., Beltrao, P., Raineri, E., *et al.* (2008) Evolvability and hierarchy in rewired bacterial gene networks. *Nature* **452**, 840–845.

Jaillon, O., Aury, J.M., Noel, B., Policriti, A., Clepet, C., Casagrande, A., *et al.* (2007) The grapevine genome sequence suggests ancestral hexaploidization in major angiosperm phyla. *Nature* **449**, 463–467.

Jeong, H., Mason, S.P., Barabasi, A.L. and Oltvai, Z.N. (2001) Lethality and centrality in protein networks. *Nature* **411**, 41–42.

Jonsson, H., Heisler, M., Reddy, G.V., Agrawal, V., Gor, V., Shapiro, B.E., *et al.* (2005) Modeling the organization of the WUSCHEL expression domain in the shoot apical meristem. *Bioinformatics* **21** (Suppl 1), i232–i240.

Kaminuma, E., Heida, N., Tsumoto, Y., Nakazawa, M., Goto, N., Konagaya, A., *et al.* (2005) Three-dimensional definition of leaf morphological traits of Arabidopsis in silico phenotypic analysis. *J Bioinform Comput Biol* **3**, 401–414.

Karp, P.D. (1996) A strategy for database interoperation. *J Comput Biol* **2** (4), 573–583.

Karwowski, R. and Prusinkiewicz, P. (2003) Design and implementation of the L+C modeling language. *Electron Notes Theor Comput Sci* **86** (2), 1–19.

Kelley, B.P., Sharan, R., Karp, R.M., Sittler, T., Root, D.E., Stockwell, B.R., *et al.* (2003) Conserved pathways within bacteria and yeast as revealed by global protein network alignment. *PNAS* **100**, 11394–11399.

Khatri, P., Draghici, S., Ostermeier, G.C. and Krawetz, S.A. (2002) Profiling gene expression using onto-express. *Genomics* **79**, 266–270.

Kim, J.Y., Song, H.R., Taylor, B.L. and Carré, I.A. (2003) Light-regulated translation mediates gated induction of the Arabidopsis clock protein LHY. *EMBO J* **22**, 935–944.

Kitano, H. (2002) Computational systems biology. *Nature* **420**, 206–210.

Kitano, H. (2004) Biological robustness. *Nat Rev Genet* **5**, 826–837.

Kitano, H. (2007) Towards a theory of biological robustness. *Mol Syst Biol* **3**, 137.

Kondo, T. (1983) Phase shifts of potassium uptake rhythm in Lemna gibba G3 due to light, dark or temperature pulses. *Plant Cell Physiol* **24**, 659–665.

Krouk, G., Tillard, P. and Gojon, A. (2006) Regulation of the high-affinity NO_3^- uptake system by a NRT1.1-mediated 'NO_3^--demand' signalling in *Arabidopsis*. *Plant Physiol* **142**, 1075–1086.

Kuromori, T., Wada, T., Kamiya, A., Yuguchi, M., Yokouchi, T., Imura, Y., *et al.* (2006) A trial of phenome analysis using 4000 Ds-insertional mutants in gene-coding regions of Arabidopsis. *Plant J* **47**, 640–651.

Lahner, B., Gong, J., Mahmoudian, M., Smith, E.L., Abid, K.B., Rogers, E.E., *et al.* (2003) Genomic scale profiling of nutrient and trace elements in Arabidopsis thaliana. *Nat Biotechnol* **21**, 1215–1221.

Lehner, B. (2007) Modelling genotype-phenotype relationships and human disease with genetic interaction networks. *J Exp Biol* **210**, 1559–1566.

Lehner, B., Crombie, C., Tischler, J., Fortunato, A. and Fraser, A.G. (2006) Systematic mapping of genetic interactions in Caenorhabditis elegans identifies common modifiers of diverse signaling pathways. *Nat Genet* **38**, 896–903.

Lehner, B. and Fraser, A.G. (2004) A first-draft human protein-interaction map. *Genome Biol* **5**, R63.

Locke, J.C., Kozma-Bognar, L., Gould, P.D., Feher, B., Kevei, E., Nagy, F., *et al.* (2006) Experimental validation of a predicted feedback loop in the multi-oscillator clock of Arabidopsis thaliana. *Mol Syst Biol* **2**, 59.

Locke, J.C., Southern, M.M., Kozma-Bognar, L., Hibberd, V., Brown, P.E., Turner, M.S., *et al.* (2005) Extension of a genetic network model by iterative experimentation and mathematical analysis. *Mol Syst Biol* **1**, 0013.

Loew, L.M. and Schaff, J.C. (2001) The virtual cell: a software environment for computational cell biology. *Trends Biotechnol* **19**, 401–406.

Lu, C., Tej, S.S., Luo, S., Haudenschild, C.D., Meyers, B.C. and Green, P.J. (2005) Elucidation of the small RNA component of the transcriptome. *Science* **309**, 1567–1569.

Ludemann, A., Weicht, D., Selbig, J. and Kopka, J. (2004) PaVESy: pathway visualization and editing system. *Bioinformatics* **20** (16), 2841–2844.

Margulies, M., Egholm, M., Altman, W.E., Attiya, S., Bader, J.S., Bemben, L.A., *et al.* (2005) Genome sequencing in microfabricated high-density picolitre reactors. *Nature* **437**, 376–380.

Matsushika, A., Makino, S., Kojima, M. and Mizuno, T. (2000) Circadian waves of expression of the APRR1/TOC1 family of pseudo-response regulators in Arabidopsis thaliana: insight into the plant circadian clock. *Plant Cell Physiol* **41** (9), 1002–1012.

McClung, C.R. (2006) Plant circadian rhythms. *Plant Cell* **18**, 792–803.

Mendes, P. (1997) Biochemistry by numbers: simulation of biochemical pathways with Gepasi 3. *Trends Biochem Sci* **22**, 361–363.

Merchant, S.S., Prochnik, S.E., Vallon, O., Harris, E.H., Karpowicz, S.J., Witman, G.B., *et al.* (2007) The Chlamydomonas genome reveals the evolution of key animal and plant functions. *Science* **318**, 245–250.

Ming, R., Hou, S., Feng, Y., Yu, Q., Dionne-Laporte, A., Saw, J.H., *et al.* (2008) The draft genome of the transgenic tropical fruit tree papaya (Carica papaya Linnaeus). *Nature* **452**, 991–996.

Mitreva, M., McCarter, J.P., Martin, J., Dante, M., Wylie, T., Chiapelli, B., *et al.* (2004) Comparative genomics of gene expression in the parasitic and free-living nematodes Strongyloides stercoralis and Caenorhabditis elegans. *Genome Res* **14**, 209–220.

Mizoguchi, T., Wheatley, K., Hanzawa, Y., Wright, L., Mizoguchi, M., Song, H.R., *et al.*

(2002) LHY and CCA1 are partially redundant genes required to maintain circadian rhythms in Arabidopsis. *Dev Cell* **2**, 629–641.

Molina, C. and Grotewold, E. (2005) Genome wide analysis of Arabidopsis core promoters. *BMC Genomics* **6**, 25.

Mueller, L.A., Zhang, P. and Rhee, S.Y. (2003) AraCyc: a biochemical pathway database for Arabidopsis. *Plant Physiol* **132**, 453–460.

Mundermann, L., Erasmus, Y., Lane, B., Coen, E. and Prusinkiewicz, P. (2005) Quantitative modeling of Arabidopsis development. *Plant Physiol* **139**, 960–968.

Oliveira, I.C. and Coruzzi, G.M. (1999) Carbon and amino acids reciprocally modulate the expression of glutamine synthetase in Arabidopsis. *Plant Physiol* **121**, 301–310.

Oliver, S.G., Winson, M.K., Kell, D.B. and Baganz, F. (1998) Systematic functional analysis of the yeast genome. *Trends Biotechnol* **16**, 373–378.

Ooi, S.L., Pan, X., Peyser, B.D., Ye, P., Meluh, P.B., Yuan, D.S., *et al.* (2006) Global synthetic-lethality analysis and yeast functional profiling. *Trends Genet* **22**, 56–63.

Ott, K.H., Aranibar, N., Singh, B. and Stockton, G.W. (2003) Metabonomics classifies pathways affected by bioactive compounds. Artificial neural network classification of NMR spectra of plant extracts. *Phytochemistry* **62**, 971–985.

Palaniswamy, S.K., James, S., Sun, H., Lamb, R.S., Davuluri, R.V. and Grotewold, E. (2006) AGRIS and AtRegNet. A platform to link cis-regulatory elements and transcription factors into regulatory networks. *Plant Physiol* **140**, 818–829.

Palenchar, P., Kouranov, A., Lejay, L. and Coruzzi, G. (2004) Genome-wide patterns of carbon and nitrogen regulation of gene expression validate the combined carbon and nitrogen (CN)-signaling hypothesis in plants. *Genome Biol* **5**, R91.

Palmer, L.E., Rabinowicz, P.D., O'Shaughnessy, A.L., Balija, V.S., Nascimento, L.U., Dike, S., *et al.* (2003) Maize genome sequencing by methylation filtration. *Science* **302**, 2115–2117.

Parker, J.S., Cavell, A.C., Dolan, L., Roberts, K. and Grierson, C.S. (2000) Genetic interactions during root hair morphogenesis in Arabidopsis. *Plant Cell* **12**, 1961–1974.

Pilgrim, M.L., Caspar, T., Quail, P.H. and McClung, C.R. (1993) Circadian and light-regulated expression of nitrate reductase in *Arabidopsis*. *Plant Mol Biol* **23**, 349–364.

Popescu, S.C., Popescu, G.V., Bachan, S., Zhang, Z., Seay, M., Gerstein, M., *et al.* (2007) Differential binding of calmodulin-related proteins to their targets revealed through high-density Arabidopsis protein microarrays. *Proc Natl Acad Sci USA* **104**, 4730–4735.

Poultney, C.S., Gutierrez, R.A., Katari, M.S., Gifford, M.L., Paley, W.B., Coruzzi, G.M., *et al.* (2007) Sungear: interactive visualization and functional analysis of genomic datasets. *Bioinformatics* **23**, 259–261.

Price, J., Laxmi, A., St Martin, S.K. and Jang, J.C. (2004) Global transcription profiling reveals multiple sugar signal transduction mechanisms in Arabidopsis. *Plant Cell* **16**, 2128–2150.

Prusinkiewicz, P. and Rolland-Lagan, A.G. (2006) Modeling plant morphogenesis. *Curr Opin Plant Biol* **9**, 83–88.

Queitsch, C., Sangster, T.A. and Lindquist, S. (2002) Hsp90 as a capacitor of phenotypic variation. *Nature* **417**, 618–624.

Rabinowicz, P.D., McCombie, W.R. and Martienssen, R.A. (2003a) Gene enrichment in plant genomic shotgun libraries. *Curr Opin Plant Biol* **6**, 150–156.

Rabinowicz, P.D., Palmer, L.E., May, B.P., Hemann, M.T., Lowe, S.W., McCombie, W.R., *et al.* (2003b) Genes and transposons are differentially methylated in plants, but not in mammals. *Genome Res* **13**, 2658–2664.

Rabinowicz, P.D., Schutz, K., Dedhia, N., Yordan, C., Parnell, L.D., Stein, L., *et al.*

(1999) Differential methylation of genes and retrotransposons facilitates shotgun sequencing of the maize genome. *Nat Genet* **23**, 305–308.

Rahayu, Y.S., Walch-Liu, P., Neumann, G., Romheld, V., von Wiren, N. and Bangerth, F. (2005) Root-derived cytokinins as long-distance signals for NO_3^--induced stimulation of leaf growth. *J Exp Bot* **56**, 1143–1152.

Raper, C.D., Jr., Thomas, J.F., Tolley-Henry, L. and Rideout, J.W. (1988) Assessment of an apparent relationship between availability of soluble carbohydrates and reduced nitrogen during floral initiation in tobacco. *Bot Gaz* **149**, 289–294.

Ravasz, E., Somera, A.L., Mongru, D.A., Oltvai, Z.N. and Barabasi, A.L. (2002) Hierarchical organization of modularity in metabolic networks. *Science* **297**, 1551–1555.

Rawat, S.R., Silim, S.N., Kronzucker, H.J., Siddiqi, M.Y. and Glass, A.D. (1999) *AtAMT1* gene expression and NH4+ uptake in roots of *Arabidopsis thaliana*: evidence for regulation by root glutamine levels. *Plant J* **19**, 143–152.

Reinhardt, D., Frenz, M., Mandel, T. and Kuhlemeier, C. (2003) Microsurgical and laser ablation analysis of interactions between the zones and layers of the tomato shoot apical meristem. *Development* **130**, 4073–4083.

Rideout, J.W., Raper, C.D., Jr. and Miner, G.S. (1992) Changes in ratio of soluble sugars and free amino nitrogen in the apical meristem during floral transition of tobacco. *Int J Plant Sci* **153**, 78–88.

Roessner, U., Willmitzer, L. and Fernie, A.R. (2001) High-resolution metabolic phenotyping of genetically and environmentally diverse potato tuber systems. Identification of phenocopies. *Plant Physiol* **127**, 749–764.

Rutherford, S.L. and Lindquist, S. (1998) Hsp90 as a capacitor for morphological evolution. *Nature* **396**, 336–342.

Salathia, N. and Queitsch, C. (2007) Molecular mechanisms of canalization: Hsp90 and beyond. *J Biosci* **32**, 457–463.

Schauer, N. and Fernie, A.R. (2006) Plant metabolomics: towards biological function and mechanism. *Trends Plant Sci* **11**, 508–516.

Schauer, N., Steinhauser, D., Strelkov, S., Schomburg, D. and Allison, G. (2005) GC-MS libraries for the rapid identification of metabolites in complex biological samples. *FEBS Lett* **579**, 1332.

Scheible, W.R., Morcuende, R., Czechowski, T., Fritz, C., Osuna, D., Palacios-Rojas, N., *et al.* (2004) Genome-wide reprogramming of primary and secondary metabolism, protein synthesis, cellular growth processes, and the regulatory infrastructure of Arabidopsis in response to nitrogen. *Plant Physiol* **136**, 2483–2499.

Schena, M., Shalon, D., Davis, R.W. and Brown, P.O. (1995) Quantitative monitoring of gene expression patterns with a complementary DNA microarray. *Science* **270**, 467–470.

Schmid, M., Davison, T.S., Henz, S.R., Pape, U.J., Demar, M., Vingron, M., *et al.* (2005) A gene expression map of Arabidopsis thaliana development. *Nat Genet* **37**, 501–506.

Scholl, R.L., May, S.T. and Ware, D.H. (2000) Seed and molecular resources for Arabidopsis. *Plant Physiol* **124**, 1477–1480.

Schwender, J., Ohlrogge, J.B. and Shachar-Hill, Y. (2003) A flux model of glycolysis and the oxidative pentosephosphate pathway in developing brassica napus embryos. *J Biol Chem* **278**, 29442–29453.

Schwikowski, B., Uetz, P. and Fields, S. (2000) A network of protein-protein interactions in yeast. *Nat Biotechnol* **18**, 1257–1261.

Siepel, A., Farmer, A., Tolopko, A., Zhuang, M., Mendes, P., Beavis W, *et al.* (2001) ISYS:

a decentralized, component-based approach to the integration of heterogeneous bioinformatics resources. *Bioinformatics* **17**, 83–94.

Shannon, P., Markiel, A., Ozier, O., Baliga, N.S., Wang, J.T., Ramage, D., *et al.* (2003) Cytoscape: a software environment for integrated models of biomolecular interaction networks. *Genome Res* **13**, 2498–2504.

Sharma, V.K., Carles, C. and Fletcher, J.C. (2003) Maintenance of stem cell populations in plants. *Proc Natl Acad Sci USA* **100** (Suppl 1), 11823–11829.

Sollars, V., Lu, X., Xiao, L., Wang, X., Garfinkel, M.D. and Ruden, D.M. (2003) Evidence for an epigenetic mechanism by which Hsp90 acts as a capacitor for morphological evolution. *Nat Genet* **33**, 70–74.

Stelling, J., Sauer, U., Szallasi, Z., Doyle, F.J., III and Doyle, J. (2004) Robustness of cellular functions. *Cell* **118**, 675–685.

Thimm, O., Blasing, O., Gibon, Y., Nagel, A., Meyer, S., Kruger, P., *et al.* (2004) MAPMAN: a user-driven tool to display genomics data sets onto diagrams of metabolic pathways and other biological processes. *Plant J* **37**, 914–939.

Turing, A.M. (1952) The chemical basis of morphogenesis. *Philos Trans R Soc Lond B Biol Sci* **237**, 37–72.

Ueda, H.R. (2006) Systems biology flowering in the plant clock field. *Mol Syst Biol* **2**, 60.

Van Leene, J., Stals, H., Eeckhout, D., Persiau, G., Van Slijke, E., Van Isterdael, G., *et al.* (2007) A tandem affinity purification-based technology platform to study the cell cycle interactome in Arabidopsis thaliana. *Mol Cell Proteomics* **6**, 1226–1238.

Waddington, C.H. (1959) Canalization of development and genetic assimilation of acquired characters. *Nature* **183**, 1654–1655.

Wagner, A. and Fell, D.A. (2001) The small world inside large metabolic networks. *Proc Biol Sci* **268**, 1803–1810.

Walch-Liu, P., Filleur, S., Gan, Y. and Forde, B.G. (2005) Signaling mechanisms integrating root and shoot responses to changes in the nitrogen supply. *Photosynth Res* **83**, 239–250.

Walch-Liu, P., Neumann, G., Bangerth, F. and Engels, C. (2000) Rapid effects of nitrogen form on leaf morphogenesis in tobacco. *J Exp Bot* **51**, 227–237.

Walhout, A.J., Sordella, R., Lu, X., Hartley, J.L., Temple, G.F., Brasch, M.A., *et al.* (2000) Protein interaction mapping in C. elegans using proteins involved in vulval development. *Science* **287**, 116–122.

Wasinger, V.C. and Corthals, G.L. (2002) Proteomic tools for biomedicine. *J Chromatogr B Analyt Technol Biomed Life Sci* **771**, 33–48.

Watanabe, T., Hanan, J.S., Room, P.M., Hasegawa, T., Nakagawa, H. and Takahashi, W. (2005) Rice morphogenesis and plant architecture: measurement, specification and the reconstruction of structural development by 3D architectural modelling. *Ann Bot (Lond)* **95**, 1131–1143.

Watts, D. and Strogatz, S. (1997) Collective dynamic of 'small-world' networks. *Nature* **393**, 440–442.

Wei, H., Persson, S., Mehta, T., Srinivasasainagendra, V., Chen, L., Page, G.P., *et al.* (2006) Transcriptional coordination of the metabolic network in Arabidopsis. *Plant Physiol* **142**, 762–774.

Wheeler, D.B., Bailey, S.N., Guertin, D.A., Carpenter, A.E., Higgins, C.O. and Sabatini, D.M. (2004) RNAi living-cell microarrays for loss-of-function screens in Drosophila melanogaster cells. *Nat Methods* **1**, 127–132.

Wilkinson, M.D., Gessler, D., Farmer, A. and Stein, L. (2003) The BioMOBY project explores open-source, simple, extensible protocols for enabling biological database interoperability. *Proc Virt Conf Genom Bioinf* **3**, 16–26 (ISSN 1547-383X).

Wilkinson, M.D. and Links, M. (2002) BioMOBY: an open source biological web services proposal. *Brief Bioinform* **3**, 331–341.

Wilson, E. (2000) *Consilience: The Unity of Knowledge* (New York, NY: Aldred A. Knopf).

Winter, D., Vinegar, B., Nahal, H., Ammar, R., Wilson, G.V. and Provart, N.J. (2007) An 'electronic fluorescent pictograph' browser for exploring and analyzing large-scale biological data sets. *PLoS ONE* **2**, e718.

Wu, M.F., Tian, Q. and Reed, J.W. (2006) Arabidopsis microRNA167 controls patterns of *ARF6* and *ARF8* expression, and regulates both female and male reproduction. *Development* **133**, 4211–4218.

Wurtele, E., Li, L., Berleant, D., Cook, D., Dickerson, J.A., Ding, J., *et al.* (2007) MetNet: systems biology software for arabidopsis. In *Concepts in Plant Metabolomics*, B.J. Nikolau and E.S. Wurtele, eds (Heidelberg: Springer), 145–158.

Yates, J.R., III (2000) Mass spectrometry. From genomics to proteomics. *Trends Genet* **16**, 5–8.

Young, N.D., Cannon, S.B., Sato, S., Kim, D., Cook, D.R., Town, C.D., *et al.* (2005) Sequencing the genespaces of Medicago truncatula and Lotus japonicus. *Plant Physiol* **137**, 1174–1181.

Yu, H., Zhu, X., Greenbaum, D., Karro, J. and Gerstein, M. (2004) TopNet: a tool for comparing biological sub-networks, correlating protein properties with topological statistics. *Nucl Acids Res* **32**, 328–337.

Yuan, Y., SanMiguel, P.J. and Bennetzen, J.L. (2003) High-Cot sequence analysis of the maize genome. *Plant J* **34**, 249–255.

Zanzoni, A., Montecchi-Palazzi, L., Quondam, M., Ausiello, G., Helmer-Citterich, M. and Cesareni, G. (2002) MINT: a molecular INTeraction database. *FEBS Lett* **513**, 135–140.

Zeeberg, B., Feng, W., Wang, G., Wang, M., Fojo, A., Sunshine, M., *et al.* (2003) GoMiner: a resource for biological interpretation of genomic and proteomic data. *Genome Biol* **4**, R28.

Zhang, X., Clarenz, O., Cokus, S., Bernatavichute, Y.V., Pellegrini, M., Goodrich, J., *et al.* (2007) Whole-genome analysis of histone H3 lysine 27 trimethylation in Arabidopsis. *PLoS Biol* **5**, e129.

Zhang, X., Yazaki, J., Sundaresan, A., Cokus, S., Chan, S.W., Chen, H., *et al.* (2006) Genome-wide high-resolution mapping and functional analysis of DNA methylation in arabidopsis. *Cell* **126**, 1189–1201.

Zhong, S., Storch, K.F., Lipan, O., Kao, M.C., Weitz, C.J. and Wong, W.H. (2004) GoSurfer: a graphical interactive tool for comparative analysis of large gene sets in Gene Ontology space. *Appl Bioinformatics* **3**, 261–264.

Zhong, W. and Sternberg, P.W. (2006) Genome-wide prediction of C. elegans genetic interactions. *Science* **311**, 1481–1484.

Zhou, M. and Cui, Y. (2004) GeneInfoViz: constructing and visualizing gene relation networks. *In Silico Biol* **4**, 0026.

Zilberman, D., Gehring, M., Tran, R.K., Ballinger, T. and Henikoff, S. (2007) Genome-wide analysis of Arabidopsis thaliana DNA methylation uncovers an interdependence between methylation and transcription. *Nat Genet* **39**, 61–69.

Zimmermann, P., Hirsch-Hoffmann, M., Hennig, L. and Gruissem, W. (2004) GENEVESTIGATOR. Arabidopsis microarray database and analysis toolbox. *Plant Physiol* **136**, 2621–2632.

Annual Plant Reviews (2009) **35**, 41–66
doi: 10.1111/b.9781405175326.2009.00002.x

www.interscience.wiley.com

Chapter 2

AN OVERVIEW OF SYSTEMS BIOLOGY

Réka Albert[1,2] and Sarah M. Assmann[3]

[1] *Department of Physics, Pennsylvania State University, University Park, PA, USA*
[2] *Huck Institutes of the Life Sciences, Pennsylvania State University, University Park, PA, USA*
[3] *Department of Biology, Pennsylvania State University, University Park, PA, USA*

Abstract: This chapter provides an overview of three crucial aspects of systems biology: constructing biological networks, analyzing and modelling the structure of biological networks and modelling the dynamics of biological networks. We describe the types of intracellular networks most often studied, and the 'omic' information available to synthesize these networks, with a special focus on plant biology. We review the computational methods used to construct or infer (reverse engineer) intracellular networks. We present the graph-theoretical measures most useful for understanding the organization of biological networks, from the single node level to the global properties of the whole network. A representative sample of biological network models is provided, ranging from static models to dynamic models that incorporate how the status of the nodes changes in time. Throughout the chapter we focus on the biological predictions possible by combining experimental, theoretical and computational methods.

Keywords: systems biology; molecular networks; interactome; network inference; graph analysis; dynamic modelling

2.1 Systems theory and biology

It is increasingly recognized that in order to understand the dynamics and function of a cell, as well as higher levels of biological organization, we need to know: (1) the components that constitute it, (2) the relations and interactions of these components and (3) their dynamic behaviour, that is, how the biological entities behave over time under various conditions (Kitano, 2002). Ultimately, this information can be combined in a model that is not only consistent with current knowledge, but provides new insights and predictions. These topics and their integration is the purview of the field of systems biology.

The origins of systems biology can be traced back to the beginning of the twentieth century to the work of Bogdanov (1922, 1980), von Bertalanffy (1968) and others. Systems thinking is used in a variety of scientific and technological fields, and concepts such as independent and dependent variables, feedback, modularity, robustness, sensitivity and control have been extensively studied in the fields of systems theory (Francois, 1999; Weinberg, 1975) and control theory (Franklin *et al.*, 2002). Systems theory is a line of inquiry based on the assumptions that (i) all phenomena can be viewed as a web of relationships among elements (i.e. a system) and (ii) all systems, whether technological or biological, have common properties and behaviours and can be handled by a common set of methods. Control theory is extensively used in designing engineered systems, and it seeks to identify and modulate the mechanisms that systematically control the state of a system to minimize malfunctions and deviations from the optimal dynamic behaviour.

In the context of biology, Biochemical Systems Theory (Voit, 2000) and Metabolic Control Theory (Heinrich and Schuster, 1996), developed between 1965 and 1975, proposed general mathematical models of biological systems at and around a steady state (equilibrium). However, these and other attempts at systems level understanding of biology suffered from inadequate data on which to base their theories and models. It was the advent of genomic technologies that brought an abundance of data on system elements, interactions and states and enabled the integration of knowledge across different levels (molecular, cellular, tissue, organ, etc.) of biological organization.

2.2 Graph elements and network attributes

Every biological system includes a network of interactions, and according to our definition, all systems biology includes network analysis. In some cases, the structure of the network and its interpretation are straightforward (e.g. a linear chain of interactions), while in other cases, more detailed analysis is required to understand how information can propagate through the network. Therefore, this chapter begins with some definitions of commonly used terms and measures in network analysis and graph theory. Many of these terms are illustrated in the hypothetical network depicted in Fig. 2.1.

A *network* (or graph) (Bollobas, 1979) is used to represent a system of elements that interact. A graph has two basic parts: the elements of the system are portrayed as *graph nodes* (also called vertices) and the interactions are portrayed as *edges*, that is, lines connecting pairs of nodes. Multi-partite graphs contain different classes of node (such as mRNA and protein). Directed edges emanate from a source (starting node) to a sink (ending node) and represent unidirectional flow of material or information. Non-directed edges are used to represent mutual interactions, for example, physical interaction between two proteins of unknown function, or interactions where the directional flow of information is not known. Signs representing activation or inhibition, or

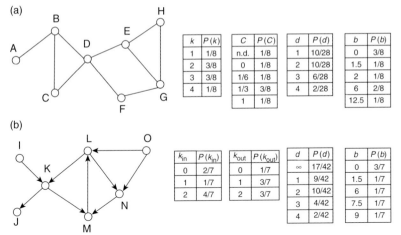

(a)

k	P(k)		C	P(C)		d	P(d)		b	P(b)
1	1/8		n.d.	1/8		1	10/28		0	3/8
2	3/8		0	1/8		2	10/28		1.5	1/8
3	3/8		1/6	1/8		3	6/28		2	1/8
4	1/8		1/3	3/8		4	2/28		6	2/8
			1	1/8					12.5	1/8

(b)

k_{in}	$P(k_{in})$		k_{out}	$P(k_{out})$		d	P(d)		b	P(b)
0	2/7		0	1/7		∞	17/42		0	3/7
1	1/7		1	3/7		1	9/42		1.5	1/7
2	4/7		2	3/7		2	10/42		6	1/7
						3	4/42		7.5	1/7
						4	2/42		9	1/7

Figure 2.1 Hypothetical networks illustrating graph terminology. The node *degree* (*k*) quantifies the number of edges that start at or end in a given node; for example, node K has both an in-degree and an out-degree of two. The *clustering coefficient* (*C*) characterizes the cohesiveness of the first neighbourhood of a node; for example, the clustering coefficient of node F is zero because its two first neighbours are not connected, and the clustering coefficient of node C is 1, indicating that it is a part of the BCD clique. The *graph distance* (*d*) between two nodes is defined as the number of edges in the shortest path between them. For example, the distance between nodes L and K is 1, the distance between nodes K and L is 2 (along the KML path) and the distance between nodes J and I is infinite because no path starting from J and ending in I exists. The *betweenness centrality* (*b*) of a node quantifies the number of shortest paths in which the node is an intermediary (not beginning or end) node. For example, the betweenness centrality of node C is zero because it is not contained in any shortest paths that do not start or end in C, and the betweenness centrality of node G is $1\frac{1}{2}$ because it is an intermediary in the FGH shortest path and in one of two alternative shortest paths between E and F (EGF and EDF). The *degree distribution* P(k) (P(k_{in}) and P(k_{out}) in directed networks) quantifies the fraction of nodes with degree k. For example, in panel (a), one node (A) has a degree of 1; three nodes (C, F, H) have a degree of 2, three nodes (B, E, G) have a degree of 3 and one node (D) has a degree of 4; the corresponding fractions are obtained by dividing by the total number of nodes (eight). The *clustering coefficient distribution* P(C) denotes the fraction of nodes with clustering coefficient C. For example, in panel (a), one node (A) has an undefined clustering coefficient because it only has a single first neighbour, one node (F) has a clustering coefficient of zero, one node (D) has a clustering coefficient of $\frac{1}{6}$, three nodes (B, D, G) have a clustering coefficient of 1/3 and two nodes (C, H) have a clustering coefficient of 1. The *distance distribution* P(d) denotes the fraction of node pairs having the distance d. The *betweenness centrality distribution* P(b) quantifies the fraction of nodes with betweenness centrality b. (a) This undirected graph is *connected*, has a range of degrees from 1 to 4, clustering coefficients between zero and 1, a range of pairwise distances from 1 to 4 and node betweenness centralities between zero and $12\frac{1}{2}$. The BCD and FGH *subgraphs* are *cliques* (completely connected subgraphs) of three nodes. (b) This directed graph contains two source nodes (I and O, both with $k_{in} = 0$), one sink node (J, with $k_{out} = 0$), one feed-forward loop (OLN; both O and L feed into N) and two feedback loops (MLN and MLK). The nodes K, L, M and N form the graph's *strongly connected* subgraph. The *in-component* of this subgraph contains the *source* nodes I and O, while its *out-component* consists of the *sink* node J.

weights indicating confidence levels or strength can be imposed on edges to enhance the information content of the network.

The organizational features of interaction graphs can be quantified by network measures denoting the importance (centrality) of individual nodes, the connectivity (reachability) among nodes and the homogeneity or heterogeneity of the network in terms of a given property. Three of the most often used network measures, in an increasing order of locality, are the node degree, the clustering coefficient and the path length.

The *degree* of a node is the number of edges pointing towards or emanating from that node (Fig. 2.1). A node's total degree is the sum of its *in-degree* and *out-degree*, which respectively quantify the number of incoming and outgoing edges of the node. In a weighted graph, one can also define a *node strength*, the sum of the weights of the edges into and out from the node. The local information on the degree of each node can be combined to yield a global description of the network known as a *degree distribution*, $P(k)$, which gives the fraction of nodes with degree k. In directed networks one can similarly define in- and out-degree distributions. A large number of cellular networks have been shown to be *scale free*, meaning that there are many different node degrees, such that one cannot validly describe the network in terms of a 'typical' node degree (reviewed in Albert and Barabási, 2002). Scale-free networks are characterized by a degree distribution that is close to a power law: $P(k) \cong Ak^{-\gamma}$, where A is a normalization constant, and where the degree exponent is typically $2 < \gamma < 3$ (Albert and Barabási, 2002).

The first *neighbourhood* of a node consists of the nodes connected to it by a single edge, and the edges among those nodes (if any). If this neighbourhood is a completely connected subgraph, it is known as a *clique*. The *clustering coefficient* of a node is the ratio of the number of edges among the first neighbours of the node and the number of edges among them if the node's first neighbourhood were a clique (Watts and Strogatz, 1998). Thus a clique has a clustering coefficient of 1; conversely, when there are no edges among the first neighbours, the clustering coefficient is zero. Large average clustering coefficients have been observed for protein–protein interaction networks (Yook et al., 2004) and metabolic networks (Wagner and Fell, 2001) indicating topological redundancy and biological cohesiveness.

Two nodes of a graph are *connected* if they are linked by a sequence of adjacent nodes and edges, a path (Bollobás, 1979). For example, a path could signify a biosynthetic pathway, or it could represent a cascade of events in a signal transduction chain. The *distance* (*path length*) between any two nodes in a network is the number of edges in the shortest path connecting those nodes (Fig. 2.1). In a *weighted network*, the distance between two nodes is the sum of edge weights along the path for which this sum is a minimum (Dijkstra, 1959). Many networks' average path length scales with the natural logarithm of the number of nodes, $d \sim \ln(N)$, indicating that path lengths of even very large networks remain small. This *small world* (Watts and Strogatz, 1998) property has been observed for metabolic, protein interaction and signal transduction networks (Jeong et al., 2000; Yook et al., 2004; Ma'ayan *et al.*, 2005) and facilitates rapid spread of information in response to inputs. Another important global property related to paths is *path redundancy*, or the availability of multiple paths between a pair of nodes (Papin and Palsson, 2004). Either by allowing multiple channels of information from input to output or as alternate routes when the preferred pathway is disrupted, path redundancy promotes the robust functioning of cellular networks by reducing reliance on individual pathways.

Networks having paths between every pair of nodes are *connected*. A directed network can be *strongly connected* if all of its node pairs are connected in both directions; alternatively, the network can have one or several strongly connected subgraphs. Each strongly connected subgraph is associated with an *in-component* (nodes that can reach the strongly connected subgraph, but that cannot be reached from it) and an *out-component* (the converse) (Fig. 2.1). The nodes of each subgraph may share a specific task within a given network. In signal transduction networks, for example, the nodes of the in-component tend to be involved in ligand sensing; the nodes of the strongly connected subgraph form a central signalling subnetwork; and the nodes of the out-component are responsible for the transcription of target genes, or for phenotypic outcomes (Ma'ayan *et al.*, 2004, 2005).

The number, directionality and strength of connections associated with a given node can be synthesized into measures of that node's *centrality*. The sources (nodes with only outgoing edges) and sinks (nodes with only incoming edges) of a directed network represent initial and terminal points of the flow of material or information. In a metabolic network describing a biosynthetic pathway, for example, the initial precursor is the source, and the final product is the sink. For nodes other than sources and sinks, the *betweenness centrality* – the number of shortest paths from node s to node t passing through the node, divided by the total number of shortest st-paths (Fig. 2.1) – indicates the importance of that node to the flow of information or materials through the network (Anthonisse, 1971; Freeman, 1977). Betweenness centrality is often, but not obligately, correlated with degree. For example, in metabolic networks of microorganisms, the most ubiquitous substrates tend to have the highest betweenness centralities but not the highest degrees (Holme *et al.*, 2003), and some low-degree metabolites are as critical to the overall network function as high-degree metabolites (Mahadevan and Palsson, 2005).

In addition to the general graph concepts and measures used to quantify the organization of biological networks, a number of specific terms are often invoked to reflect their functional constraints.

2.2.1 Hubs

In scale-free networks, small-degree nodes are most common, and the node degrees are highly heterogeneous such that the highest degree nodes have degrees that are orders of magnitude higher than the average degree. Such highest degree nodes are commonly referred to as *hubs*, although there is no explicit definition of the boundary between hubs and non-hubs.

Consequently in scale-free networks random node disruptions do not lead to a major loss of connectivity, whereas the loss of the hubs causes the breakdown of the network into isolated clusters (Albert and Barabási, 2002). This point has been experimentally verified in *Saccharomyces cerevisiae*, where the severity of a gene knockout has been shown to correlate with the number of interactions in which the gene's products participate (Jeong *et al.*, 2001; Said *et al.*, 2004). Indeed, as much as 73% of the *S. cerevisiae* genes are non-essential, that is, the knockout has no phenotypic effects (Giaever *et al.*, 2002), and this confirms cellular networks' robustness in the face of random disruptions. Conversely, the likelihood that a gene is essential (i.e. that its knockout is lethal) is greater for high-degree nodes (Jeong *et al.*, 2001; Said *et al.*, 2004). This confirms the intuitive prediction that cell viability is more likely to be compromised by loss of highly interactive nodes such as hubs.

2.2.2 Modularity

Cellular networks are expected to be decomposable to subnetworks (pathways) corresponding to specific biological functions (Hartwell *et al.*, 1999). These subnetworks or modules should be distinguishable within interaction networks by the fact that they have dense intra-module connectivity but sparse inter-module connectivity. Methods to identify functional modules are based on the physical location or function of network components (Rives and Galitski, 2003), the topology of the interaction network (Spirin and Mirny, 2003; Guimera and Amaral, 2005), or the evolutionary conservation of the nodes (von Mering *et al.*, 2003; Sharan *et al.*, 2005). The demonstrated overlap and crosstalk between pathways (Han *et al.*, 2004) present a challenge to module-detecting algorithms, as is the observation of hierarchical modularity, in which modules are made up of smaller and more cohesive modules (Ravasz *et al.*, 2002).

2.2.3 Motifs

Cellular networks contain recurring interaction motifs, which are small subgraphs that have well-defined topologies. Interaction motifs such as autoregulation (usually a negative feedback) and feed-forward loops have a higher abundance in transcriptional regulatory networks than expected based on the degree distribution alone (Shen-Orr *et al.*, 2002; Balázsi *et al.*, 2005). In general, protein interaction motifs such as small cliques are both abundant (Giot *et al.*, 2003) and evolutionarily conserved (Wuchty *et al.*, 2003), partly because many of them represent subgraphs of protein complexes. Feed-forward loops, positive and negative feedback loops and triangles of scaffolding (protein) interactions are over-represented in signal transduction networks (Ma'ayan *et al.*, 2005). Interaction motifs are proposed to form functionally separable building blocks of cellular networks (Mangan and Alon, 2003). For example, the abundance of negative feedback loops in the early steps of signal transduction networks and of positive feedback loops at later steps suggest mechanisms to filter weak or short-lived signals and to amplify strong and persistent signals (Ma'ayan *et al.*, 2005).

2.3 Building biological networks: identifying nodes and mapping interactions

Genome-level information concerning cellular networks is often described using five 'omes': genome, transcriptome, proteome, metabolome and interactome. During the last decade, the respective omics approaches have produced an incredible quantity of expression and interaction data, providing extensive, albeit still incomplete, knowledge regarding the nodes and edges of biological networks (Pandey and Mann, 2000; Caron *et al.*, 2001).

2.3.1 Identifying nodes

Transcriptome data convey the identity of each expressed gene and its level of expression for a given cell type, tissue, organ or organism. High-throughput mRNA data are obtained by serial analysis of gene expression (Wang, 2007), microarrays (DeRisi *et al.*, 1997; Cho *et al.*, 1998), massively parallel signature sequencing (Reinartz *et al.*, 2002; Rensink and Buell, 2005) and 454 sequencing (Kanehisa and Goto, 2000; Margulies *et al.*, 2005) among other methods. For model plant species such as Arabidopsis, much of the microarray information has been compiled in databases such as AtGenExpress at The Arabidopsis Information Resource (TAIR) and in NASCArrays at the European Arabidopsis Stock Centre (NASC), as well as at the Gene Expression Omnibus (GEO) site at NCBI (see Table 2.1 for a summary of the websites mentioned in this chapter). These databases provide genome-wide information on transcript levels in individual tissues or organs, or under specific stress or developmental conditions, and thus provide information on which transcripts are co-expressed and potentially co-regulated. Genevestigator and the Botany Array Resource (BAR) both provide pictorial summaries of spatial patterns of gene expression.

Information about transcription factor binding motifs is available from the Transcription Factor Database (TRANSFAC) (Wingender *et al.*, 1996), the Regulon Database (RegulonDB) (Huerta *et al.*, 1998) and the Kyoto Encyclopedia of Genes and Genomes (KEGG). To identify novel motifs, one can download promoter sequences from TAIR and evaluate them using motif-finding algorithms such as AlignACE (Hughes *et al.*, 2000), MEME (Grundy *et al.*, 1996), MotifSampler (Aerts *et al.*, 2003) and Weeder (Pavesi *et al.*, 2004), the latter two of which provide the option of using background models derived from Arabidopsis-specific datasets.

Proteome data provide information on global protein expression patterns in the sample of interest. The multitude of possible post-translational modifications dictates that the complexity of the proteome is much greater than that of the transcriptome, and proteomics methods are still rapidly evolving (Glinski and Weckwerth, 2006). In addition, currently there are fewer proteomics studies than transcriptomics studies (Rossignol *et al.*, 2006). For these reasons, proteo͏͏ ͏ ͏ ͏ein data (as opposed to predict͏ ͏complete, and this is

Table 2.1 URL addresses of databases cited in this chapter

Database	URL
The Arabidopsis Information Resource (TAIR)	www.arabidopsis.org
The European Arabidopsis Stock Centre	http://arabidopsis.info
AtGenExpress	http://www.arabidopsis.org/info/expression/ATGenExpress.jsp
NASCArrays	http://affymetrix.arabidopsis.info/narrays/experimentbrowse.pl
Gene Expression Omnibus (GEO)	http://www.ncbi.nlm.nih.gov/geo
Arabidopsis Gene Expression Database	http://www.arexdb.org
Genevestigator	https://www.genevestigator.ethz.ch/at
Botany Array Resource	http://bbc.botany.utoronto.ca
TRANSFAC	http://www.gene-regulation.com
RegulonDB	http://regulondb.ccg.unam.mx
Kyoto Encyclopedia of Genes and Genomes (KEGG)	http://www.genome.jp/kegg
Aligns Nucleic Acid Conserved Elements (AlignACE)	http://arep.med.harvard.edu/mrnadata/mrnasoft.html
MEME	http://meme.sdsc.edu/meme/meme.html
MotifSampler	http://homes.esat.kuleuven.be/~thijs/BioDemo/MotifSampler.html
Weeder	http://159.149.109.16:8080/weederWeb/howto.html
ExPASy	http://ca.expasy.org
GABIPD	http://gabi.rzpd.de/projects/Arabidopsis_Proteomics
SUBA	http://www.suba.bcs.uwa.edu.au
MetaCyc	http://metacyc.org
AraCyc	http://www.arabidopsis.org/biocyc/index.jsp
MoTo Database	http://appliedbioinformatics.wur.nl
The Golm Metabolome Database	http://csbdb.mpimp-golm.mpg.de/csbdb/gmd/gmd.html
Metabolomics Fiehn Lab	http://fiehnlab.ucdavis.edu/projects
Database of Interacting Proteins	http://dip.doe-mbi.ucla.edu
Munich Information Center for Protein Sequences	http://mips.gsf.de
Human Protein Reference Database	http://www.hprd.org
Search Tool for the Retrieval of Interacting Proteins	http://string.embl.de

especially true for multicellular organisms. Plant proteomics databases are available at NASC, at the Arabidopsis page in the ExPASy (Expert Protein Analysis System) proteomics server of the Swiss Institute of Bioinformatics, and at GABIPD, although they are currently sparsely populated. The Subcellular Proteomic database (SUBA) hosts information on subcellular localization of plant proteins, based on GFP tagging and proteomics methods. Other, more specific databases, are described in (Komatsu, 2005).

Metabolome data. Metabolome analysis includes the study of both metabolites and the enzymes that catalyze their production, and methods for metabolome analysis include gas chromatography–mass spectrometry, liquid chromatography–mass spectrometry and nuclear magnetic resonance (Hall, 2006). Information on metabolic pathways is available at KEGG and MetaCyc. It has been estimated that plants produce between 100 000 and 200 000 secondary metabolites (Oksman-Caldentey and Inze, 2004). Primary and secondary metabolites have highly diverse functions, for example as enzyme cofactors, hormones and signalling agents. Although only a fraction of all the plant metabolites has been characterized, many of the known secondary metabolites have important functions in human biology, serving as medicines, flavourings, perfumes, colorants, etc. A metabolic database for Arabidopsis is available at AraCyc in TAIR and a database for the secondary metabolites of tomato fruits, MoTo DB, was recently developed (Moco *et al.*, 2006). Metabolite databases with an emphasis on primary metabolites are available at the Golm Metabolome Database and at http://fiehnlab.ucdavis.edu/projects.

2.3.2 Mapping interactions

Advancements in molecular biology techniques have increased the resolution, type and scale of interactions that can be detected. These improvements promise to change the focus of biology from an understanding of local, binary interactions to an understanding of the integration of these interactions into a functional system. At the genome level, transcription factors and chromatin structure promote or inhibit gene transcription. Since transcription factors and chromatin modulating agents are themselves products of genes, the ultimate effect is that genes regulate each other's expression as part of gene-regulatory networks. Proteins participate in diverse post-translational processes as well as binding to form protein scaffolds, modulate subcellular localization or alter enzyme activity; the totality of these processes is called a protein–protein interaction network. The biochemical reactions of cellular metabolism can likewise be integrated into a metabolic network whose fluxes are regulated by enzymes catalyzing the metabolic reactions. Often these different interaction types are intertwined, as occurs when an external signal triggers a cascade of interactions that involve biochemical reactions and metabolite fluxes, as well as transcriptional regulation.

Transcriptional regulatory maps link two types of nodes – transcription factors and mRNAs – and have two types of directed edges, corresponding to transcriptional regulation and translation (Lee *et al.*, 2002), where the regulatory edges can have two types of signs, corresponding to activation or repression. Transcriptional regulatory maps exist for *Escherichia coli* (Shen-Orr *et al.*, 2002) and *S. cerevisiae* (Guelzim *et al.*, 2002; Lee *et al.*, 2002; Luscombe *et al.*, 2004). A given transcription factor usually regulates multiple genes, and this is reflected in the approximately scale-free out-degree distribution of these maps. Unidirectional regulation is prevalent, thus these networks do not have large strongly connected components, in contrast to protein interaction maps (see below).

In Arabidopsis, a transcriptional regulatory map has been created for cold signalling mediated by the ICE1 transcription factor (Benedict *et al.*, 2006). The source–sink distances are small (4–5 edges) in this network as well as in the *E. coli* and yeast networks, suggesting that this may be a common feature of transcriptional regulatory maps.

In *protein interaction graphs*, the nodes are proteins, and two proteins are connected by a directed edge if the direction of information flow during their interaction is known. Two proteins are connected by a non-directed edge if there is strong evidence of their physical interaction or association without, however, evidence for a directionality of interaction. Non-directional protein interaction networks are commonly generated by assays of interaction in yeast-based systems (yeast two hybrid and split ubiquitin assays), and in assessment of co-immunoprecipitation and mass spectrometry-based identification of the composition of protein complexes. There are many protein interactome databases, including the Database of Interacting Proteins (DIP) (Xenarios *et al.*, 2002), yeast and mammalian protein interaction databases at the Munich Information Center for Protein Sequences (MIPS) (Mewes *et al.*, 2004) and the Human Protein Reference Database (HPRD) (Peri *et al.*, 2004). Plant-specific global interaction databases derived from wet bench data were not available at the time of this writing. However, the Search Tool for Interacting Proteins (STRING) will, upon user query, output predicted interactions based on orthology, as well as known interactions.

Protein–protein interaction maps have been constructed for a variety of prokaryotes and eukaryotes (McCraith *et al.*, 2000; Uetz *et al.*, 2000; Rain *et al.*, 2001; Giot *et al.*, 2003; Li *et al.*, 2004; Rual *et al.*, 2005). Although these maps are incomplete, and some assays for protein interaction have a high rate of false positives (Hart *et al.*, 2006), extant maps nevertheless are revealing common attributes, including an approximately scale-free degree distribution (Jeong *et al.*, 2001; Giot *et al.*, 2003; Yook *et al.*, 2004) and a large connected subgraph with short distances (Giot *et al.*, 2003; Yook *et al.*, 2004). This latter finding suggests why pleiotropy is commonly observed, since perturbations of a single gene or protein can propagate through the network, and have seemingly unrelated effects.

In plants, interaction maps have been experimentally defined for homo- and heterodimerization within two large classes of transcription factors: the MADS (**M**CM1, **A**gamous, **D**eficiens, **S**RF) box transcription factors (Immink *et al.*, 2003; de Folter *et al.*, 2005) and the MYB (myeloblastosis)

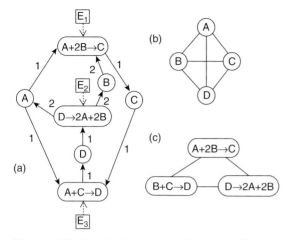

Figure 2.2 Hypothetical metabolic network illustrating different possible representations. (a) Directed and weighted tripartite graph representation whose three types of node are metabolites (circles), reactions (ovals) and enzymes (squares), and whose two types of edge represent mass flow (solid lines) and catalytic regulation (dashed lines), respectively. Mass flow edges connect reactants to reactions and reactions to products, and are marked by the stoichiometric coefficients of the metabolites; enzymes catalyzing the reactions are connected by regulatory edges to the nodes signifying the reaction. (b) Metabolite (substrate) graph, whose nodes are metabolites, joined by a non-directed edge if they occur in the same chemical reaction (Wagner and Fell, 2001). (c) Reaction graph, whose nodes are reactions, connected by a non-directed edge if they share at least one metabolite.

transcription factor family (Zimmermann *et al.*, 2004). In addition, based on known protein–protein interactions in other species, interaction of the homologous proteins in Arabidopsis has been predicted on a global scale (Yu *et al.*, 2004a). A database of these 'interologs' is available at http://interolog.gersteinlab.org/.

Metabolic networks link two types of nodes, metabolites and enzymes, by edges that represent enzyme-catalyzed chemical reactions. An idealized metabolic network and its simplified representations as a metabolite graph or a reaction graph (Wagner and Fell, 2001) are given in Fig. 2.2. Because some nodes of metabolic networks, for example ATP (adenosine triphosphate), participate in a multitude of reactions while other nodes, for example most enzymes, participate only in one specific reaction (Franklin *et al.*, 2002), there is a range of node degrees in metabolic networks (Arita, 2004), resulting in broad-tailed or scale-free degree distributions (Jeong *et al.*, 2000; Wagner and Fell, 2001; Tanaka, 2005). The path length of known metabolic networks is small (Jeong *et al.*, 2000; Wagner and Fell, 2001) although its value depends on the network representation used (Ma and Zeng, 2003).

In plants, metabolite profiling has revealed characteristic metabolomes of different Arabidopsis ecotypes (Keurentjes *et al.*, 2006) and has illustrated that metabolite signatures can be used in a predictive fashion to identify recombinant inbred Arabidopsis lines that accumulate biomass more rapidly (Meyer *et al.*, 2007). Another major emphasis of plant metabolomics currently is in the use of combined metabolomic and transcriptomic datasets (Oksman-Caldentey and Saito, 2005; Glinski and Weckwerth, 2006). Maps have been created, for example, for carbon nitrogen regulatory networks (Gutierrez *et al.*, 2007) and responses to sulfur and nitrogen deficiency (Hirai *et al.*, 2004). A combination of metabolomics and transcriptomics has also been used to predict functions of previously uncharacterized genes involved in metabolic processes, including sulfur metabolism (Hirai *et al.*, 2005), anthocyanin synthesis (Tohge *et al.*, 2005) and volatile formation (Guterman *et al.*, 2002; Goossens *et al.*, 2003; Mercke *et al.*, 2004).

Signal transduction networks connect signals to effectors with (typically) directed edges, and can encompass all the node types discussed above: RNAs, proteins and metabolites. The largest reconstructed signal transduction network to date, for a brain neuronal cell type, consists of 545 nodes and 1259 interactions (Ma'ayan *et al.*, 2005); yet the average source-to-sink path length is only four, suggesting a rapid response capability, which is also supported by the observation that 60% of the nodes are strongly connected. By contrast, the largest reconstructed cellular signal transduction network to date for plants, for stomatal closure induced by the plant hormone abscisic acid, has only ∼40 nodes (Li *et al.*, 2006), indicating that plant biologists are still amassing the types of information required to build comprehensive cellular network models.

Finally, information on gene co-expression (Stuart *et al.*, 2003), gene co-occurrence (Valencia and Pazos, 2002) or genetic interactions (Tong *et al.*, 2004) can be used to construct *networks of gene functional relationships*. For example, such analyses in Arabidopsis have yielded information on co-regulation of genes involved in cell wall biosynthesis (Persson *et al.*, 2005) and primary and secondary metabolism (see above references). Genetic interactions and functional relationships are often complementary and only sometimes overlapping with physical interactions (Tong *et al.*, 2004). Functional relationships and associations between genes are often inferred from gene expression information, and methods of inference are briefly reviewed in the next section.

2.4 Building biological networks: computational methods for network inference

Computational inference (also referred to as reverse engineering) is an approach to infer causal relationships within and between the transcriptome, proteome and metabolome when direct experimental determination of these causal relationships has not been performed.

Data-mining schemes typically extract relationships between two entities based on their statistical co-occurrence, for example their shared inclusion in journal articles (Marcotte *et al.*, 2001; Stapley and Benoit, 2000). Algorithms of this nature have been used extensively to infer protein–protein interactions based on their genes' co-occurrence in the same chromosomal neighbour-hood (Snel *et al.*, 2000), shared evolutionary pattern (Pellegrini *et al.*, 1999) or co-expression. Search tools such as the Search Tool for the Retrieval of Inter-acting Proteins (STRING) employ data-mining methods for the inference of protein–protein interactions in eukaryotes and prokaryotes (Snel *et al.*, 2000; von Mering *et al.*, 2006).

Inference of functional relationships among gene products based on their mRNA, protein or metabolite expression profiles frequently invokes *statistical methods* such as clustering (Qian *et al.*, 2001; Dougherty *et al.*, 2002) and Bayesian networks (Drawid and Gerstein, 2000; Jansen *et al.*, 2003). *Clustering* aims to find groups of genes that respond in a similar manner to vary-ing conditions, and that might therefore be co-regulated (Alon *et al.*, 1999; Qian *et al.*, 2001; Gargalovic *et al.*, 2006; Horvath *et al.*, 2006). In clustering algorithms, genes or proteins with statistically similar expression profiles (Qian *et al.*, 2001) are grouped using hierarchical clustering algorithms (Wen *et al.*, 1998), self-organizing maps (Toronen *et al.*, 1999), K-mean clustering (Tavazoie *et al.*, 1999) or topological measures of the inferred networks (Gupta *et al.*, 2006; Horvath *et al.*, 2006). Because of the strong evidence of correlations between co-expression of a pair of genes and interactions among the pair of proteins encoded by these two genes (Ge *et al.*, 2001; Qian *et al.*, 2001), clus-tering methods have also been used to augment protein–protein interaction networks (Ge *et al.*, 2001). *Bayesian networks* aim to infer a directed, acyclic graph that summarizes the dependency relationships among variables in the system, and a set of local joint probability distributions that statistically convey these relationships (Friedman *et al.*, 2000). The initial links of the dependency graph are established either randomly or based on prior knowl-edge, and the network is refined by an iterative search-and-score algorithm in which multiple candidate networks are scored against experimental ob-servations and against one another (Yu *et al.*, 2004b). Bayesian networks have been employed to sort yeast proteins into functional groupings (Drawid and Gerstein, 2000) and to infer protein interactions in the *S. cerevisiae* cell cycle (Friedman *et al.*, 2000).

Several methods proposed for the *inference of gene-regulatory networks* from time-course gene expression data seek to relate the change in the expression level of a given gene with the levels of other genes' transcripts in the network by describing it as a differential equation (Chen *et al.*, 1999; Gupta *et al.*, 2005) or a discrete relationship (Smolen *et al.*, 2000; Shmulevich *et al.*, 2002). The resulting system of equations is typically underdetermined because there are more unknowns than experimental time points; for this reason additional assumptions (e.g. maximum parsimony) are also invoked (Gupta *et al.*, 2005). Deterministic methods based on systems of linear differential equations have, for example, been used to infer gene-regulatory networks in *Bacillus subtilis* (Gupta *et al.*, 2005) and in the rat central nervous system (Chen *et al.*, 1999). De-terministic Boolean methods approximate gene expression levels with binary variables (e.g. by using a threshold), and describe gene regulation with logical functions (using the Boolean operators 'and', 'or' and 'not') (Liang *et al.*, 1998;

Shmulevich *et al.*, 2002). Each gene's logical function is found by a systematic search for the minimum set of regulator nodes whose combined expressions explain the experimentally observed state of the gene. Probabilistic Boolean methods (Shmulevich *et al.*, 2002; Dougherty and Shmulevich, 2003) incorporate uncertainty and fluctuations in expression levels by assigning several alternative logical functions to each gene (Shmulevich *et al.*, 2002; Dougherty and Shmulevich, 2003); a machine learning algorithm (Dougherty *et al.*, 2000; Dougherty and Shmulevich, 2003) then selects the most probable logical function at each time point. This method was used to infer the regulatory networks involved in embryonic segmentation and muscle development in *Drosophila melanogaster* (Zhao *et al.*, 2006).

Metabolic pathway reconstruction from known stoichiometric information is usually performed by constraint-based deterministic methods (Famili *et al.*, 2003) such as Flux Balance Analysis (Famili and Palsson, 2003; Reed *et al.*, 2003) or S-systems (Irvine and Savageau, 1990). Recently, a linear optimization strategy that first selects a subset of a predetermined set of possible metabolic reactions, and then optimizes the metabolic flux distribution, was proposed (Burgard *et al.*, 2003) and used to find the changes in an *E. coli* genome-scale metabolic model that are needed to minimize the discrepancy between model predictions and experimentally measured flux data (Herrgard *et al.*, 2006a).

The majority of network inference methods presented in this section use node-level (expression) information to infer causal relationships. There also exists a complementary approach of *reverse inference*: inferring interactions from indirect causal relationships. Indeed, experimental information about the involvement of a protein in a process is often restricted to evidence of differential responses to a stimulus in the wild-type organism versus an organism in which the respective protein's expression or activity is disrupted. These observations can be incorporated as two intersecting paths (denoting the stimulus-response and protein-response indirect relationships) in an incompletely mapped interaction network. The inference algorithm must integrate indirect and direct evidence to find a network consistent with all experimental observations (Li *et al.*, 2006). This inference approach is incorporated in the software NET-SYNTHESIS (Albert *et al.*, 2007; Kachalo *et al.*, 2008), and we expect it will play an increasing role when integrating information from disparate data sources.

2.5 Biological network models: data integration

Once a network has been derived or inferred, it is encapsulated in a graph. The graph measures and analysis techniques described in Section 2.2 can then be used to quantify the global connectivity (reachability) among nodes, the densely interconnected clusters (modules) and the importance (centrality) of individual nodes. These graph measures, alone or combined with additional information regarding the network nodes (such as the functional annotation of the corresponding genes/proteins), provide testable biological predictions on several scales, from single interactions to functional modules.

The predictive power of biological network reconstructions can be substantially enhanced by integrating several types of interactions and functional associations. Composite networks superpose protein–protein and protein–DNA interactions (Yeger-Lotem *et al.*, 2004), protein–protein interactions, genetic interaction, transcriptional regulation, sequence homology and expression correlation (Zhang *et al.*, 2005) or metabolic reactions and transcriptional regulation of metabolic genes (Covert *et al.*, 2001; Herrgard *et al.*, 2006b). Alternatively, gene–gene linkages can be defined as probabilistic summaries of physical and functional associations such as protein interaction, mRNA co-expression, synthetic lethal interactions and comparative genomics (Lee *et al.*, 2004).

The *functions of unannotated proteins* can be inferred on the basis of the annotation of their first neighbours in the protein interaction network (Vazquez *et al.*, 2003; Lee *et al.*, 2004). *New protein interactions* can be predicted based on the presence of conserved interaction motifs within the network (Albert and Albert, 2004). New protein functions and interactions can be inferred through global alignment between protein interaction networks in different species (Kelley *et al.*, 2004). Conversely, the relatively well-mapped protein interaction networks of *S. cerevisiae* and *D. melanogaster* allowed the determination of the closest pair of functionally orthologous proteins, one in *S. cerevisiae* and one in *Drosophila*, by modelling the orthology relation as a probabilistic function of the orthology relations of the immediate network neighbours of each member of the pair (Bandyopadhyay *et al.*, 2006).

The *S. cerevisiae* multi-networks allowed the identification of *regulatory themes* supported by several data types such as two interacting transcription factors regulating the same target gene or one transcription factor regulating several genes whose protein products are members of the same protein complex (Yeger-Lotem *et al.*, 2004; Zhang *et al.*, 2005). There is great interest in identifying high confidence (multi-)network subgraphs corresponding to components working towards a particular cellular function or within a common *pathway*. Frequently encountered subgraphs include linear or branching paths of interaction (suggestive of pathways), densely connected clusters (suggestive of functional protein complexes) and parallel clusters in which the proteins in one cluster are associated with the proteins in the other cluster by orthology or genetic interactions (Mak *et al.*, 2007). The connected subgraphs of the probabilistic gene–gene linkage network have been used to identify highly connected *gene clusters* (modules). The highly coherent functional annotation of genes within each cluster allowed the annotation of unknown proteins that are part of a cluster (Lee *et al.*, 2004).

Finally, the integrated transcriptional and metabolic network allowed global predictions of growth phenotypes and qualitative gene expression changes in *E. coli* (Covert *et al.*, 2004) and yeast (Herrgard *et al.*, 2006b). The creative (modelling) aspects of the definition of these (multi-)networks and the variety of predictions enabled by them demonstrate that they are not simply compilations of data but *qualitative biological network models*. A recently developed database, CellCircuits, offers a repository of network models spanning yeast, worm, fly, *Plasmodium falciparum* and human, and four types of interaction (Mak *et al.*, 2007). It was proposed that qualitative network models may develop into something akin to a Biological Information System, incorporating components, interactions and causal relationships described by a small group of verbs such as 'promote', 'inhibit', 'bind' etc. (Endy and Brent, 2001).

2.6 Biological network models: from network structure to dynamics

The nodes of cellular interaction networks represent populations of proteins or other molecules. The abundances of these populations can range from a few copies of an mRNA to hundreds or thousands of molecules per cell, and they vary in time and in response to external or internal stimuli. To capture these changes, the interaction network needs to be augmented by quantitative variables indicating the expression, concentration or activity – in short, the state – of each node, and by a set of equations indicating how the state of each node changes in response to changes in the state of its regulators.

Dynamic network models have as input information (i) the interaction network summarizing the regulatory relationships among components, (ii) the equations indicating how the state of a node depends on the state of its regulators and (iii) the initial state of each component in the system. The model's output is the time evolution of the state of the system, for example the system's response to the presence or absence of a given signal. A dynamic model that correctly captures experimentally observed normal behaviour allows researchers to track the changes in the system's behaviour due to perturbations. It is easier to use a model to search for perturbations that have a significant effect on system behaviour that it is to perform comparable experiments on the living system; for example, models can predict multiple small perturbations that produce large effects when combined. In the following, we briefly outline the main mathematical frameworks for modelling the dynamics of cellular networks, and give examples of their applications.

Continuous and deterministic models characterize node states by concentrations and describe the rate of production or decay of all components by differential equations based on mass-action (or more general) kinetics (Irvine and Savageau, 1990). With sufficiently thorough knowledge of the elementary biochemical reactions and fluxes comprising a system and the associated reaction rates, it is possible to accurately reproduce the system's dynamics and to explore the effect of perturbations. For example, a differential equation-based model of an 11-node signalling network responsible for programmed cell death after infection of *Arabidopsis thaliana* with *Pseudomonas syringae* led to significant refinement of the signalling circuitry (by discounting two previously proposed negative feedback loops) and of the kinetic parameters (Agrawal *et al.*, 2004). The *stochasticity* (non-determinism) of biological processes is usually taken into account by appending stochastic (noise) terms to differential equations. *Discrete events* (such as the initiation of transcription) and low abundances for certain molecules are incorporated by characterizing the node states by the copy number of each molecule and describing the time evolution of the probabilities of each of a system's possible states (Morton-Firth and Bray, 1998; Rao *et al.*, 2002; Andrews and Arkin, 2006). A recent model of the ethylene signalling pathway and its gene response in *A. thaliana* combines mass-action kinetics for signalling proteins with a probabilistic description of the target genes' states (Diaz and Alvarez-Buylla, 2006). This model reproduces the experimentally observed differential responses to different ethylene concentrations and predicts that the pathway filters rapid stochastic fluctuations in ethylene availability.

Boolean models characterize network nodes by one of two binary states corresponding to, for example, an expressed or not expressed gene, open or closed ion channel or above-threshold or below-threshold concentration of a molecule. The change in state of each regulated node is usually described by a logical function having as inputs the state of the node's regulators. Boolean models predict dynamic trends in the absence of detailed kinetic parameters; for example, such models have been used successfully to describe wild type and mutant behaviour in the development of floral organs (Mendoza and Alvarez-Buylla, 1998) and the process of abscisic acid-induced stomatal closure (Li *et al.*, 2006) in *A. thaliana*. *Hybrid dynamic models* meld a Boolean description of combinatorial regulation with continuous synthesis and decay by describing each node with both a continuous variable (akin to a concentration) and a Boolean variable (akin to activity) (Glass and Kauffman, 1973; Yuh *et al.*, 2001; Chaves *et al.*, 2006). For example, a hybrid model of the transcriptional regulation of the Endo16 sea urchin gene revealed that its spatial control during embryonic development is mediated by a cis-regulatory switch (Yuh *et al.*, 2001).

2.7 Perspectives

Systems biology develops through an ongoing dialogue and feedback among experimental, computational and theoretical approaches. High-throughput experiments reveal, or allow the inference of, the edges of global interaction networks. Graph-theoretical analysis of these networks enables general insight into the topological and functional organization of cellular regulation. Comparative network analysis feeds back to network inference (Albert and Albert, 2004; Gupta *et al.*, 2006; Horvath *et al.*, 2006; Christensen *et al.*, 2007) and expands the tools of graph theory to incorporate the diversity of molecular interactions. While genome-level interaction maps help us in understanding regulatory design features, dynamic modelling of systems with a less than genome-wide scope and specified inputs and outputs allows the identification of key regulatory components or parameters.

Biochemical reactions within and between cells take place on timescales spanning several orders of magnitude (Papin *et al.*, 2005), and these timescales are modulated by the spatial aspects of the interactions, the types of biomolecules or complexes that are interacting, and the environmental conditions to which the system is subjected (Han *et al.*, 2004; Balázsi *et al.*, 2005). Although the use of extensive dynamic modelling is limited by the incomplete availability of detailed transfer functions and kinetic parameters, emergent qualitative and hybrid modelling techniques that map the propagation of signals through a network (Ma'ayan *et al.*, 2005; Li *et al.*, 2006) give hope that even when exhaustive knowledge of parameters is unreachable, predictive modelling of biological processes will still be possible. Similarly augmenting the currently available directionless interactome networks with information regarding the sources (signals) and outputs of the network and the cause-and-effect (directional) relationships along the edges will significantly enhance their functional information content.

In addition to the dynamic changes in the state of network nodes, the topology of biological networks itself is shaped by dynamic events whose impact occurs on multigenerational (e.g. imprinting) and evolutionary (e.g. gene duplications and point mutations) timescales. Integration of epigenetic and evolutionary aspects with transcriptional, metabolic and signal transduction networks represents the 'final frontier' of systems biology.

It is arguably of primary importance to the systems biologist to discern whether, and how, the graph properties of biological networks reflect their functional and evolutionary constraints. However, the general architectural features of many biological networks described so far have been found to be shared to a large degree by other complex systems, ranging from technological networks to social networks. This universality is intriguing and should make systems biology of great interest to other fields, ranging from engineering to sociology.

References

Aerts, S., Thijs, G., Coessens, B., Staes, M., Moreau, Y. and De Moor, B. (2003) Toucan: deciphering the cis-regulatory logic of coregulated genes. *Nucleic Acids Res* **31**, 1753–1764.

Agrawal, V., Zhang, C., Shapiro, A.D. and Dhurjati, P.S. (2004) A dynamic mathematical model to clarify signaling circuitry underlying programmed cell death control in Arabidopsis disease resistance. *Biotechnol Prog* **20**, 426–442.

Albert, I. and Albert, R. (2004) Conserved network motifs allow protein–protein interaction prediction. *Bioinformatics* **20**, 3346–3352.

Albert, R. and Barabási, A.L. (2002) Statistical mechanics of complex networks. *Rev Mod Phys* **74**, 47–97.

Albert, R., DasGupta, B., Dondi, R., Kachalo, S., Sontag, E., Zelikovsky, A., *et al.* (2007) A novel method for signal transduction network inference from indirect experimental evidence. *J Comput Biol* **14**, 927–949.

Alon, U., Barkai, N., Notterman, D.A., Gish, K., Ybarra, S., Mack, D., *et al.* (1999) Broad patterns of gene expression revealed by clustering analysis of tumor and normal colon tissues probed by oligonucleotide arrays. *Proc Natl Acad Sci USA* **96**, 6745–6750.

Andrews, S.S. and Arkin, A.P. (2006) Simulating cell biology. *Curr Biol* **16**, R523–R527.

Anthonisse, J.M. (1971) The rush in a directed graph. Technical Report BN 9/71 Stichting Mathematicsh Centrum, Amsterdam.

Arita, M. (2004) The metabolic world of Escherichia coli is not small. *Proc Natl Acad Sci USA* **101**, 1543–1547.

Balázsi, G., Barabási, A.L. and Oltvai, Z.N. (2005) Topological units of environmental signal processing in the transcriptional regulatory network of Escherichia coli. *Proc Natl Acad Sci USA* **102**, 7841–7846.

Bandyopadhyay, S., Sharan, R. and Ideker, T. (2006) Systematic identification of functional orthologs based on protein network comparison. *Genome Res* **16**, 428–435.

Benedict, C., Geisler, M., Trygg, J., Huner, N. and Hurry, V. (2006) Consensus by democracy. Using meta-analyses of microarray and genomic data to model the cold acclimation signaling pathway in Arabidopsis. *Plant Physiol* **141**, 1219–1232.

Bogdanov, A. (1980) *Essays in Tektology: The General Science of Organization* (Seaside, CA: Intersystems Publications).

Bogdanov, A. (1922) *Tektologiya: Vseobschaya Organizatsionnaya Nauka* (Berlin: Grez'bin).

Bollobás, B. (1979) *Graph Theory: An Introductory Course* (New York: Springer).

Burgard, A.P., Pharkya, P. and Maranas, C.D. (2003) Optknock: a bilevel programming framework for identifying gene knockout strategies for microbial strain optimization. *Biotechnol Bioeng* **84**, 647–657.

Caron, H., van Schaik, B., Van Der Mee, M., Baas, F., Riggins, G., van Sluis, P., *et al.* (2001) The human transcriptome map: clustering of highly expressed genes in chromosomal domains. *Science* **291**, 1289–1292.

Chaves, M, Sontag, E.D. and Albert, R. (2006) Methods of robustness analysis for Boolean models of gene control networks. *Syst Biol (Stevenage)* **153**, 154–167.

Chen, T., He, H.L. and Church, G.M. (1999) Modeling gene expression with differential equations. *Pac Symp Biocomput* **4**, 29–40.

Cho, R.J., Campbell, M.J., Winzeler, E.A., Steinmetz, L., Conway, A., Wodicka, L., *et al.* (1998) A genome-wide transcriptional analysis of the mitotic cell cycle. *Mol Cell* **2**, 65–73.

Christensen, C., Gupta, A., Maranas, C.D. and Albert, R. (2007) Inference and graph-theoretical analysis of Bacillus subtilis gene regulatory networks. *Physica A* **373**, 796–810.

Covert, M.W., Knight, E.M., Reed, J.L., Herrgard, M.J. and Palsson, B.O. (2004) Integrating high-throughput and computational data elucidates bacterial networks. *Nature* **429**, 92–96.

Covert, M.W., Schilling, C.H. and Palsson, B. (2001) Regulation of gene expression in flux balance models of metabolism. *J Theor Biol* **213**, 73–88.

de Folter, S., Immink, R.G., Kieffer, M., Parenicová, L., Henz, S.R., Weigel, D., *et al.* (2005) Comprehensive interaction map of the Arabidopsis MADS box transcription factors. *Plant Cell* **17**, 1424–1433.

DeRisi, J.L., Iyer, V.R. and Brown, P.O. (1997) Exploring the metabolic and genetic control of gene expression on a genomic scale. *Science* **278**, 680–686.

Diaz, J. and Alvarez-Buylla, E.R. (2006) A model of the ethylene signaling pathway and its gene response in Arabidopsis thaliana: pathway cross-talk and noise-filtering properties. *Chaos* **16**, 023112.

Dijkstra, E.W. (1959) A note on two problems in connection with graphs. *Numerische Math* **1**, 269–271.

Dougherty, E.R., Barrera, J., Brun, M., Kim, S., Cesar, R.M., Chen, Y., *et al.* (2002) Inference from clustering with application to gene-expression microarrays. *J Comput Biol* **9**, 105–126.

Dougherty, E.R., Kim, S. and Chen, Y. (2000) Coefficient of determination in nonlinear signal processing. *Signal Process* **80**, 2219–2235.

Dougherty, E.R. and Shmulevich, I. (2003) Mappings between probabilistic Boolean networks. *Signal Process* **83**, 799–809.

Drawid, A. and Gerstein, M. (2000) A Bayesian system integrating expression data with sequence patterns for localizing proteins: comprehensive application to the yeast genome. *J Mol Biol* **301**, 1059–1075.

Endy, D. and Brent, R. (2001) Modelling cellular behaviour. *Nature* **409**, 391–395.

Famili, I., Forster, J., Nielsen, J. and Palsson, B.O. (2003) Saccharomyces cerevisiae phenotypes can be predicted by using constraint-based analysis of a genome-scale reconstructed metabolic network. *Proc Natl Acad Sci USA* **100**, 13134–13139.

Famili, I. and Palsson, B.O. (2003) The convex basis of the left null space of the stoichiometric matrix leads to the definition of metabolically meaningful pools. *Biophys J* **85**, 16–26.

Francois, C. (1999) Systemics and cybernetics in a historical perspective. *Syst Res Behav Sci* **16**, 203–219.

Franklin, G., Powell, J. and Emami-Naeimi, A. (2002) *Feedback Control of Dynamic Systems* (Englewood Cliffs, NJ: Prentice Hall).

Freeman, C.L. (1977) A set of measures of centrality based on betweenness. *Sociometry* **40**, 35.

Friedman, N., Linial, M., Nachman, I. and Pe'er, D. (2000) Using Bayesian networks to analyze expression data. *J Comput Biol* **7**, 601–620.

Gargalovic, P.S., Imura, M., Zhang, B., Gharavi, N.M., Clark, M.J., Pagnon, J., *et al.* (2006) Identification of inflammatory gene modules based on variations of human endothelial cell responses to oxidized lipids. *Proc Natl Acad Sci USA* **103**, 12741–12746.

Ge, H., Liu, Z., Church, G.M. and Vidal, M. (2001) Correlation between transcriptome and interactome mapping data from Saccharomyces cerevisiae. *Nat Genet* **29**, 482–486.

Giaever, G., Chu, A.M., Ni, L., Connelly, C., Riles, L., Veronneau, S., *et al.* (2002) Functional profiling of the Saccharomyces cerevisiae genome. *Nature* **418**, 387–391.

Giot, L., Bader, J.S., Brouwer, C., Chaudhuri, A., Kuang, B., Li, Y., *et al.* (2003) A protein interaction map of Drosophila melanogaster. *Science* **302**, 1727–1736.

Glass, L. and Kauffman, S.A. (1973) The logical analysis of continuous, non-linear biochemical control networks. *J Theor Biol* **39**, 103–129.

Glinski, M. and Weckwerth, W. (2006) The role of mass spectrometry in plant systems biology. *Mass Spectrom Rev* **25**, 173–214.

Goossens, A., Hakkinen, S.T., Laakso, I., Seppanen-Laakso, T., Biondi, S., De Sutter, V., *et al.* (2003) A functional genomics approach toward the understanding of secondary metabolism in plant cells. *Proc Natl Acad Sci USA* **100**, 8595–8600.

Grundy, W.N., Bailey, T.L. and Elkan, C.P. (1996) ParaMEME: a parallel implementation and a web interface for a DNA and protein motif discovery tool. *Comput Appl Biosci* **12**, 303–310.

Guelzim, N., Bottani, S., Bourgine, P. and Kepes, F. (2002) Topological and causal structure of the yeast transcriptional regulatory network. *Nat Genet* **31**, 60–63.

Guimera, R. and Nunes Amaral, L.A. (2005) Functional cartography of complex metabolic networks. *Nature* **433**, 895–900.

Gupta, A., Maranas, C.D. and Albert, R. (2006) Elucidation of directionality for co-expressed genes: predicting intra-operon termination sites. *Bioinformatics* **22**, 209–214.

Gupta, A., Varner, J.D. and Maranas, C.D. (2005) Large-sale inference of the transcriptional regulation of Bacillus subtilis. *Comput Chem Eng* **29**, 565–576.

Guterman, I., Shalit, M., Menda, N., Piestun, D., Dafny-Yelin, M., Shalev, G., *et al.* (2002) Rose scent: genomics approach to discovering novel floral fragrance-related genes. *Plant Cell* **14**, 2325–2338.

Gutierrez, R.A., Lejay, L.V., Dean, A., Chiaromonte, F., Shasha, D.E. and Coruzzi, G.M. (2007) Qualitative network models and genome-wide expression data define carbon/nitrogen-responsive molecular machines in Arabidopsis. *Genome Biol* **8**, R7.

Hall, R.D. (2006) Plant metabolomics: from holistic hope, to hype, to hot topic. *New Phytol* **169**, 453–468.

Han, J.D., Bertin, N., Hao, T., Goldberg, D.S., Berriz, G.F., Zhang, L.V., *et al.* (2004) Evidence for dynamically organized modularity in the yeast protein–protein interaction network. *Nature* **430**, 88–93.

Hart, G.T., Ramani, A.K. and Marcotte, E.M. (2006) How complete are current yeast and human protein-interaction networks? *Genome Biol* **7**, 120.

Hartwell, L.H., Hopfield, J.J., Leibler, S. and Murray, A.W. (1999) From molecular to modular cell biology. *Nature* **402**, C47–C52.

Heinrich, R. and Schuster, S. (1996) *The Regulation of Cellular Systems* (New York: Chapman & Hall).

Herrgard, M.J., Fong, S.S. and Palsson, B.O. (2006a) Identification of genome-scale metabolic network models using experimentally measured flux profiles. *PLoS Comput Biol* **2**, e72.

Herrgard, M.J., Lee, B.S., Portnoy, V. and Palsson, B.O. (2006b) Integrated analysis of regulatory and metabolic networks reveals novel regulatory mechanisms in Saccharomyces cerevisiae. *Genome Res* **16**, 627–635.

Hirai, M.Y., Klein, M., Fujikawa, Y., Yano, M., Goodenowe, D.B., Yamazaki, Y., *et al.* (2005) Elucidation of gene-to-gene and metabolite-to-gene networks in Arabidopsis by integration of metabolomics and transcriptomics. *J Biol Chem* **280**, 25590–25595.

Hirai, M.Y., Yano, M., Goodenowe, D.B., Kanaya, S., Kimura, T., Awazuhara, M., et al. (2004) Integration of transcriptomics and metabolomics for understanding of global responses to nutritional stresses in Arabidopsis thaliana. *Proc Natl Acad Sci USA* **101**, 10205–10210.

Holme, P., Huss, M. and Jeong, H. (2003) Subnetwork hierarchies of biochemical pathways. *Bioinformatics* **19**, 532.

Horvath, S., Zhang, B., Carlson, M., Lu, K.V., Zhu, S., Felciano, R.M., *et al.* (2006)

Analysis of oncogenic signaling networks in glioblastoma identifies ASPM as a molecular target. *Proc Natl Acad Sci USA* **103**, 17402–17407.

Huerta, A.M., Salgado, H., Thieffry, D. and Collado-Vides, J. (1998) RegulonDB: a database on transcriptional regulation in Escherichia coli. *Nucleic Acids Res* **26**, 55–59.

Hughes, J.D., Estep, P.W., Tavazoie, S. and Church, G.M. (2000) Computational identification of cis-regulatory elements associated with groups of functionally related genes in Saccharomyces cerevisiae. *J Mol Biol* **296**, 1205–1214.

Immink, R.G.H., Ferrario, S., Busscher-Lange, J., Kooiker, M., Busscher, M. and Angenent, G.C. (2003) Analysis of the petunia MADS-box transcription factor family. *Mol Genet Genomics* **268**, 598–606.

Irvine, D.H. and Savageau, M.A. (1990) Efficient solution of nonlinear ODE's expressed in S-system canonical form. *SIAM J Numer Anal* **27**, 704–735.

Jansen, R., Yu, H., Greenbaum, D., Kluger, Y., Krogan, N.J., Chung, S., *et al.* (2003) A Bayesian networks approach for predicting protein–protein interactions from genomic data. *Science* **302**, 449–453.

Jeong, H., Mason, S.P., Barabási, A.L. and Oltvai, Z.N. (2001) Lethality and centrality in protein networks. *Nature* **411**, 41–42.

Jeong, H., Tombor, B., Albert, R., Oltvai, Z.N. and Barabási, A.L. (2000) The large-scale organization of metabolic networks. *Nature* **407**, 651–654.

Kachalo, S., Zhang, R., Sontag, E., Albert, R. and DasGupta, B. (2008) NET-SYNTHESIS: a software for synthesis, inference and simplification of signal transduction networks. *Bioinformatics* **24**, 293–295.

Kanehisa, M. and Goto, S. (2000) KEGG: Kyoto encyclopedia of genes and genomes. *Nucleic Acids Res* **28**, 27–30.

Kelley, B.P., Yuan, B., Lewitter, F., Sharan, R., Stockwell, B.R. and Ideker, T. (2004) PathBLAST: a tool for alignment of protein interaction networks. *Nucleic Acids Res* **32**, W83–W88.

Keurentjes, J.J., Fu, J., de Vos, C.H., Lommen, A., Hall, R.D., Bino, R.J., *et al.* (2006) The genetics of plant metabolism. *Nat Genet* **38**, 842–849.

Kitano, H. (2002) Systems biology: a brief overview. *Science* **295**, 1662–1664.

Komatsu, S. (2006) Plant proteomics databases: their status in 2005. *Curr Bioinformatics* **1**, 33–36.

Lee, I., Date, S.V., Adai, A.T. and Marcotte, E.M. (2004) A probabilistic functional network of yeast genes. *Science* **306**, 1555–1558.

Lee, T.I., Rinaldi, N.J., Robert, F., Odom, D.T., Bar-Joseph, Z., Gerber, G.K., *et al.* (2002) Transcriptional regulatory networks in Saccharomyces cerevisiae. *Science* **298**, 799–804.

Li, S., Assmann, S.M. and Albert, R. (2006) Predicting essential components of signal transduction networks: a dynamic model of guard cell abscisic acid signaling. *PLoS Biol* **4**, e312.

Li, S., Armstrong, C.M., Bertin, N., Ge, H., Milstein, S., Boxem, M., *et al.* (2004) A map of the interactome network of the metazoan C. elegans. *Science* **303**, 540–543.

Liang, S., Fuhrman, S. and Somogyi, R. (1998) Reveal, a general reverse engineering algorithm for inference of genetic network architectures. *Pac Symp Biocomput* **3**, 18–29.

Luscombe, N.M., Babu, M.M., Yu, H.Y., Snyder, M., Teichmann, S.A. and Gerstein, M. (2004) Genomic analysis of regulatory network dynamics reveals large topological changes. *Nature* **431**, 308–312.

Ma, H.W. and Zeng, A.P. (2003) The connectivity structure, giant strong component and centrality of metabolic networks. *Bioinformatics* **19**, 1423–1430.

Ma'ayan, A., Blitzer, R.D. and Iyengar, R. (2004) Toward predictive models of mammalian cells. *Annu Rev Biophys Biomol Struct* **34**, 319–349.

Ma'ayan, A., Jenkins, S.L., Neves, S., Hasseldine, A., Grace, E., Dubin-Thaler, B., *et al.* (2005) Formation of regulatory patterns during signal propagation in a mammalian cellular network. *Science* **309**, 1078–1083.

Mahadevan, R. and Palsson, B.O. (2005) Properties of metabolic networks: structure versus function. *Biophys J* **88**, L07–L09.

Mak, H.C., Daly, M., Gruebel, B. and Ideker, T. (2007) CellCircuits: a database of protein network models. *Nucleic Acids Res* **35**, D538–D545.

Mangan, S. and Alon, U. (2003) Structure and function of the feed-forward loop network motif. *Proc Natl Acad Sci USA* **100**, 11980–11985.

Marcotte, E.M., Xenarios, I. and Eisenberg, D. (2001) Mining literature for protein–protein interactions. *Bioinformatics* **17**, 359–363.

Margulies, M., Egholm, M., Altman, W.E., Attiya, S., Bader, J.S., Bemben, L.A., *et al.* (2005) Genome sequencing in microfabricated high-density picolitre reactors. *Nature* **437**, 376–380.

McCraith, S., Holtzman, T., Moss, B. and Fields, S. (2000) Genome-wide analysis of vaccinia virus protein–protein interactions. *Proc Natl Acad Sci USA* **97**, 4879–4884.

Mendoza, L. and Alvarez-Buylla, E.R. (1998) Dynamics of the genetic regulatory network for Arabidopsis thaliana flower morphogenesis. *J Theor Biol* **193**, 307–319.

Mercke, P., Kappers, I.F., Verstappen, F.W., Vorst, O., Dicke, M. and Bouwmeester, H.J. (2004) Combined transcript and metabolite analysis reveals genes involved in spider mite induced volatile formation in cucumber plants. *Plant Physiol* **135**, 2012–2024.

Mewes, H.W., Amid, C., Arnold, R., Frishman, D., Guldener, U., Mannhaupt, G., *et al.* (2004) MIPS: analysis and annotation of proteins from whole genomes. *Nucleic Acids Res* **32**, D41–D44.

Meyer, R.C., Steinfath, M., Lisec, J., Becher, M., Witucka-Wall, H., Torjek, O., *et al.* (2007) The metabolic signature related to high plant growth rate in Arabidopsis thaliana. *Proc Natl Acad Sci USA* **104**, 4759–4764.

Moco, S., Bino, R.J., Vorst, O., Verhoeven, H.A., de Groot, J., van Beek, T.A., *et al.* (2006) A liquid chromatography-mass spectrometry-based metabolome database for tomato. *Plant Physiol* **141**, 1205–1218.

Morton-Firth, C.J. and Bray, D. (1998) Predicting temporal fluctuations in an intracellular signalling pathway. *J Theor Biol* **192**, 117–128.

Oksman-Caldentey, K.M. and Inze, D. (2004) Plant cell factories in the post-genomic era: new ways to produce designer secondary metabolites. *Trends Plant Sci* **9**, 433–440.

Oksman-Caldentey, K.M. and Saito, K. (2005) Integrating genomics and metabolomics for engineering plant metabolic pathways. *Curr Opin Biotechnol* **16**, 174–179.

Pandey, A. and Mann, M. (2000) Proteomics to study genes and genomes. *Nature* **405**, 837–846.

Papin, J.A., Hunter, T., Palsson, B.O. and Subramaniam, S. (2005) Reconstruction of cellular signalling networks and analysis of their properties. *Nat Rev Mol Cell Biol* **6**, 99–111.

Papin, J.A. and Palsson, B.O. (2004) Topological analysis of mass-balanced signaling networks: a framework to obtain network properties including crosstalk. *J Theor Biol* **227**, 283–297.

Pavesi, G., Mereghetti, P., Mauri, G. and Pesole, G. (2004) Weeder Web: discovery of transcription factor binding sites in a set of sequences from co-regulated genes. *Nucleic Acids Res* **32**, W199–W203.

Pellegrini, M., Marcotte, E.M., Thompson, M.J., Eisenberg, D. and Yeates, T.O. (1999) Assigning protein functions by comparative genome analysis: protein phylogenetic profiles. *Proc Natl Acad Sci USA* **96**, 4285–4288.

Peri, S., Navarro, J.D., Kristiansen, T.Z., Amanchy, R., Surendranath, V., Muthusamy, B., *et al.* (2004) Human protein reference database as a discovery resource for proteomics. *Nucleic Acids Res* **32**, D497–D501.

Persson, S., Wei, H., Milne, J., Page, G.P. and Somerville, C.R. (2005) Identification of genes required for cellulose synthesis by regression analysis of public microarray data sets. *Proc Natl Acad Sci USA* **102**, 8633–8638.

Qian, J., Dolled-Filhart, M., Lin, J., Yu, H. and Gerstein, M. (2001) Beyond synexpression relationships: local clustering of time-shifted and inverted gene expression profiles identifies new, biologically relevant interactions. *J Mol Biol* **314**, 1053–1066.

Rain, J.C., Selig, L., De Reuse, H., Battaglia, V., Reverdy, C., Simon, S., *et al.* (2001) The protein–protein interaction map of Helicobacter pylori. *Nature* **409**, 211–215.

Rao, C.V., Wolf, D.M. and Arkin, A.P. (2002) Control, exploitation and tolerance of intracellular noise. *Nature* **420**, 231–237.

Ravasz, E., Somera, A.L., Mongru, D.A., Oltvai, Z.N. and Barabási, A.L. (2002) Hierarchical organization of modularity in metabolic networks. *Science* **297**, 1551–1555.

Reed, J.L., Vo, T.D., Schilling, C.H. and Palsson, B.O. (2003) An expanded genome-scale model of Escherichia coli K-12 (iJR904 GSM/GPR). *Genome Biol* **4**, R54.

Reinartz, J., Bruyns, E., Lin, J.Z., Burcham, T., Brenner, S., Bowen, B., *et al.* (2002) Massively parallel signature sequencing (MPSS) as a tool for in-depth quantitative gene expression profiling in all organisms. *Brief Funct Genomic Proteomic* **1**, 95–104.

Rensink, W.A. and Buell, C.R. (2005) Microarray expression profiling resources for plant genomics. *Trends Plant Sci* **10**, 603–609.

Rives, A.W. and Galitski, T. (2003) Modular organization of cellular networks. *Proc Natl Acad Sci USA* **100**, 1128–1133.

Rossignol, M., Peltier, J.B., Mock, H.P., Matros, A., Maldonado, A.M. and Jorrin, J.V. (2006) Plant proteome analysis: a 2004–2006 update. *Proteomics* **6**, 5529–5548.

Rual, J.F., Venkatesan, K., Hao, T., Hirozane-Kishikawa, T., Dricot, A., Li, N., *et al.* (2005) Towards a proteome-scale map of the human protein–protein interaction network. *Nature* **437**, 1173–1178.

Said, M.R., Begley, T.J., Oppenheim, A.V., Lauffenburger, D.A. and Samson, L.D. (2004) Global network analysis of phenotypic effects: protein networks and toxicity modulation in Saccharomyces cerevisiae. *Proc Natl Acad Sci USA* **101**, 18006–18011.

Sharan, R., Suthram, S., Kelley, R.M., Kuhn, T., McCuine, S., Uetz, P., *et al.* (2005) Conserved patterns of protein interaction in multiple species. *Proc Natl Acad Sci USA* **102**, 1974–1979.

Shen-Orr, S.S., Milo, R., Mangan, S. and Alon, U. (2002) Network motifs in the transcriptional regulation network of Escherichia coli. *Nat Genet* **31**, 64–68.

Shmulevich, I., Dougherty, E.R., Kim, S. and Zhang, W. (2002) Probabilistic Boolean Networks: a rule-based uncertainty model for gene regulatory networks. *Bioinformatics* **18**, 261–274.

Smolen, P., Baxter, D.A. and Byrne, J.H. (2000) Mathematical modeling of gene networks. *Neuron* **26**, 567–580.

Snel, B., Lehmann, G., Bork, P. and Huynen, M.A. (2000) STRING: a web-server to retrieve and display the repeatedly occurring neighbourhood of a gene. *Nucleic Acids Res* **28**, 3442–3444.

Spirin, V. and Mirny, L.A. (2003) Protein complexes and functional modules in molecular networks. *Proc Natl Acad Sci USA* **100**, 12123–12128.

Stapley, B.J. and Benoit, G. (2000) Biobibliometrics: information retrieval and visualization from co-occurrences of gene names in Medline abstracts. *Pac Symp Biocomput* **5**, 529–540.

Stuart, J.M., Segal, E., Koller, D. and Kim, S.K. (2003) A gene-coexpression network for global discovery of conserved genetic modules. *Science* **302**, 249–255.

Tanaka, R. (2005) Scale-rich metabolic networks. *Phys Rev Lett* **94**, 168101.

Tavazoie, S., Hughes, J.D., Campbell, M.J., Cho, R.J. and Church, G.M. (1999) Systematic determination of genetic network architecture. *Nat Genet* **22**, 281–285.

Tohge, T., Nishiyama, Y., Hirai, M.Y., Yano, M., Nakajima, J., Awazuhara, M., *et al.* (2005) Functional genomics by integrated analysis of metabolome and transcriptome of Arabidopsis plants over-expressing an MYB transcription factor. *Plant J* **42**, 218–235.

Tong, A.H., Lesage, G., Bader, G.D., Ding, H., Xu, H., Xin, X., *et al.* (2004) Global mapping of the yeast genetic interaction network. *Science* **303**, 808–813.

Toronen, P., Kolehmainen, M., Wong, G. and Castren, E. (1999) Analysis of gene expression data using self-organizing maps. *FEBS Lett* **451**, 142–146.

Uetz, P., Giot, L., Cagney, G., Mansfield, T.A., Judson, R.S., Knight, J.R., *et al.* (2000) A comprehensive analysis of protein–protein interactions in Saccharomyces cerevisiae. *Nature* **403**, 623–627.

Valencia, A. and Pazos, F. (2002) Computational methods for the prediction of protein interactions. *Curr Opin Struct Biol* **12**, 368–373.

Vazquez, A., Flammini, A., Maritan, A. and Vespignani, A. (2003) Global protein function prediction from protein–protein interaction networks. *Nat Biotechnol* **21**, 697–700.

Voit, E.O. (2000) *Computational Analysis of Biochemical Systems* (Cambridge: Cambridge University Press).

von Bertalanffy, L. (1968) *General System Theory: Foundations, Development, Applications* (New York: George Braziller).

von Mering, C., Jensen L., Kuhn, M., Chaffron, S., Doerks, T., Kruger, B., *et al.* (2006) STRING 7 – recent developments in the integration and prediction of protein interactions. *Nucleic Acids Res* **35**, D358–D362.

von Mering, C., Zdobnov, E.M., Tsoka, S., Ciccarelli, F.D., Pereira-Leal, J.B., Ouzounis, C.A., *et al.* (2003) Genome evolution reveals biochemical networks and functional modules. *Proc Natl Acad Sci USA* **100**, 15428–15433.

Wagner, A. and Fell, D.A. (2001) The small world inside large metabolic networks. *Proc R Soc Lond Ser B Biol Sci* **268**, 1803–1810.

Wang, S.M. (2007) Understanding SAGE data. *Trends Genet* **23**, 42–50.

Watts, D. and Strogatz, S.H. (1998) Collective dynamics of 'small-world' networks. *Nature* **393**, 440–442.

Weinberg, G. (1975) *An Introduction to General Systems Thinking* (New York: Wiley-Interscience).

Wen, X., Fuhrman, S., Michaels, G.S., Carr, D.B., Smith, S., Barker, J.L., *et al.* (1998)

Large-scale temporal gene expression mapping of central nervous system development. *Proc Natl Acad Sci USA* **95**, 334–339.

Wingender, E., Dietze, P., Karas, H. and Knuppel, R. (1996) TRANSFAC: a database on transcription factors and their DNA binding sites. *Nucleic Acids Res* **24**, 238–241.

Wuchty, S., Oltvai, Z.N. and Barabasi, A.L. (2003) Evolutionary conservation of motif constituents in the yeast protein interaction network. *Nat Genet* **35**, 176–179.

Xenarios, I., Salwinski, L., Duan, X.J., Higney, P., Kim, S.M. and Eisenberg, D. (2002) DIP, the Database of Interacting Proteins: a research tool for studying cellular networks of protein interactions. *Nucleic Acids Res* **30**, 303–305.

Yeger-Lotem, E., Sattath, S., Kashtan, N., Itzkovitz, S., Milo, R., Pinter, R.Y., *et al.* (2004) Network motifs in integrated cellular networks of transcription-regulation and protein–protein interaction. *Proc Natl Acad Sci USA* **101**, 5934–5939.

Yook, S.H., Oltvai, Z.N. and Barabási, A.L. (2004) Functional and topological characterization of protein interaction networks. *Proteomics* **4**, 928–942.

Yu, H., Luscombe, N.M., Lu, H.X., Zhu, X., Xia, Y., Han, J.D., *et al.* (2004a) Annotation transfer between genomes: protein–protein interologs and protein–DNA regulogs. *Genome Res* **14**, 1107–1118.

Yu, J., Smith, V.A., Wang, P.P., Hartemink, A.J. and Jarvis, E.D. (2004b) Advances to Bayesian network inference for generating causal networks from observational biological data. *Bioinformatics* **20**, 3594–3603.

Yuh, C.H., Bolouri, H. and Davidson, E.H. (2001) Cis-regulatory logic in the endo16 gene: switching from a specification to a differentiation mode of control. *Development* **128**, 617–629.

Zhang, L.V., King, O.D., Wong, S.L., Goldberg, D.S., Tong, A.H., Lesage, G., *et al.* (2005) Motifs, themes and thematic maps of an integrated Saccharomyces cerevisiae interaction network. *J Biol* **4**, 6.

Zhao, W., Serpedin, E. and Dougherty, E.R. (2006) Inferring gene regulatory networks from time series data using the minimum description length principle. *Bioinformatics* **22**, 2129–2135.

Zimmermann, I.M., Heim, M.A., Weisshaar, B. and Uhrig, J.F. (2004) Comprehensive identification of Arabidopsis thaliana MYB transcription factors interacting with R/B-like BHLH proteins. *Plant J* **40**, 22–34.

Annual Plant Reviews (2009) **35**, 67–136
doi: 10.1111/b.9781405175326.2009.00003.x

Chapter 3

Prokaryotic Systems Biology

Thadeous Kacmarczyk,[1] Peter Waltman[2] and
Richard Bonneau[1,2]

[1] *Department of Biology and Center for Genomics and Systems Biology, New York University, New York, NY, USA*
[2] *Computer Science Department, Courant Institute for Mathematical Sciences, New York University, New York, NY, USA*

Abstract: This chapter will briefly review the most common technologies employed in prokaryotic systems biology projects. When possible, we will highlight the relevance to prokaryotic biology and the role prokaryotic systems might have had in the inception of the technology overall. We will then walk the reader through four functional genomics projects. Although this review cannot cover all details of the full, field-wide, efforts directed at even a single organism, we do think it is instructive to attempt to outline the efforts directed to each of these systems. We wish to highlight the biological insights gleaned from these studies and demonstrate that these studies illustrate a very productive interface between systems biology, computational biology and more traditional (small numbers of genes) molecular biology that is already in operation in the studies we review.

Keywords: systems biology; network inference; Halobacterium NRC-1; *Halobacterium salinarum*; *Bacillus subtilis*; *Caulobacter crescentus*; *Escherichia coli*; microarray; proteomics

3.1 Introduction

Recent advances in systems biology have dramatically accelerated the rate at which biologists can acquire data on all informational levels of the cell (genome sequence, RNA, protein, protein modification, metabolites, etc.). Concurrent advances in computational biology have begun to allow for large multi-group efforts to generate predictive dynamical models of whole cells. Several groups have produced first drafts of global models of regulation and metabolism, and these groups are taking next steps to significantly improve the comprehension and accuracy of these models by working to validate aspects of these models and include additional data types (e.g. metabolomics and more accurate proteomics). In this review, we will discuss in detail several prokaryotic functional genomics projects that employ systems biology approaches. We will show that, although many challenges remain, we are beginning to cross critical milestones in our efforts to learn systems-wide quantitative models of prokaryotic cells and their interactions with their environments.

3.1.1 The importance of microbes

Bacteria and archaea are abundant, diverse and important organisms. Many currently relevant human pathogens are prokaryotic. Microbes have been used for fermentation of foodstuffs for aeons and more recently have been used in engineering and synthesis applications spanning the full range of human activities (e.g. bacteria can serve as platforms for the synthesis of drugs, vitamins, food additives). Prokaryotes play critical roles in our environment and are central to efforts to mitigate the human impact associated with solid waste/sewage, industrial toxic waste and agriculture. Prokaryotic biology is critical to our understanding the history of our environment. Prokaryotes have traditionally provided biologists useful tools for molecular and cell biology across all systems.

3.1.2 Experimental advantages of prokaryotic systems biology

Archaea and bacterial systems offer a distinct advantage in complexity. Although they have all the properties of life that warrant our awestruck admiration, such as self-assembly, robustness, reproducible autonomous decision making, they are orders of magnitude less complex than eukaryotes, they often allow for collection of larger amounts of material in the lab. Prokaryotes are often synchronizable (as are many eukaryotic systems) and amenable/robust to the manipulations needed for single-celled measurements (Alon, 2007). Often the genetics of a given prokaryotic system will allow for rapid construction of knock out and/or overexpression strains that can be used to directly query the global result of specific genetic perturbation (this is the case for all organisms described herein). Unfortunately, these experimental advantages do not extend to all organisms and several prokaryotes participate in complex communities that currently elude even laboratory culture, and are thus only now coming into focus via metagenomic sequencing directly from the environment (Handelsman, 2004). In this review, we will focus on systems biological approaches used in organisms that are amenable to genetics, culture and have full genome sequence.

3.2 Types of questions

Before we begin our discussion, we need to discuss the types of questions one might answer with prokaryotic systems biology.

3.2.1 Core biology

The first and most fundamental question one might ask is: 'how do all systems components interact to form core aspects of biology with components and/or strategies common to many systems?' For example, we might study the cell cycle in several organisms and compare common themes in an attempt to reveal the functional requirements or ancestral progenitor of cell cycle control in different niches/organisms. Systems biology becomes essential in answering this type of question due to the shear number of genes involved in many core processes. So the fact that much of the cell is involved makes techniques based on global measurements a natural fit to the question. So-called master regulators (hubs) are prevalent in biology and determining the targets and control of such master regulators is more directly accomplished via global techniques (such as ChIP-chip, yeast one hybrid, microarray measurement following a genetic perturbation to the gene, etc.).

3.2.2 Environmental

Another case where global measurements are key is in the deciphering of an organisms response to its environment. A typical structure for such a study involves the use of genomics techniques to identify key players in a physiological response to a given cell environment, followed by more focused studies to investigate/validate the role or necessity of the discovered proteins/genes. Many of the earliest studies employing microarrays in prokaryotic systems were designed to characterize cells genome-wide/transcriptome response to environmental stress. In these studies, we look for novel regulation of known processes that have been discovered, novel associations between proteins of unknown function with known environmental responses.

3.2.3 Disease-related pathogens

In cases where the prokaryote of interest is also a human pathogen, our question is: 'how can intervention maximally disrupt the pathogen, disrupt its interaction with the human host or vector or otherwise mitigate its effect on human health?' In this study we will focus less on this type of study, as the interaction with the human host often requires as much study as the internal workings of the pathogen of interest. This prokaryote–host interaction is, although currently the focus of several systems biology efforts, beyond the scope of this review.

3.2.4 Engineering

Genome-wide models will inevitably be required if we are to rationally engineer microbial systems. Reasons for engineering microbial systems span human efforts and include: bioenergy, remediation of industrial waste sites and production of difficult to synthesize compounds.

3.3 A typical prokaryotic systems biology project

Figure 3.1 outlines a typical functional genomics platform. Different aspects of this schematized project will be described in greater detail (with reference to the biological questions being asked) below, as each project will require dramatically different paths through this simplified view. This review is intended to be a review about systems biology of prokaryotes, thus many of the projects and efforts we discuss are centred on global measurements. One goal of this work is to illustrate that when we approached the biology of these organisms, we found that the global studies, at least for these systems, were well integrated with countless other more traditional molecular biology studies. Thus, if one looks at any single group one might incorrectly see a divide between systems biology and biology as a whole. However, looking across all studies for a single organism, one sees that hypotheses generated by global studies have permeated field-wide and, in a corresponding manner, high-confidence single-gene results from traditional reductionist biology commonly guide the design of global studies. Thus, we optimistically see the field of prokaryotic biology evolving into a happily balanced field where studies focusing on small numbers of genes are carried out alongside global studies in a highly complimentary manner (loosely coupled via the literature, i.e. in a manner in no way different than how this scientific field has operated for the past century). We find the global tools and studies described herein, and the idea of systems biology becoming cost effective and commonplace, exciting. We do not, however, think that these new global tools nucleate a pedagogical or philosophical crisis. The literature supports the notion that these new tools produce an acceleration of, but *do not* produce a pedagogical rift or fundamental change in the types of questions approached by biologists.

3.4 Global models

3.4.1 Emergent properties

Emergent properties are properties of a system that cannot be trivially traced back to properties of any single component of the system. Simple examples of emergent properties abound in nature such as flock behaviour, the decisions and patterns of ant and termite colonies, dramatic trends in human economies, a tabby cat's stripes, spiral waves in heart defibrillation, etc. When we refer to the meaningful properties of highly complex systems as emergent in this review, it is simply a compact way of describing the simple notion that if large complex systems have many inter-component interactions then only by modelling the global system can we hope to recapitulate or model the overall system behaviour. Systems that involve interactions on multiple scales, interactions between components that involve loops (such as feedback loops) and nonlinear effects such as saturation, recovery and auto excitation all contribute to the degree to which systems are likely to have difficulty in predicting emergent properties.

Nearly, all biological systems exhibit complex phenotypes and physiologies that are not attributable to single subsystems or genes, and all biological systems are large, complex systems involving all the interaction types typically leading to systems dominated by emergent behaviour. Thus, we must view important properties of living systems as interdependent, emergent or at least highly epigenetic phenomena. Regardless of our diction, we rapidly arrive at the conclusion that highly interconnected phenomena like metabolism, signalling and regulation require modelling at the global, genome-wide, scale if we are to construct predictive models of cellular behaviour.

3.4.2 Global models require the new approaches to experimental design, technologies and analysis

This motivation for global measurement and modelling of biology has lead to prokaryotic biologists, working on several systems, to adopt some aspect of genomic (genome-wide) experimentation and analysis. In the end this has led to many successes and many mistakes, as the field wrestles with technical and computational challenges generated by high-throughput methods. After a decade of systems biology, many biologists feel a bit unclear, pedagogically, as to the state of modern biology. Many people incorrectly feel that biology is currently a disjoint field, with labs that perform systems biology/global studies existing in a sub-field separate from those biologists that perform one-gene-at-a-time studies. One point we hope to convey by reviewing several functional genomics projects below is that many of the most interesting results are from work where more focused studies of subsystems and small numbers of genes are embedded in or guided by global analysis.

3.5 Comparative functional genomics of prokaryotes

Although not the explicit focus of this chapter, the comparison of the results from multiple functional genomics projects devoted to different organisms, offers a look into the evolution of not just sequences but subnetworks, networks and biomodules across bacterial and archaeal clades. This possibility is made particularly exciting by recent advances in the reconstruction of phylogenetic histories of microbes that explicitly model lateral gene transfer. Lateral gene transfer makes the prospect of comparative genomics particularly exciting, as even very distantly related organisms will share very similar modules (regulation and strategies) for some cellular processes and employ dramatically different modules (strategies and regulation) for other cellular processes (Price et al., 2006, 2008; Shapiro and Alm, 2008). Uncovering these relationships at the module and network level (in addition to the sequence level) is possible given the scale of prokaryotic systems; in fact several metagenomics projects already exist and multi-species functional genomics projects and analysis are beginning to show results as we speak. Given the large number of prokaryotic functional genomics projects, multi-species analysis (inferring networks and modules over multiple species data sets) is one of the next major challenges, as prokaryotic systems rarely exist in clonal isolation (consortia of microbes inhabiting ecological niches are the relevant system to study in many cases).

3.6 Review of core technologies for prokaryotic systems biology

Here, we will briefly review the core technologies found in a typical prokaryotic functional genomics pipeline as discussed throughout the paper. We will place emphasis on these techniques in our discussion of four specific functional genomics projects below. This section illustrates that many of these technologies, although found throughout studies of eukaryotes as well, were first developed in prokaryotic systems.

3.6.1 Genomics

The sequenced genome is an essential pre-requisite to determining the parts list for an organism, encoding its RNA transcripts, proteins, as well as several patterns and properties beyond our current understanding. The field of genomics has expanded during the past decade from the static study of DNA sequences, annotation and structure to dynamic studies of functional and comparative genomics, but all rests squarely on our ability to determine complete genome sequences for organisms in a cost-effective manner. The process and capabilities of genome sequencing has dramatically changed since the first complete genome in 1995 of *Haemophilus influenzae Rd.* (Fleischmann *et al.*, 1995) with innovations in cloning and high-throughput DNA sequencing technology. The Sanger (Sanger *et al.*, 1977), or chain termination method, is still the primary method for DNA sequencing (although new technologies are most certainly poised to overtake it as the most commonly used method). Sanger sequencing has seen many optimizations and improvements since the laboratory of Leroy Hood first automated the process in the mid 1980s. These advancements include advances in fluorescent labels and detection, capillary electrophoresis and microfluidics, automation, informatics and computational power, and now typically produce ∼100 kbp per run of a typical capillary sequencing machine. There are two new promising technologies: 454 Life Sciences sequencing can produce 30 Mbp per run by utilizing a sequence-by-synthesis (SBS) approach which integrates pyrosequencing, massively parallel sequencing and microfabricated picolitre reactors (Margulies *et al.*, 2005). This technology has already resulted in economically sequenced genomes (in combination with Sanger for longer reads). Solexa sequencing technology also uses SBS and massively parallel technology on a clonal single molecule array, and is working towards 1G bp per run. Emerging methods based on other technologies such as, sequence-by-hybridization, mass spectroscopy and single molecule nanopore sequencing are also being investigated. Regardless of which technique wins the race, it is clear that sequencing hundreds of prokaryotic genomes by single groups with modest funding is on the horizon. Even without these new technologies high-throughput sequencing is pouring out raw data at a fantastic rate. This along with new techniques for protein annotation has allowed us to compile a very large compendium of gene and protein families that greatly facilitate our management of the complexity of any given proteome (Tatusov *et al.*, 1997; Finn *et al.*, 2006). Comparative genomics can illustrate genetic programmes that are global properties of organisms as well as properties specific to a species. This sequencing power offers opportunities into the natural microbial world.

Much of the earth's biomass comprises microorganisms that participate in tightly interconnected microbial communities. In many cases, these communities are too complex or adapted to a very particular microenvironment to culture. This inability to culture a large number of microbes important to the environment under standard laboratory conditions has motivated the development of metagenomics (sequencing microbial communities directly from the environment, for example host tissue or soil, to study dynamic species interactions and diversity is being called environmental genomics or metagenomics (Tringe and Rubin, 2005)). Recent studies emphasize the insights to be gained from metagenomic studies. Assembly of environmental microbial sequences from acid mine drainage biofilms is one of several recent metagenomic projects that illustrates that microbial community genomes can be reconstructed to high completeness given sufficient coverage (Tyson *et al.*, 2004). Sogin *et al.* (2006) surveyed the deep sea to show that current sequence databases represent only as small fraction of global microbial diversity. The Sargasso Sea metagenomics survey revealed a substantial amount of phylogenetic diversity and complexity, identified 1.2 million genes and sampled from an estimated 1800 bacterial species (Venter *et al.*, 2004). Three new investigations from the Sorcerer II Global Ocean Sampling expedition have enhanced this data set, which now includes 6.3 billion base pairs (Rusch *et al.*, 2007; Yooseph *et al.*, 2007). Metagenomics shows us environmentally relevant protein frequency of occurrence and diversity and that, when we consider the planet-wide diversity of microbial ecology, we have just scratched the surface, with respect to diversity, of microbial genomes and proteomes.

3.6.2 Proteome annotation

Give the genome, the next step is to predict proteins, functional RNAs and other transcribed regions; we will only discuss annotation briefly. Methods for annotating proteomes are still evolving, but generally rely on a mix of sequence-similarity, protein-domain or protein family searches (such as COG (clusters of orthologous genes) and Pfam). Structure prediction-based methods for genome annotation are emerging (Bonneau *et al.*, 2004; Malmstrom *et al.*, 2007) and rely on fold recognition and de novo structure prediction to extend the reach of our ability to detect distant homology (structure similarity is conserved across a greater evolutionary distance than sequence similarity). Methods for solving protein structures experimentally remain costly and a mix of experimental structural biology and computational structure biology are likely going to lead to prokaryotic genomes characterized at the protein 3D (three dimensional) structure level to high levels of completeness. Another promising note is that as more sequences are added to the databases our ability to find sequence-based homology via intervening sequences (e.g. via multiple iterations of position specific iterative BLAST (PSI-BLAST)) also increases.

3.6.3 Transcriptomics

Transcriptomics is the measurement and study of the properties and dynamics of all mRNA transcripts in the cell (the transcriptome). There are a variety of tools used to measure transcriptomes, the most common being the microarray. All such tools are high-throughput methods for detecting and measuring the expression level, or relative abundance of mRNA transcripts, for every gene within the cell, and result in a snapshot of all the genes present at one time in the cell for a given condition. The methods measure the abundance of RNAs, which is a convolution of the rate of synthesis, transport and degradation. Two common goals of transcriptomics are to identify genes that are differentially expressed and recognize patterns in gene expression that correlate with the phenotype. The main technologies used to explore this are DNA microarrays and serial analysis of gene expression (SAGE). SAGE quantifies transcript levels by sequencing and counting cDNAs converted from small unique tags of samples of RNA (Velculescu *et al.*, 1995). Parallel gene expression analysis is typically done by either one-colour oligonucleotide arrays from Affymetrix (GeneChip) or NimbleGen, or by two-colour spotted/printed arrays that can be oligonucleotides, cDNAs or expressed sequence tags printed onto a glass slide. The physical microarray consists of probes (complementary to the RNA being measured), the oligonucleotide or cDNA, printed (in the case of cDNA arrays) or built (for oligo arrays) onto a glass slide or silicon chip. The array is perfused with the cell extracts of RNA tagged with a fluorescent dye (Cy3, Cy5); labelled RNAs thus hybridize to the DNA probes. Ideally, it is the specificity guaranteed (excepting cross-hybridization) by reverse complementarity that is core to all microarray technologies (alas, nothing similar to reverse complementarity exists for proteins). Lastly, fluorescent intensities are read to measure the relative abundance. It is therefore important to design the experiment correctly for the comparison to be made. Data are collected by exciting the fluorescent dye-tagged RNA and scanning the image. Array scanners usually have software that automatically scans the image, locates the spots and computes the intensities. The intensity data are converted into numerical data that can then be further analyzed statistically to identify differentially expressed genes.

Expression profiles can be compared among different cells or tissues (e.g. cancerous versus non-cancerous), time points and perturbations. Clustering microarray data were an important development (hierarchical clustering, k-means, self-organizing maps) for identifying patterns of co-expressed genes.

3.6.4 Proteomics

Proteomics is a very large and ever expanding field with a large diversity of aims and corresponding techniques. Recent advances have allowed identification and quantification of all of the proteins that exist in a cell, their abundance, post-translational modifications, interactions, localization and modifications. However, determination of an organism's proteome is difficult due to the complexity of the large number of proteins and their modifications (Bray, 1995). We focus on studies that aim to complete the characterization of the proteome by identifying and quantifying all of the proteins encoded by the genome. The main methods for quantifying, characterizing and profiling proteins and complexes are: two-dimensional gel electrophoresis (2DE), mass spectroscopy (MS), matrix-assisted laser desorption–ionization time-of-flight and other combinations of MS (LC-MS, GC-MS, etc.) (Aggarwal and Lee, 2003). Advanced protein array technology can assay protein activity as well as identifying protein–protein and protein–DNA interactions, here we focus on MS-based proteomics (Poetz *et al.*, 2005; Vemuri and Aristidou, 2005).

Recent developments have included new methods for measuring relative protein levels (proteome-wide) by incorporating stable isotopically labelled reagents into multiple samples (by cell culture, in SILAC (stable isotope labelling by amino acids in cell culture), and by labelling with reagents in ITRAQ (isobaric tag for relative and absolute quantitation), ICAT (isotope coded affinity tag)) (Gygi *et al.*, 1999; Zhang *et al.*, 2006). In these experiments each sample is labelled with a reagent containing different numbers of stably incorporated heavy isotopes and MS is simultaneously performed on multiple samples. These methods (e.g. SILAC, ICAT, ITRAQ) promise to provide proteome-wide measurements analogous to multi-colour microarrays. Many technical challenges remain, but mass spectrometry-based proteomics is currently central to many functional genomics projects, and with inbound improvements in resolution, reliability and cost (as well as improvements in surrounding methods such as reagents and fractionation steps) we will only see the increase in importance of these technologies.

3.6.5 Techniques for measuring protein–DNA and protein–protein interactions

Proteins function as networks of interconnected components, involving networks composed of protein–protein and protein–DNA and protein–RNA interactions for the cell overlaid to form an overall network for a given organism (Ge *et al.*, 2003). Techniques for measuring such interactions are thus highly relevant to prokaryotic functional genomics projects.

High-throughput interaction mapping methods have been developed for measuring all three of these interaction networks. For example, yeast 2-hybrid (Y2H) (Walhout and Vidal, 2001) and chromatin immunoprecipitation (ChIP-chip) assays are methods for identifying protein–DNA interactions and co-immunoprecipitation (co-IP) is used to identify protein complexes from cell extracts. Chromatin immunoprecipitation (ChIP-chip) assays aim to identify the specific regions of the genome a given protein binds. Proteins that interact with DNA will, by this procedure, enrich segments containing high-affinity-binding sites for these proteins. Introduced in 2000 and 2001 by three papers that reported its first successful use, the general goal of ChIP-chip is to use chromatin immunoprecipitation to help identify the upstream-binding sites for a given transcription factor. To accomplish this, the general strategy is as follows. Once the transcription factor protein under consideration has been bound either in vivo or in vitro to its DNA target, it is cross-linked to the DNA target, often with formaldehyde, which can easily be unlinked with heat. After cross-linking, the DNA is sheared, usually by sonication and the protein–DNA complex is then immunoprecipitated using an antibody specific to the transcription factor being studied, allowing the cross-linked protein–DNA complex to be isolated. After unlinking the transcription factor from the DNA, the DNA fragments are *polymerase chain reaction* (PCR) amplified and labelled before finally being evaluated with microarrays to identify enriched regions of the genome that correspond to binding regions for the transcription factor.

3.6.6 Metabolomics

Metabolites are low molecular weight (compared to the other large biopolymers) molecules and chemicals such as: hormones, carbohydrates, amino acids, fatty acids, vitamins, lipids, etc. that are the building blocks of the cell. We refer to methods that aim to simultaneously measure many metabolites as metabolomics. Generally these techniques are able to measure hundreds of metabolites, but are quite new and the field is still evolving rapidly. The diversity of metabolites makes their global measurement challenging; metabolites have various and wide-ranging chemical properties like polarity and volatility that make quantifying all of them simultaneously difficult for any one analysis or technology. An intensely growing field termed metabolomics, and a related field metabonomics (which focuses on disease and therapeutic metabolite fluctuations), has emerged to identify and quantify all of the metabolites in a biological system (Dunn and Ellis, 2005). Metabolites have been shown to be important influences on key cellular processes; metabolites have interactions with every informational level in the cell. For example, metabolites can affect transcription factors, riboswitches, enzymes, lipids and protein complexes. Since it is exceptionally difficult to simultaneously quantify multiple metabolites, metabolic target analysis or profiling is usually performed (Wang *et al.*, 2006). The main technologies and methods for metabolomic research revolve around a wide variety of analytical techniques including but not limited to: MS vibrational spectroscopy; gas chromatography (GC-MS), liquid chromatography (LC-MS) and high performance-LC, capillary electrophoresis (CE-MS), nuclear magnetic resonance (NMR) spectroscopy and Fourier Transform InfraRed spectroscopy. Metabolic profiling generally involves the analysis of multiple metabolites in a pathway, in several pathways, or resolution by a specific technology (Wang *et al.*, 2006). Target analysis is the quantification of a specific metabolite or few metabolites related to a specific system. Metabolites are essentially the final products of gene expression in an information flow from genome to transcriptome to proteome to metabolome (Fiehn, 2002). It has been shown that the concentration of metabolites in a cell can change significantly while the genes and proteins remain fairly constant for some environmental or genetic perturbations (Wang *et al.*, 2006). Phenotype is dependent upon genotype but can manifest itself as disproportions of metabolites. Thus, metabolomics is fundamental to understanding the function of the cell. One advantage of studying the metabolome is that there are fewer metabolites than genes or proteins, so there is less complexity (Dunn and Ellis, 2005). The true potential of metabolomics is to complement transcriptomics and proteomics. Integration of these data sets can help resolve mechanisms of biological processes and cellular functions by reconstructing these networks to provide insight at a systems level. This can help to identify gene functions, regulatory mechanisms and to focus research.

3.7 *Caulobacter crescentus*

The non-pathogenic oligotroph *Caulobacter crescentus* is a gram-negative α-proteobacterium that lives in aquatic environments; for the remainder of this section, we will refer to it as *Caulobacter*. Morphologically, *Caulobacter* exhibits three distinct phenotypes. The first, referred to as swarmer cells (SW cells), are motile, rod-like cells that have in one pole both a flagellum, as well as two type IV pili adjacent to the flagellum. Due to an as of yet unknown signal, an SW cell will metamorphose into a stalked cell (ST cell), during which the pili are retracted and the flagellum is ejected, replaced by a 'stalk' that is formed from a thin extension of the cell wall that can help serve as an anchor for the new ST cell (Ausmees and Jacobs-Wagner, 2003; Skerker and Laub, 2004; Holtzendorff *et al.*, 2006).

At the same time as this SW to ST cell transformation, chromosomal replication, which had been repressed in the SW cell, is initiated from a single origin of replication and the cell enters S phase. Following the completion of the chromosomal replication, the two copies are sequestered to the two polar halves of the pre-divisional (PD) cell, a new flagellum and pili are generated on the pole opposite of the stalk, and a diffusion barrier develops separating the two polar halves. Once cell division is complete, yielding both an SW and an ST cell, chromosomal replication is reinitiated in the ST cell, while the new SW cell will relocate via chemotaxis, with chromosomal replication inhibited in it until it differentiates into an ST cell and the entire process reinitiates.

The key observations to draw from the *Caulobacter* cell cycle are: (1) its asymmetrical nature – as it yields two morphologically different daughter cells, and (2) the replication process yields exactly two daughter cells (Skerker and Laub, 2004). In contrast, *Escherichia coli* cell division in logarithmic phase can replicate the genome up to four times before cell division occurs (Skerker and Laub, 2004). As cell division in *Caulobacter* yields exactly two daughter cells, it exhibits a periodicity that lends itself well to the examination of the bacterial cell cycle. In addition, the asymmetric nature of its cell cycle allows researchers the opportunity to study bacterial cell differentiation – an aspect shared with many other bacteria such as *Bacillus subtilis* (Skerker and Laub, 2004). However, as this asymmetry is accomplished via asymmetric localization of proteins, a good portion of the current research directed at deciphering this process uses lab techniques directed at study of single genes (such as localization studies using green fluorescent protein, GFP).

We will outline/review both systems-wide studies employing microarrays, proteomics and ChIP-chip (as is the mandate of this chapter) alongside studies aimed at determining the function of small numbers/single genes. Thus, our goal is to illustrate how systems-level techniques have been used alongside these more focused studies to successfully identify the regulation of the cell cycle in *Caulobacter*.

3.7.1 A first application of genome-wide expression profiling to *Caulobacter*

The first systems-level examination of *Caulobacter's* RNA expression during its cell cycle was reported by Laub *et al.* (2000). Interestingly, this was reported in advance of the publishing of *Caulobacter's* complete genome which was published 3 months later by Nierman *et al.* (2001). As such, the cDNA microarrays they used did not cover the entire set of ORFs in the *Caulobacter* genome, however they did represent 2966 predicted ORFs, corresponding to nearly 80% of the 3767 that would be reported by Nieman *et al.* Sampling every 15 min over the complete 150 min cell cycle progression from SW cell to ST cell and final asymmetric cell division, Laub *et al.* (2000) identified 553 cell cycle-regulated genes, 72 of which had been previously identified using earlier genetic techniques. Clustering these cell cycle-regulated genes using self-organizing maps, Laub *et al.* (2000) discovered that these were organized into sets of functionally associated genes that were induced in synchronization with the various events of the cell cycle. These included coordinated sets of genes involved in DNA replication and cell division, protein synthesis and polar morphogenesis. Significant among these included homologues of the *E. coli* cell division genes *ftsI, ftsW, ftsQ, ftsA* and *ftsZ*, the gene for the tubulin-like GTPase, FtsZ, an essential protein for cell division. Additionally, 16 histidine kinases were among these cell cycle-regulated genes, of which only 4 at the time had been characterized, these being CheA, DivJ, CckA and PleC.

3.7.2 Laub *et al.* (2000) – probing the CtrA regulon

In addition to this time-course expression profile, Laub *et al.* (Laub *et al.*, 2000) also explored the regulon of CtrA, a member of the two-component response regulators that had already been identified using earlier genetic techniques to be a master regulator of the *Caulobacter* cell cycle (Ausmees and Jacobs-Wagner, 2003). This was accomplished by comparing the expression profiles of wild-type *Caulobacter* with those of a temperature-sensitive mutant, revealing 144 differentially expressed gene transcripts as a result of CtrA expression. To identify which of these were directly regulated by CtrA, Laub *et al.* (2000) used MEME (Bailey and Elkan, 1994) to construct a consensus profile of known CtrA-binding sites and then used this profile in conjunction with the expression data to identify several previously unknown genes under direct CtrA regulation, including *divK*, a single-domain response regulator. Finally, Laub *et al.* (2000) compared the mRNA expression of wild-type *Caulobacter* with another that contained an allele that produces a form of CtrA that is both proteolysis-resistant and constitutively active, resulting in cell cycle to arrest at the G_1 (SW) stage. From these assays, they were able to identify a nearly 70% overlap with those genes differentially expressed in the temperature-sensitive mutant.

These findings were partially validated in a 2002 paper where CtrA targets were identified by performing chromatin immunoprecipitation with microarrays (a.k.a. ChIP-chip or ChIP-on-chip). Using this, then new, ChIP-chip method, Laub *et al.* (2002) identified 138 regions enriched for CtrA binding; the 196 genes flanking these regions were then considered likely targets of CtrA. Of these, 116 had been assayed by the microarray expression profiling reported by their earlier paper, as well as new expression profiling they performed of a *ctrA* temperature-sensitive mutant over a 4 h time period (longer than the 2.5 h cell cycle) that was aimed at identifying CtrA-dependent genes (including those not involved in the cell cycle). Combining these three data sets together allowed Laub *et al.* (2002) to identify 55 CtrA-binding sites that corresponded to 34 individual genes and 21 putative operons yielding a total of 95 genes. Among these included five genes involved in cell division and cell wall metabolism, 14 regulatory genes and 29 polar morphogenesis genes, with the remaining 47 either unknown (25) or not discussed (22). Notably, these also included *ccrM*, a methyltransferase previously known to be under CtrA regulation, as well the gene responsible for producing SAM (S-adenosylmethionine), the substrate used by CcrM for methylation. In addition, they also confirmed other prior results including those that showed CtrA had multiple binding sites in the origin of replication, as well as directly regulated a number of the main genes responsible for cell division, including *ftsA*, *ftsQ*, *ftsW* and *ftsZ*.

3.7.3 DivK

Soon after these global characterizations of CtrA effects, Hung and Shapiro (2002) described the impact of the single-domain response regulator, DivK, on the *Caulobacter* cell cycle by using a cold sensitive, *divK-cs*, strain. They discovered that when grown at the restrictive temperature, the *divK-cs* strain developed into long, filamentous stalk-like cells. A return to the permissive temperature allowed these cells to recover morphologically, as cell division was permitted to proceed, indicating the cell cycle of the *divK-cs* strain had been halted at the G_1-S stage by the restrictive temperature. To further explore this behaviour Hung and Shapiro next used cDNA microarrays to characterize the mRNA expression profiles of the *divK-cs* strain during growth in both the restrictive and permissive temperatures. From these, they discovered that many of the *Caulobacter* cell cycle genes, including those involved in DNA replication, as well as pili and flagellar synthesis, were repressed during growth in the restrictive temperature, but became induced following the return to the permissive temperature. Combining these new results with the prior understanding that CtrA must be proteolyzed in order for DNA replication to initiate, they next performed a series of immunoblot and pulse-chase analyses to examine CtrA quantities in the *divK-cs* strain. From these experiments, they discovered that at the restrictive temperature, the *divK-cs* strain failed to proteolyze CtrA, thus preventing the initiation of DNA replication, leading Hung and Shapiro to conclude that DivK is requisite for CtrA proteolysis. While the exact mechanism by which DivK mediated CtrA proteolysis was still unclear, Hung and Shapiro, noting that *divK* had been shown to be part of the CtrA regulon, further concluded that the two participate in a regulatory circuit with each other.

3.7.4 Dissection of CckA's global effect

Shortly following these reports on the role of DivK in the *Caulobacter* cell cycle, Jacobs *et al.* (2003) described the results of a series of experiments performed to elucidate the effects of CckA, a histidine kinase, upon the phosphorylation of the CtrA response regulator. As phosphorylation of CtrA is one of the mechanisms by which CtrA activity is regulated and earlier studies had indicated CckA has a role in phosphorylating CtrA, the goal of their study was to explore CckA's role in regulating CtrA activity. As their initial step, Jacobs *et al.* (2003) used microarrays and gel electrophoresis to compare the RNA and protein expression profiles of a *ctrA* temperature-sensitive mutant strain with those of a temperature-sensitive mutant strain for *cckA*. Discovering that RNA and protein expression was virtually identical in both strains, Jacobs *et al.* (2003) next used ^{32}P radiolabelling and immunoprecipitation with a *Caulobacter* wild-type strain to illustrate that phosphorylated CtrA and CckA (CtrA~P and CckA~P) possessed nearly matching patterns of expression during the cell cycle. Subsequent viability studies illustrated that while a $\Delta cckA$ mutant strain was unviable, it could be rescued via a phosphorylation-independent *ctrA* mutation, providing evidence that suggested CckA was crucial for providing

CckA~P-mediated phosphorylation of CtrA. A final test comparing RNA expression of a $\Delta ctrA\Delta cckA$ double mutant strain with that from a $\Delta cckA$ strain, revealed nearly identical expression of the cell cycle-regulated genes for both strains. As such, Jacobs *et al.* (2003) concluded from all these tests that CckA is a required regulator for CtrA phosphorylation and subsequent activation, though they were unsure of what the exact mechanism for this regulation is.

3.7.5 The cell cycle circuit circa 2004

Thus, from the results of these systems-level experiments, along with those from other non-systems-level studies of *Caulobacter* proteomic localization, a regulatory circuit centred on CtrA that governed *Caulobacter's* cell cycle gradually began to emerge by early 2004. For example, it was understood that CtrA was expressed at high levels during the SW cell (or G_1) stage, but was quickly proteolyzed by a ClpXP-dependent process during the transition to an ST cell. As a result of the decrease in CtrA in the cell, it was understood that the CtrA-controlled inhibition of DNA replication is released, allowing for replication to begin. Additionally, it was also understood that expression of *ctrA* was induced shortly following the initiation of replication, however, there was still confusion about the transcription machinery driving this (Skerker and Laub, 2004).

Specifically, by 2004 it was understood that as the levels of CtrA increase in the cell, CtrA acts to repress transcription from a weak upstream promoter, CtrAP1, while also activating expression from a stronger upstream promoter CtrAP2. It was still unclear, though, what exact mechanism was behind the expression of either of these two promoters. For example, it was understood that *ctrAP1* could only be expressed during the short window of replication when the new daughter strand is unmethylated. Furthermore, it had been discovered that newly expressed CtrA is quickly phosphorylated into its active form, CtrA~P, which subsequently induces expression of the CcrM methyltransferase that methylates the daughter strand. In doing so, CtrA~P inhibits further activity from the *ctrAP1* promoter. However, it was not yet clear what transcription factor induces the transcription from *ctrAP1* (Skerker and Laub, 2004).

It was also understood that the newly produced CtrA was phosphorylated (CtrA~P) and that in the ST portion of the PD cell CtrA was again proteolyzed by a ClpXP-dependent process, allowing DNA replication to continue. However, while the phosphorylation of CtrA was understood to be related to CckA phosphorylation, as described above, it was still unclear how the two were related. Furthermore, the mechanism that allowed for the localized degradation of CtrA within the ST end of the PD cell was still unknown, though, it was suspected that it was related to the localization to the ST end of DivJ, a DivK kinase, which as described above, will induce CtrA proteolysis (Skerker and Laub, 2004).

3.7.6 Holtzendorff's GcrA model

The next major step in the exploration of *Caulobacter's* cell cycle was provided by Holtzendorff *et al.* (2004) who reported that they had identified GcrA as a second master regulator of the *Caulobacter* cell cycle. In their findings, Holtzedndorff *et al.* (2004) discovered that GcrA participates in a regulatory circuit with CtrA where in the first step of this circuit *gcrA* is transcriptionally repressed by CtrA. However, in the next step of the circuit, the proteolysis of CtrA upon entry into S phase releases both the CtrA-mediated inhibition of DNA replication, as well as CtrA's repression of *gcrA* expression. This subsequently allows GcrA to induce *ctrA* expression from the CtrA P1 promoter during the short period while *ctrAP1* is still in its hemi-methylated state on the daughter strand. The circuit is closed when the resulting CtrA~P expression from the activation of the CtrA P1 promoter consequently re-represses *gcrA* transcription, thereby indirectly repressing the activation of the P1 promoter.

While the majority of the methods Holtzendorff *et al.* (2004) applied to identify the role of GcrA were not systems-level techniques, such as β-galactosidase assays and immunoblotting, they also performed expression profiling to characterize its regulon once its role had been identified. Using oligoarrays that contained probe sets for 3761 predicted ORFs, Holtzendorff examined the expression profile of a Δ*gcrA* mutant strain in which a copy of *gcrA* was added under the control of a xylose-inducible promoter. From the expression profile of this strain, Holtzendorff discovered 125 known cell cycle genes that were GcrA dependent. Of these 125 genes, however, only 8 overlapped with the CtrA regulon that had been identified previously by Laub *et al.* (2000, 2002). Moreover, the fact that the two regulons for CtrA and GcrA consisted of only 30% of the 553 cell cycle-regulated genes Laub *et al.* (2000) identified led Holtzendorff *et al.* (2004) to conclude that there were likely to exist additional proteins regulating *Caulobacter's* cell cycle.

3.7.7 Global exploration of the effects of DnaA

The next such cell cycle-regulating protein to be identified was DnaA, the DNA replication initiation factor. At the time, it was already well established that DnaA played a major role in the initiation of DNA replication whereby binding to specific binding motifs within the origin of replication, called DnaA boxes, it 'melts' the hydrogen bonds holding together the double-stranded DNA, allowing polymerases to access the individual strands. However, in 2005 Hottes *et al.* (2005) published results that indicated, similar to both *E. coli* and *B. subtilis*, DnaA also functioned as a transcription factor in *Caulobacter*. Using a *dnaA*-inducible strain (*dnaA* under control of a xylose-inducible promoter), Hottes *et al.* (2005) performed expression profiling to identify 40 genes that were DnaA-dependent, 10 of which were known to be GcrA induced. They next used the *in silico* motif-prediction tool, MEME, to identify DnaA boxes within the upstream regions, of 13 of these, including *gcrA*, *ftsZ* and *podJ* which Hottes *et al.* (2005) verified by using electrophoretic mobility-shift assays. Given these results, Hottes *et al.* (2005) concluded that these 13 genes comprised a regulon under the direct transcriptional control of DnaA, with DnaA serving as a promoter for GcrA, FtsZ and PodJ.

3.7.8 Holtzendorff's model of the cell cycle control circuit

Thus, by this point, we had an emerging model involving three master regulators. Starting with active CtrA~P, the dephosphorylation and proteolysis of CtrA release its repression of DNA replication as well as both the *gcrA* promoter (P_{gcrA}) and its own weak P1 promoter (*ctrA*P1). The release of this CtrA-mediated repression consequently allows DnaA to induce expression of GcrA. In kind, GcrA induces expression of *ctrA* via expression of CtrA's weak P1 promoter, see Fig. 3.1. However, as illustrated in Fig. 3.1, this newly expressed and phosphorylized CtrA (CtrA~P) subsequently further accelerates its own induction by simultaneously repressing its P1 promoter, while inducing expression of its stronger P2 promoter (*ctrA*P2), with this repression of its P1 promoter occurring via two mechanisms. The first mechanism was direct repression of *ctrA*P1 by the binding CtrA~P upstream of the P1 promoter. The second occurring when CtrA-induced expression of the CcrM methyltransferase methylates the newly generated daughter strand, and thereby completely suppresses further expression of the P1 promoter by GcrA (Holtzendorff *et al.*, 2006). However, still left unanswered by this model are questions such as what is the mechanism by which phosphorylated CckA (CckA~P) controls the phosphorylation (and, thus activity) of CtrA. Another is the question of what is the mechanism by which phosphorylated DivK (DivK~P) induces the dephosphorylation and proteolysis of CtrA. A recent work by Biondi *et al.* (2006b) addresses many of these questions; however, before discussing this paper, we need to make a brief detour to describe the underlying methods and motivation of the work.

3.7.9 Skerker *et al.*'s phosphotransfer method

In their paper, Biondi *et al.* (2006b) utilized a biochemical phosphotransfer mapping method that had been developed in their lab and described by Skerker *et al.* (2005) which they named phosphotranfer profiling. In this phosphotransfer profiling technique, a soluble kinase domain of a histidine kinase is autophosphorylated with radiolabelled adenosine triphosphate ($[\gamma^{-32}]$ATP) and then incubated in separate in vitro experiments with each individual full-length response regulator. Using an added autophosphorylated histidine kinase as a reference, phosphotranfer reactions between the kinase domain and their specific response regulators can be identified when the radiolabel is either depleted from the histidine kinase band or is transferred to the response regulator (which can be identified as a band that corresponds to its molecular weight). Therefore with this method, researchers can systematically examine the complete complement of response regulators of a given genome for phosphotransfer reactions with a given kinase.

With this phosphotransfer method, Skerker *et al.* (2005) identified a signalling pathway between the cell envelope proteins CenK and CenR, and soon after, Biondi working with Skerker and others used the method to identify a signalling pathway involved in stalk biogenesis between ShkA and TacA (Biondi *et al.*, 2006b). Later, noting these open questions regarding CckA and DivK and their relationships with CtrA, Biondi *et al.* (2006a) set out to determine their roles in *Caulobacter's* cell cycle. Their first step was to definitively determine whether or not CckA had the capacity to phosphorylize CtrA, which they accomplished by using phosphotransfer profiling. From these tests, Biondi *et al.* (2006a) determined that while CckA could autophosphorylate via the phosphorylation of its reciever domain (CckA-RD) by its histidine kinase domain (CckA-HK), CckA had no direct role in the phosphorylation of CtrA. Given these results, they suspected there existed a histidine phosphotransferase (HPT) which served as an intermediary between CckA~P and CtrA, as Jacobs *et al.* (2003) had speculated in their initial exploration of CckA's relationship with CtrA.

3.7.10 Identifying the key histidine phosphotransferase

However, none of the predicted genes in the *Caulobacter* genome were annotated as being an HPT. Therefore, using common characteristics of HPTs as criteria, along with the requirement that any such gene must have an orthologue in another genome that also contained orthologues for CckA and CtrA as well, Biondi *et al.* (2006a) identified a single candidate that they subsequently named ChpT. To validate this hypothesis, they next performed viability as well as expression profiling experiments of a *chpT* deletion strain containing a plasmid with a xylose-inducible copy of *chpT*. From these tests, Biondi *et al.* (2006a) discovered that in a glucose-only environment this strain was virtually identical to the *ctrA^{ts}* and *cckA^{ts}* strains that Jacobs *et al.* (2003) had used when grown at the restrictive temperature, strongly indicating a connection between the three genes. Given these results, Biondi *et al.* (2006a) next returned to the phosphotransfer profiling method to examine the relationship between these three genes. From this method, Biondi *et al.* (2006a) ascertained that, indeed, ChpT serves as the HPT bridge between CckA and CtrA.

Furthermore, they also discovered that while CckA is ChpT's only input, ChpT can phosphorylate both CtrA as well as the single-domain response regulator, CpdR, which had only just recently been shown by Iniesta *et al.* (2006) to be critical to the localization of CtrA's protease, ClpXP, to the ST cell pole during the SW to ST transition. Though not discussed in detail here as it was primarily a non-systems-level study, this earlier work had demonstrated that CpdR while in its unphosphorylated state controls the localization of ClpXP to the ST cell pole, thereby facilitating CtrA proteolysis by ClpXP. Moreover, they too had demonstrated that Cck~P was responsible for CpdR phosphorylation, resulting in ClpXP delocalization from the pole. Thus, by demonstrating that ChpT served as the HPT between both CtrA and CpdR, Biondi *et al.* had shown the mechanism by which CckA both activated and prevented its proteolysis.

3.7.11 DivK's role in CtrA regulation

With these results indicating a clear phosphotransfer CckA-ChpT-CtrA pathway, Biondi *et al.* (2006a) turned their attention to DivK and its role in the dephosphorylation and proteolysis of CtrA. Combining their results along with those of Hung and Shapiro who had shown that a *divK^{ts}* mutant strain was phenotypically similar to a constitutive expressing CtrA strain led Biondi *et al.* (2006a) to hypothesize that phosphorylated DivK (DivK~P) inhibited CtrA activity by inhibiting activity of CckA. To verify their theory, they compared the CckA~P levels within a *divK^{ts}* mutant strain with those of a wild-type strain, finding a fourfold increase of CckA~P in the *divK^{ts}* mutant strain, giving evidence that DivK~P inhibited CckA~P.

3.7.12 *divK* localization impacts cckA

However, as Jacobs *et al.* (2003) had illustrated that CckA~P was also dynamically localized during the cell cycle, Biondi *et al.* (2006a) performed a long series of GFP localization experiments to determine the mechanisms driving this. In their earlier work, Jacobs *et al.* (2003) had shown that CckA was localized to the SW pole during G$_1$ phase, but was subsequently delocalized during the G$_1$-S phase transition before becoming localized to both poles of the PD cell and then later delocalized in the new ST cell. While not discussed in detail, Biondi *et al.* (2006a) used GFP localization experiments to illustrate that indeed, DivK~P triggers CckA to delocalize and inactivate, resulting in a consequent inactivation of CtrA. Furthermore, as previous studies had shown that DivJ, a DivK kinase, localized to the ST pole, while PleC, a DivK~P phosphatase, localized to the SW pole, Biondi *et al.* (Biondi *et al.*, 2006a) hypothesized that cell division was crucial for DivK~P-induced delocalization of CckA which they also verified using GFP localization experiments. Finally, using a constitutively expression DivK strain, they also demonstrated that the timing of DivK expression, normally mediated by CtrA, was necessary for normal or wild-type cell cycle progression.

3.7.13 The current model

Thus, from the results of this work have emerged two more new feedback loops that drive the *Caulobacter* cell cycle, both of which involve a phosphotransfer cascade that starts with CckA and its activated form, CckA~P, and are determined by the proteomic localization within the cell. In this circuit, as is illustrated in Fig. 3.1, CtrA~P induces expression of DivK, which when phosphorylated by its kinase, DivJ, will cause delocalization and proteolysis of CckA, preventing CckA from initiating this cascade. In contrast, when DivK~P is inactivated by its phosphatase, PleC, into its inactive form DivK, this repression of CckA is lifted, allowing CckA to initiate a phosphotranser cascade that passes through the HPT, ChpT. In turn, ChpT~P both deactivates CpdR-mediated proteolysis of CtrA by phophorylating it into its inactive form, CpdR~P, as well as phosphorylating CtrA into its active form, CtrA~P, thereby completing the loop. Significant to understanding this regulatory circuit is to recognize the role that localization plays in determining the activity and inactivity of DivK. Specifically, as illustrated in Fig. 3.2, DivK's kinase and phosphatase, DivJ and PleC, respectively, are located in the two opposing poles of a late PD cell, with the PleC phosphatase in the SW pole and DivJ in the ST pole. As such, with PleC in the SW pole inhibiting DivK~P activity, CtrA~P is left unencumbered to repress further DNA replication, while the opposite is the case in the ST pole, where DivJ-induced phosphorylation of DivK and subsequent proteolysis of CtrA~P allows replication to reinitiate.

3.7.14 Future *Caulobacter* work

Thus far, the bulk of the current research has focused on the regulatory relationships of the *Caulobacter* cell cycle. However, as Biondi *et al.* identified, the temporal dynamics of expression will need to be an area of further study. Additionally, given its crucial role in the organism, localization and the mechanisms driving this deserve further attention. On this last point, effort has focused on the polar organelle development protein, PodJ, which has been associated with PleC localization, though, neither the exact relationship and mechanism is known at this time (Jacobs-Wagner, 2004) nor is that which determines DivJ localization. Further mapping of *Caulobacter's* stress response and metabolism also present areas for further research as well.

3.8 Bacillus subtilis

B. subtilis is one of the best studied model organisms in biology today. *B. subtilis* is a robust, non-pathogenic, aerobic, rod-shaped bacterium in the division Firmicutes; it is a member of the class Bacilli that includes other gram-positive genera such as *Staphylococcus, Streptococcus, Enterococcus* and *Clostridium*. As a model organism, *B. subtilis* has been studied for over a century (happily predating the earliest PubMed article), it was chosen as the best representative of the gram-positive bacteria, and studying it can help us understand the biology of these organisms. The importance of the Bacillus genus spans biomedicine (with several pathogenic spore forming closely related species), industry (with several economically critical syntheses carried out in Bacillus species) and agriculture (members of the genus are insect pathogens that are used as a bioinsecticide). Bacilli are commonly found in soil, water sources and in association with plants (Kunst *et al.*, 1997). *B. subtilis* can be manipulated with relative ease since much of its genetics, biochemistry and physiology are well established. Other important properties that make *B. subtilis* useful to study are: it is naturally competent, can form endospores, contains systems for motility, has a highly diversified set of two-component signal transduction pathways, quorum sensing and a protein secretion system useful for expression of engineered proteins.

B. subtilis plays an important role in industrial and medical fields and has been used as a platform for the biosynthesis of small molecules and proteins because it is one of several bacteria that can secrete enzymes at gram per litre concentrations directly into the medium (Kunst *et al.*, 1997). It is known specifically for producing proteases and amylases and is currently being developed as a vaccine development platform (Kunst *et al.*, 1997; Ferreira *et al.*, 2005). Importantly, its secretion system is more compact (has fewer components) than that of *E. coli* (Yamane *et al.*, 2004).

Bacteria commonly use a two-component signal transduction mechanism to respond to changing environmental conditions (Fabret *et al.*, 1999). These phosphotransfer systems contain two components, a histidine protein kinase that autophosphorylates, and a response regulator protein that elicits a specific response (as described above) (Stock *et al.*, 2000; Mascher *et al.*, 2006). Homologous versions of this system in several organisms have been shown to initiate and direct various processes such as sporulation, chemotaxis, aerobic and anaerobic respiration and competence (Fabret *et al.*, 1999; Ogura *et al.*, 2007).

Several species of Bacillus also produce and release chemical signals, called autoinducers or pheromones, which act as cell–cell signalling molecules between bacteria (Miller and Bassler, 2001). As population density increases so do these signals, until a threshold is reached and gene expression is modulated. This process is called quorum sensing and controls responses such as competence, sporulation, motility, biofilm formation and others (Miller and Bassler, 2001). Quorum sensing is an active area of research, as biofilm formation is critical to several biomedical and bioindustrial applications.

3.8.1 Genome sequence and annotation

The complete genome sequence of *B. subtilis* became available in 1997 revealing a sequence of 4.21 Mbp containing about 4106 protein coding genes (Kunst *et al.*, 1997). Bioinformatics approaches revealed other properties of the genome such as, a large family of putative ABC transporters, a variable G+C ratio of 43.5%, repetitive elements, and an average predicted protein size of 890 bp (Kunst *et al.*, 1997). The *B. subtilis* genome is similar in size to that of *E. coli* (4.6 Mbp) and shares roughly 1000 orthologous genes. Comparing these two genomes, which diverged about one billion years ago, will facilitate evolutionary studies of core genes, while comparisons of *B. subtilis* to other more closely related genomes, such as *Bacillus anthracis*, may provide information about conserved promoter structure and aid in diverse bioinformatics techniques from biclustering to gene finding.

3.8.2 Initial forays into transcriptomics

Exploration of whole genome expression profiles in *B. subtilis* began in 2000 by Fawcett *et al.* (2000), who were able to assign a number of genes to the sporulation process by using nylon-substrate macroarrays, covering ~96% of predicted ORFs, and Hidden Markov models to study the transcriptional profile of early to middle stages of sporulation. Ye *et al.* (2000), using two-colour glass slide arrays, compared mRNA levels from aerobic and anaerobic conditions. The results of these initial genome-wide investigations revealed complex expression patterns, including many genes of unknown function with highly different expression under the measured conditions, indicating that much still remained to be learned about the control of spore formations and spore induction/control.

3.8.3 Bacillus stress responses

A number of investigations have focused on the cellular response to stress at the transcriptome level in *B. subtilis* (this so-called stress response is a key focus of several prokaryotic functional genomics projects). Yoshida *et al.* (2001), studied glucose repression by a combined approach of microarray and 2DE, with a focus on the genes dependent on catabolite control protein, CcpA. Helmann *et al.* (2001) investigated the general stress response to heat shock in order to establish its profile thus allowing it to be compared to other stress response profiles. Nakano *et al.* (2003) described the role of Spx as a global transcriptional regulator of disulfide stress conditions. Ren *et al.* (2004) observed the induction of stress response genes by investigating the growth inhibition mechanism of a natural brominated furanone. Also in the search of new antibiotics, Lin *et al.* (2005) determined *B. subtilis* expression profiles in response to treatment with subinhibitory amounts of chloramphenicol, erythromycin and gentamicin. Hayashi *et al.* (2005) determined that there is a direct interaction, during H_2O_2 oxidative stress, between PerR, a stress response regulator, and srfA, an operon involved in surfactin biosynthesis. Allenby *et al.* (2005) characterized the phosphate starvation, PhoP, regulon, identifying some new members and a connection to the sigB general stress regulon. Ogura and Fujita (2007) investigated the role of RapD, one of 11 Rap proteins that typically inhibit response regulators, and found it to be a negative regulator, in conjunction with SigX and RghR, of the ComA regulon.

Overall these genomic studies helped to bring in many key proteins that would have been missed, including several proteins never before linked to a known process. Once these proteins are discovered by genomic techniques they are quickly validated and integrated into the aggregate picture of stress response. Furthermore, identification of key genes and proteins has enabled the construction of networks between the various pathways and processes within the cell.

3.8.4 Exploration of Bacillus two-component regulatory systems

As described above, two-component regulatory systems are characterized by a sensor protein (e.g. kinase) and a response regulator protein (e.g. DNA-binding protein). Ogura *et al.* (2001) began using whole genome microarray analysis in order to identify the target genes of the response regulators DegU, ComA and PhoP. Using the same strategy as Ogura *et al.* (2001), overexpressing the response regulator in mutants for their sensor kinase, Kobayashi *et al.* (2001) further analyzed 24 different two-component regulatory systems. These studies greatly expanded our knowledge of kinase target-gene specificity, and interestingly, the role of crosstalk between these sensory systems. For example, they identified many new genes regulated by ComK along with some previously known genes and identified a cellular state they called, the K-state, as a time for the cell to rest and recover from stress that is separate from sporulation (Berka *et al.*, 2002). This work was quickly followed up by Ogura *et al.* (2002), who then explored the roles of many ComK-regulated genes, in order to better understand competence. Britton *et al.* (2002), performed a genome-wide analysis of sigmaH, which is involved mainly in transitioning from growth to stationary phase, but is also involved in initiation into sporulation and competence. Hamon *et al.* (2004) investigated genes involved in biofilm formation that are regulated by AbrB whose results led to the discovery of two non-transcription factor gene products, a signal peptidase and a secreted protein, that play an essential role in biofilm formation. Serizawa *et al.* (2005) studied the YvrGHb two-component system and found it to control the maintenance of the cell surface and its proteins, as well as being involved in preventing autolysis. Keijser *et al.* (2007) investigated the regulatory process and outlined key events of spore germination and outgrowth by microscopy, genome-wide expression profiles and metabolite analysis.

3.8.5 Other uses for microarrays

Several reports have focused on RNAs other than mRNA, such as tRNA (transfer RNA), untranslated RNAs and RNAs involved in the processing of other RNAs. Ohashi *et al.* (2003) examined the modulation of the translation machinery during sporulation, finding in accordance with previous reports that there tends to be a dramatic global decrease in RNA, but that certain ribosomal rRNA and mRNA genes either remain the same or can increase. Dittmar *et al.* (2004) aimed to quantify tRNA transcription, processing and degradation levels on a genomic scale and developed specifically for tRNAs, a microarray and method of selectively labelling them. Silvaggi *et al.*

(2006) investigated the small non-translated RNAs involved in sporulation by microarray analysis with a microarray of intergenic regions as probes and a comparative computational analysis that predicts conserved RNA secondary structures.

Earl *et al.* (2007) examined 17 *B. subtilis* strains in order to quantify their diversity and identify regions of variability by microarray-based comparative genomic hybridization (M-CGH). M-CGH results in a measure of gene presence or absence by quantifying the relative hybridization efficiencies from two differently labelled bacterial strains. As bacterial genomes are dynamic, they found the gene content of their collection of strains to have at least 28% variability, meaning the genes could either have diverged or are missing.

3.8.6 Probing Bacillus with ChIP-chip

ChIP-chip in combination with transcriptional profiling and gel electrophoretic mobility-shift assays has been performed to identify 103 additional genes regulated by Spo0A, the master regulator for entry into sporulation (Molle *et al.*, 2003a) and many new targets of CodY, a guanosine-5′-triphosphate (GTP)-activated repressor of early stationary genes in *B. subtilis* (Molle *et al.*, 2003b). Also, a centromere-like element in *B. subtilis* was defined by mapping the binding sites for RacA, a chromosome remodelling and anchoring gene, and identifying 25 high selectivity-binding sites (Ben-Yehuda *et al.*, 2005).

3.8.7 The *B. subtilis* proteome

The global study of proteomes (e.g. using mass spectroscopy coupled with multiple separation strategies) lags behind transcriptome studies in reproducibility, cost and accuracy. Studying the dynamic proteome is confounded by several factors, for example: (1) there is a lack of cost-effective methods for designing high-affinity, high-specificity capture agents for all proteins in a given genome, and (2) several post-translational modifications of a protein can complicate its identification and quantification. The genome of *B. subtilis* contains more than 4100 genes and therefore we expect at least on the order of 4100 gene products. The proteome of *B. subtilis* has been studied for more than 20 years starting with explorations of heat shock proteins (Streips and Polio, 1985). Then with the sequencing of the genome, establishment of on-line databanks, and advances in MS and two-dimensional polyacrylamide gel electrophoresis (2D-PAGE) technology, proteome-wide characterizations became possible. In the cytosol of vegetatively growing cells, Buttner *et al.* (2001), first identified over 300 proteins, then Eymann *et al.* (2004) identified 876 proteins. Tam Le *et al.* (2006) identified over 200 proteins in cells under stress or starvation conditions. Finally, Wolff *et al.* (2006, 2007) has increased the number of identified proteins to 1395, thus covering over one-third of the *B. subtilis* proteome. Clearly, with slightly more than a third of the *B. subtilis* proteome identified, dynamical characterization of the proteome (both levels of proteins and protein modifications) will reveal a great deal of novel biological information (sequence-specific degradation and translational control, specificity and dynamics of modification, etc.).

3.8.8 Yeast 2-hybrid investigation of the Bacillus protein interaction network

As described above, Y2H analysis is a widely used method for detecting protein–protein interactions and screens can scale to test whole genomes (Fields and Song, 1989). Noirot-Gros *et al.* (2002), made an initial Y2H analysis of DNA replication components in *B. subtilis* identifying 69 proteins with 91 interactions. Their investigation yielded several interesting results that connect DNA replication to diverse cellular processes, including membrane and signalling pathways.

Predictions from the work of Noirot-Gros *et al.* (2002) influenced Meile *et al.* (2006) to perform a larger scale semi-systematic protein localization study for over 100 proteins in *B. subtilis*. To accomplish this, they developed a new approach for the rapid construction of GFP fusion constructs. In their study, 110 ORFs were selected, 50 chosen from known DNA replication components identified by previous Y2H screens. The remaining 60 selections were from various functional categories, including some of unknown function, from different functional categories based on annotations from Subtilist, Swiss-Prot and National Center for Biotechnology Information (NCBI). Overall, 90% of the proteins they studied were tagged with GFP with 78% tagged on both the N- and C-ends. In summary, they were able to identify interesting localization patterns for 85 previously unlocalized proteins, and thus identified new proteins associated with DNA-replication machinery. The locations of all proteins in the cell, under various conditions, will need to be compiled before there can be a clear picture of the organism at the systems level.

3.8.9 Investigating metabolome changes during sporulation

Clearly, the levels of metabolites are important to microbial biology, but methods for measuring the metabolome are much less widely adopted than methods for measuring the transcriptome and proteome. CE-MS is a powerful, quantitative tool for the direct and sensitive global analysis of metabolites. Soga *et al.* (2003) were able to determine a total of 1692 metabolites by splitting sample using three purification schemes (one each for cationic metabolites, anionic metabolites and nucleotides/coenzyme A compounds) in parallel to separate and subsequently identify metabolites (Soga *et al.*, 2003). To detect as many metabolites as possible they used an instrument wide range of approximately 70–1000 m/z. Their novel strategy was lengthy, 16 h per run, with several runs required, but is highly automated. Soga *et al.* (2003) used their metabolomic approach to profile metabolites before and during sporulation. They characterized unknown peaks by combining CE-MS results with bioinformatics and made headway into determining the (partially characterized prior) link between sporulation in B. subtilis and the metabolic network. Thus, revealing possible functional links from some uncharacterized metabolites. The power of their approach was nicely demonstrated by the ability to simultaneously monitor glycolytic, pentose phosphate and tricarboxylic acid pathway sporulation metabolite responses consistent with previous data. The study showed that metabolite concentrations cannot be accurately resolved by transcriptome analysis and revealed significant changes in metabolites during B. subtilis sporulation important for deciphering this important process.

3.8.10 A systems approach to reconstruction of the sporulation control circuit

As is commonly known, multi-cellular organisms contain many different types of cells. The mechanism of cellular differentiation is a fundamental problem in biology. Various developmental processes such as cell growth, morphogenesis and cell death occur in bacteria, with sporulation being a prime example. Sporulation can be considered a developmental process, albeit a simple one, as it is the process by which an organism differentiates from a vegetative cell type into a completely different cell type, the spore. The fate of each cell type is due to both its particular developmental gene expression programme, as well as its interaction with the cell's environment. B. subtilis, like many gram-positive, low G+C content, bacteria is known to undergo this transformation, and is among the best studied in this area. Inhospitable environmental conditions cause B. subtilis to begin the sporulation process, but it is typically induced in the laboratory by low nutrient conditions, for example the removal of a carbon, nitrogen or phosphorus source (Piggot and Hilbert, 2004). In the beginning of sporulation, a septum forms asymmetrically, near one end of the cell, dividing it into two cells, the larger mother cell and the smaller forespore; the forespore is to become the mature spore. Immediately following septum formation, the two cells have identical genomes but asymmetric gene expression programmes. In the next stage, the forespore is completely engulfed by the mother cell in a phagocytic-like process. The mother cell then nurtures the endospore surrounding it with proteins that form a spore cortex, and a spore coat. Finally, the mother cell lyses to release the fully developed and remarkably resilient spore (the spore is resistant to heat, ultraviolet (UV) and γ radiation, and various chemicals and enzymes). When nutrients are again sensed in the environment, the spore can germinate and flourish as a vegetative cell (Setlow, 2003).

Various independent transcriptome analyses have elucidated, on a genome-wide level, many of the relationships between genes, including a catalogue for sporulation the process of at least 600 genes (Fawcett et al., 2000; Britton et al., 2002; Eichenberger et al., 2003; Feucht et al., 2003; Molle et al.,

2003a). Eichenberger *et al.* (Eichenberger *et al.*, 2004) utilized an elegant microarray strategy in conjunction with computational, biochemical and in vivo analyses attempting to take transcriptome analysis a step further. Their systems-level investigation comprehensibly illustrated a regulatory circuit by integrating data from transcriptomics and genomics approaches thus characterizing the mechanism controlling the cell's decision to sporulate, and the timing of the process by which the spore is assembled.

Transcription in bacteria is mediated by sigma (σ) factors (general transcription factors involved in a large fraction of bacterial transcription initiations). Sigma factors bind to specific promoter regions, and in Bacillus have been shown to be master regulators with sequence-specific affinity for separate promoters. There are at least 17 sigma factors in *B. subtilis* but only 6 have a notable role in sporulation (Moszer, 1998; Moszer *et al.*, 2002). Gene expression during sporulation is coordinated by four sigma factors σ^E, σ^F, σ^G and σ^K. The regulatory cascade in the forespore is initiated by σ^F; it includes 48 genes organized in 36 transcription units whose products govern spore morphogenesis and germination properties (Wang *et al.*, 2006). After engulfment, σ^G regulates transcription of genes involved in chromosome condensation and equipping the spore for germination. In the mother cell, σ^E begins the cascade and turns on 262 genes (Zheng and Losick, 1990; Eichenberger *et al.*, 2003, 2004). Two of the targets of σ^E are DNA-binding proteins, SpoIIID and GerR (Kunkel *et al.*, 1989; Stevens and Errington, 1990; Tatti *et al.*, 1991; Errington, 2003; Eichenberger *et al.*, 2004). The function of GerR was previously unknown and now has a role as a negative regulator, switching off genes in the σ^E regulon. SpoIIID is interesting in that it acts as a repressor for some genes activated by σ^E and activates additional genes in conjunction with σ^E. SpoIIID is important for activating many coat proteins and especially the genes for an inactive proprotein, pro-σ^K, that ultimately converts to mature σ^K upon reception of an intercellular signal governed by forespore specific σ^G. This signal is important for keeping the separate mother cell and forespore programmes coordinated during the morphogenesis (Errington, 2003; Hilbert and Piggot, 2004). The σ^K regulon includes sets of genes for the spore cortex, structural components of the spore coat and germination (Steil *et al.*, 2005), and importantly GerE. Last in the mother cell line hierarchy, GerE, a DNA-binding protein, activates a final set of 36 genes and represses about half of the genes activated by σ^K. For example, two cell wall hydrolases are activated that play a role in lysis of the mother cell when spore morphogenesis is complete.

Eichenberger *et al.* (2004) compared RNA from mutants in transcriptional regulators suspected/known to control sporulation; using prior knowledge of the sporulation process they were able to construct near-optimal experimental designs for measuring the effects of these perturbed transcription factors. As a result of their transcriptional profiling strategy, two DNA-binding proteins, SpoIIID and GerR, turned on by σ^E were found to have significant effects on the σ^E regulon. SpoIIID extensively affects the σ^E-regulated transcription pattern, influencing over half of the σ^E regulon. This seems to be accomplished by direct interaction, as evidenced by assaying the promoter regions of the modulated genes. Evidence for direct interaction with the promoter regions was obtained first by identifying SpoIIID-binding sites with gel electrophoresis mobility-shift assays and DNAse I footprinting. Their application of in vivo ChIP-chip revealed many regions on the chromosome that SpoIIID bounds that did not include genes not known to be under its control, and some sites were located within protein coding regions, possibly indicating an architectural role for SpoIIID. Finally, computational-binding site sequence analysis was used to find putative conserved motifs in the upstream region of genes regulated by SpoIIID. Analysis of GerR by transcriptional profiling found that no genes that were dependent upon GerR for activation, but many genes were inhibited by GerR. Following SpoIIID in the cascade, the σ^K regulon was delineated by transcriptional profiling and further resolved by computational sequence analysis to identify a conserved motif in the promoters of the σ^K-regulated genes. The last regulator in this cascade, another DNA-binding protein, GerE, was found to inhibit the expression of slightly over half of the σ^K regulon and activates at least 36 additional genes at the end of the mother cell line of gene expression.

A comprehensive programme of the mother cell line of gene expression can be drawn from these results together, see Fig. 3.3. The resulting model consists of a hierarchical regulatory cascade of three DNA-binding proteins (SpoIIID, GerR and GerE) and two general transcription factors (sigma factors σ^E and σ^K); σ^E begins the cascade by activating transcription of 262 genes. SpoIIID and GerR repress many genes of the σ^E regulon and SpoIIID and σ^E activate 10 additional genes. σ^K activates 75 more genes, and finally, GerE, represses over half of the σ^K regulon and activates 36 more genes. Eichenberger *et al.* (2004) compiled these results into a transcriptional network composed of a linked series of five type 1 feed forward loops (FFLs) (Milo *et al.*, 2002; Shen-Orr *et al.*, 2002; Mangan *et al.*, 2003). Two of the FFLs are coherent and have the property of being persistence detectors (low pass filters); these may be used to minimize the effect of high frequency noise (Mangan *et al.*, 2003). Three of the FFLs are incoherent and have the property of producing pulses of gene transcription (Mangan and Alon, 2003).

Finally, they performed comparative analyses to determine possible conservation of this spore formation circuit in other endospore-forming bacteria. There are differences in the presence of certain regulatory proteins, for example *Bacillus* and *Clostridium* contain orthologues for σ^E, σ^K and SpoIIID including conserved sequence recognition domains, but *Clostridium* is missing GerE and GerR. Also, there is variation in the composition of each individual regulon among species, for example: 75% of the *B. subtilis* σ^E regulon have orthologues in *B. anthracis* and *Bacillus cereus* whereas only 40% have orthologues in *Clostridium*, and 50% of the *B. subtilis* σ^K regulon have orthologues in *B. anthracis* and *B. cereus* compared to 20% that have orthologues in *Clostridium*. They show that this pattern of conservation is consistent with the fact that the σ^K regulon contains many components of the spore's outer surface and that spore surfaces of *B. subtilis*, *B. anthracis* and *B. cereus* are known to be quite different (Chada *et al.*, 2003), the low level of conservation among σ^K regulons may be due to adaptation to an ecological niche. Thus, the sporulation circuit (the regulatory control of the decision to sporulate and the subsequent control of spore assembly) is more conserved than the target protein components (the spore coat proteins). Finally, Wang *et al.* (2006) extended this work by investigating the forespore line of gene expression and synthesized a single model summarized in Fig. 3.3.

3.9 Escherichia coli

3.9.1 Early systems-wide studies and initial genome sequence

Discovered in 1886 by Theodore Escherich, *E. coli* is a gram-negative species of bacteria that inhabit the mammalian gut, specifically the colon or lower intestines. As one of the best studied organisms of the pre-genomic era, *E. coli*, like *B. subtilis* was an early target for sequencing and in 1997, the complete sequence for the K-12 (MG1655) strain, consisting of 4 639 221 base pairs, was completed and reported by Blattner *et al.* (1997). Later that same year, the first two genome-wide microarray studies of *Sacromyces cerevisiae* were reported. The first by DeRisi *et al.* (1997) used spotted cDNA arrays to profile the expression changes of yeast during diauxic shift, and then later Wodicka *et al.* (1997) used 25-mer oligonucleotide arrays from Affymetrix to profile the expression differences of yeast grown on rich versus minimal media. Closely following these initial studies, two early projects were performed to develop microarrays for *E. coli*. The first of these, described by Tao *et al.* (1999), was a microarray that used nylon membranes and radiolabels for the cDNA; making the experiment essentially a genome-wide Northern blot. In contrast, the second project by Wei *et al.* (2001) used the technique developed by Pat Brown to develop a two-colour, spotted cDNA microarray on glass slides. Shortly following these initial studies, Richmond *et al.* (1999) compared these two micorarray technologies by comparing the expression profiles reported for two well-studied environmental responses. Specifically, in their comparison, they used both technologies to explore the RNA expression profiles of *E. coli's* heat shock response, as well as exposure to the *lac* operon inducer, isopropyl-b-D-thiogalactopyranoside (IPTG). In their results, the authors reported that both microarray varieties indicated expression differences for genes in both the *lac* and melibiose operons for the IPTG tests, both of which were expected given previous published experimental work. A sizeable intersection between the genes that the two technologies reported as being induced during the heat shock response was found; 62 of the 77 genes reported by the nylon membrane microarrays were also identified as being induced by the glass cDNA arrays. In contrast, the authors reported little overlap between the genes the two technologies identified as being down-regulated. Despite this discrepancy, the authors concluded that glass microarrays were more reproducible and therefore recommended it as the preferred method.

3.9.2 Overview of early *E. coli* microarray studies

Shortly after these initial projects, Selinger *et al.* (2000) introduced the first Affymetrix chips designed for *E. coli* in a paper that compared the expression profiles of *E. coli* during logarithmic growth and stationary phases (on a rich medium). In addition to probes for the 4290 predicted ORFs in the *E. coli* genome, these new chips also contained probe sets for non-coding RNAs such as tRNAs and ribosomal rRNAs. While there was some discussion of results of the biological findings of their experiment, the focus of the paper, not surprisingly, was on the technology and the advantages offered by using short oligos, rather than whole cDNAs. The primary advantage being lower cross-hybridization. However, it is important to also note that as these were still the early days of microarray design, these chips had the design flaw of failing to randomize the location of the probes on the chip. For example, the top half of the array contained all the probe sets for ORFs and untranslated RNAs, while all the tRNA and rRNA probe sets were all located along the bottom edge of the chip. As described by Qian and Kluger (Qian *et al.*, 2003), chips that manifest such a linearity in probe location are prone to biasing the expression levels reported when there is an uneven distribution of RNA in the solution that is hybridized to the chip.

Following the announcements of these new *E. coli*-specific genome-wide microarrays, they were quickly adopted by researchers who began applying them in systems-wide studies of various environmental and metabolic responses. While many of these responses had already been the subject of earlier studies using previously existing genetic techniques, for most this was the first time they had been studied at a genome, or systems-wide level. Early examples include explorations of the 'SOS response (Courcelle *et al.*, 2001), metal-ion tolerance (Brocklehurst and Morby, 2000), osmostress (Weber and Jung, 2002) and adaptation to acetate and propionate (Polen *et al.*, 2003). More recent examples of stress-response examinations include inhibition of cell division (Arends and Weiss, 2004), anti-microbial peptides (Hong *et al.*, 2003; Tomasinsig *et al.*, 2004) and cadmium toxicity (Wang and Crowley, 2005).

These early studies were primarily descriptive in nature but were also key in motivating the development of several analysis techniques suited to these genome-wide measurement technologies. In this sense, they can be viewed as foundational as they reported systems-wide expression differences, from which new hypotheses could be drawn that could be validated and further explored in later studies. For example, it was shown by Barbosa and Levy

(2000), and later partially validated by Pomposiello *et al.* (2001) that there was a previously unknown overlap between the multiple antibiotic resistance and oxidative stress regulons (MarRA and SoxRS, respectively), a finding that would not have been easily identifiable using previous experimental methodologies. Another example would be the results reported by Zheng *et al.* (2001) who discovered an additional overlap for the SoxRS response regulon with that of the OxyR response regulon.

3.9.3 System-level studies of regulatory interactions governing the glutamate-dependent acid response (AR)

One example of how systems-level biology assisted in the study of *E. coli* focused on and helped elucidate a complex network of regulatory interactions governing its glutamate-dependent acid resistance or response. While the ability of *E. coli* to develop acid resistance was first observed over 50 years ago, it was not until 1995, during the pre-genomic era, that it was discovered that there exist four distinct systems within *E. coli* for acquiring AR (Lin *et al.*, 1995; Foster, 2004). These include one system that is repressed by glucose (and only functions in its absence), another that is dependent on arginine, as well as one more that is dependent upon lysine, and finally a fourth that is glutamate dependent (with the last three functioning in environments that include glucose). Of these, the glutamate-dependent system is the most effective, the best studied, and the one upon which systems biology has had the most impact.

The first systems-level foray into the understanding of *E. coli*'s glutamate-dependent AR was performed by Hommais *et al.* (2001), though the original goal of the work was to explore the role of *E. coli*'s nucleoid-associated protein, H-NS. Using nylon membrane microarrays to compare the RNA expression profiles of wild type and an *hns* mutant strain, Hommais *et al.* (2001) identified expression differences for genes involved in processes including those that were then known to be involved in osmolarity and acid resistance. Note, for the majority of the observed gene expression differences, the expression was induced or elevated in the Δ*hns* strain, leading them to conclude that H-NS was a repressor of gene regulation. Among the genes upregulated in the Δ*hns* strain included *evgA*, the regulator from the EvgAS two-component system, as well as *gadA* and *gadB*, the two glutamate decarboxlyases known to be required for acid resistance, as well as *gadC*, the GABA/glutamate antiporter required by AR. Noting the induction of the genes involved in acid resistance, Hommais *et al.* (2001) next explored the impact of the Δ*hns* upon acid resistance. Comparing the effects of arginine, lysine and glutamate acid stress upon both the Δ*hns* and the wild-type strains, Hommais *et al.* (2001) discovered that the Δ*hns* strain only conferred a resistance when in the presence of glutamate. Based on these results, Hommais *et al.* (2001) used plasmid-induced overexpression strains to identify *yhiX* (later renamed to *gadX*) as a gene whose overexpression will impart acid resistance, leading them to conclude that it was likely to be a transcription factor necessary for glutamate-dependent AR.

A year after Hommais *et al.* (2001) published their results, Masuda and Church (2002) set out to explore the regulon of the EvgA response regulator protein in the EvgAS two-component signalling system, with the hope that characterizing the response would help identify EvgA's functional role. To accomplish this, they used *E. coli* specific chips from Affymetrix to compare the expression profiles of EvgA knockout and overexpressing (via a transfected plasmid) strains to identify potential target genes of the EvgA regulon. Now, as EvgA's functional role was still unclear at the time, they also developed a similar set of strains from an *acrAB* knockout strain, as it had been reported by Nishino and Yamaguchi (2001) that EvgA overexpression would bestow antibiotic resistance to this strain. Comparing the expression profiles of all these strains, Masuda and Church were able to identify 79 genes with induced expression as well as another 24 that were repressed or reduced.

3.9.4 Exploring the genes necessary for acid resistance

Of these, they noted that several genes were known to be involved in conferring acid resistance, motivating their exploration of the effect of EvgA overexpression upon the organism's response to acid stress. Thus, to verify their hypothesis, they performed survivability tests for *E. coli* in a low pH environment and discovered that, as they suspected, EvgA overexpressing strains were, indeed, acid resistant. Given this validation, the authors then performed another series of survivability experiments using knockout strains for each of the genes most strongly induced by EvgA overexpression. From these tests, Masuda and Church were able to identify three genes, *ydeO*, *ydeP* and *yhiE* (later renamed to *gadE*) that were required for the AR of *E. coli* in logarithmic growth, while also discovering that *gadE* is key to the organism's AR while in stationary phase.

3.9.5 Identifying EvgA's role in acid resistance

Along with their findings for the acid stress response, Masuda and Church also performed a similar set of experiments to explore the drug resistance that was induced by EvgA overexpression in the Δ*acr* strains. In doing so, they were able to identify the YhiUV efflux pump and the TolC outer membrane channel proteins as being key to the Δ*acr* strain's drug resistance during EvgA overexpression. However, in later tests, they also observed that EvgA overexpression could not confer drug resistance for strains without this Δ*acr* deletion. For this reason, combining their observations about both the drug and acid shock response, Masuda and Church concluded that EvgA's primary role is not in coordinating the organism's drug response, but instead its acid shock response.

3.9.6 Expanding the list of AR regulators

In a similar project, performed nearly concurrently with that done by Masuda and Church, Nishino *et al.* (2003) partially validated Masuda and Church's findings. For example, they too recognized the induced expression of genes known to be involved in the organism's AR and thus tested the effects of EvgA overexpression on the survivability of the organism. While they also observed an increased resistance to acid shock, they however did not pursue this further and thus did not identify the critical roles of *ydeO*, *ydeP* and *gadE* in its AR.

In contrast to the Masuda and Church (2002) and Nishino *et al.* (2003) investigations, the goal of Tucker *et al.* (2002) was specifically to explore *E. coli*'s glutamate-dependent AR. To accomplish this, they used nylon membrane chips to compare the expression profiles of *E. coli* during logarithmic growth in glucose-rich media of varying pH, with pHs of 7.4, 5.5 and 4.5. Of the genes they identified as being induced were six genes that were either known or suspected of being transcription factors, including four in the *hdeA-gadA* region with these being *yhiF*, *gadE*, *gadX* and *gadW*[1]. Similar to Masuda and Church, to further explore the roles of the induced genes, Tucker *et al.* (2002) generated gene knockout strains and performed survivability tests on these. Focusing on seven genes in the *hdeA-gadA* region, they discovered that only one, *gadE*, was critical for the organism to become acid resistant and for this reason, they concluded that it likely was an AR transcription factor.

Following up their initial study, Masuda and Church (2003) developed a set of deletion and overexpression *E. coli* strains for each of the *ydeO*, *ydeP* and *gadE* genes they identified in their earlier study. From the results of a series of susceptibility tests for these strains, they hypothesized that there exists a set of cascading regulatory interactions where EvgA induces YdeO which subsequently induces GadE. To validate this, they used a combination of in vitro and *in silico* systems-level methods. Specifically, via the expression profiles of a new set of deletion mutants for the *ydeO* and *evgA* genes, individually and in combination, Masuda and Church identified two distinct regulons. One was induced directly by EvgA expression (including YdeO), while the other was indirectly induced by EvgA via YdeO. To further validate EvgA induction of YdeO, Masuda and Church used the *in silico* motif discovery tool, ALIGNACE (Roth *et al.*, 1998), to identify a putative 18 bp binding motif in the upstream regions of the genes that they predicted to be induced directly by EvgA. Next, the putative-binding sites in the upstream regions of *ydeP* and *b1500* (a gene upstream of *ydeO* that they suspected formed an operon with it) were mutated in a new set of *E. coli* strains that were subsequently

---

[1] Note, in the text the last three genes are referred to as *yhiE*, *yhiX* and *yhiW*, but were later renamed using the *gad* prefix once they were recognized as being members of the glutamate AR regulon. For the sake of clarity and consistency, we use their current naming scheme rather than those used in the original text.

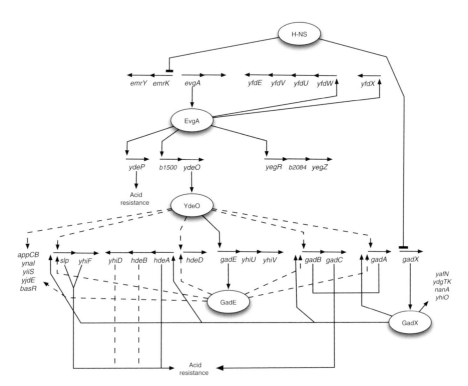

Figure 3.4   *E. coli* glutamate dependent acid resistance circuit. Masuda and Church's model for *E. coli's* glutamate dependent acid response, adapted from Masuda and Church (2002). Solid lines represent confirmed regulatory relationships, while dotted lines represent relationships that were unclear. In this model, H-NS serves to repress evgA and gadX expression, while EvgA expression induces expression of ydeO and ydeP, both necessary for acid resistance, with YdeO expression inducing gadE expression. Additionally, dotted lines are used to connect both YdeO and GadE with gadA, hdeD and the slp-yhiF, hdeAB-yhiD and gadBC operons to reflect uncertainty as to whether these were under YdeO's direct control or via GadE. Reflecting their conclusion that GadX did not induce GadE expression, GadX is shown to induce some of the acid response genes and operons, but not gadE-yhiUV.

subjected to acid resistance tests. The results from these experiments indicated that the putative EvgA-binding sites were, as they suspected, necessary for acid resistance. Combining their latest results with those of Hommais *et al.* (2001), Masuda and Church postulated a regulatory cascade with H-NS repressing EvgA, while EvgA induces YdeO. As described above, YdeO was proposed to induce GadE, the transcription factor responsible for inducing acid resistance, with their complete model summarized in Fig. 3.4.

It is important to note, however, that in the network they proposed, Masuda and Church argued that GadX – an AraC-like protein identified by Hommais *et al.* (2001) as being induced in *hns* mutants – did not induce GadE. In contrast, earlier studies had concluded that GadX was part of a complex regulatory circuit involving another AraC-like protein GadW, the stress sigma factor, σ^S, and CRP (cAMP receptor protein). Masuda and Church based their argument on a comparison of the expression profiles of *gadX* deletion and *gadX* over-expression strains during exponential growth, which did not show *gadE* to be differentially expressed. In contrast, in a nearly concurrent study, Tucker *et al.* (2003) compared the expression profiles of wild type and deletion strains for *gadX* and *gadW* during stationary phase which they believed indicated a regulatory relationship between GadX and GadE. However, Tucker *et al.* (2003) argued that GadX works with GadW to integrate signals from other sources, though the exact mechanism of this is as yet unknown. Regardless, this inconsistency reminds us that co-expression is one facet of a highly interconnected system (one on many informational levels including protein, protein modification, etc.), but do not diminish the pioneering contributions made by these first global studies.

3.9.7 Global computational models of *E. coli* metabolism and regulation

In addition to allowing researchers to study *E. coli* on a genome-wide scale, the completion of the sequencing of the *E. coli* genome also opened the door for the first genome-scale *in silico* models. In fact, *in silico* models of *E. coli* have been around since as early as 1990 (Majewski and Domach, 1990); however, these were limited in both scale and complexity, usually comprising a small set of genes and modelling only a few processes. In contrast, the more recent models of regulation and metabolism contain thousands of genes involved in a nearly comprehensive number of processes (Covert *et al.*, 2004).

We will describe these models and the ways they are being used in greater detail below, but first we need to cover a few basics.

3.9.8 Data-driven models of the *E. coli* regulatory network

Generally speaking, the full spectrum of *in silico* research can be divided into two distinct classes consisting of: (1) regulatory network inference and (2) modelling of the full metabolic network and its interactions with a subset of the regulatory network. While there are networks that have been generated via manual collation and collection of experimentally validated interactions from published literature (Ogata *et al.*, 1998; Karp *et al.*, 2000; Salgado *et al.*, 2006), it is expensive and time consuming to create and maintain these networks as they require expert knowledge and extensive experimentation (full field times 50 years) to be generated. As a result, there have been a number of *in silico* methods developed that attempt to infer regulatory relationships from genome-wide experimental data such as microarray expression and ChIP-chip data. Often, these methods use computational learning algorithms that have been adapted to work specifically with biological data

(D'Haeseleer *et al.*, 1999; Friedman *et al.*, 2000; Vanet *et al.*, 2000; van Someren *et al.*, 2000, 2002; Segal *et al.*, 2001, 2003; Bar-Joseph *et al.*, 2003; Stuart *et al.*, 2003; Hashimoto *et al.*, 2004; Bonneau *et al.*, 2006; Slonim *et al.*, 2006; Faith *et al.*, 2007). In addition to offering the prospect of a cheaper and less costly solution, these automatic methods also have the possibility of identifying previously unknown protein interactions, providing quantitative means for experimental design, and a means for inferring the roles of genes of unknown function. The development of these algorithms is still in the nascent stage and currently there exists no 'gold standard' data set or known interaction map that can be used to gauge their performance. We will return to the second class (models including flux through metabolic networks) shortly, but first let us discuss one network inference project that has recently been applied to *E. coli*.

While, as mentioned above, there have been a number of efforts reported in recent years to infer the regulatory networks of various organisms and other systems, the first such effort for *E. coli* has only recently been reported in early 2007 by Faith *et al.* (2007). Faith *et al.* (2007) used an unsupervised network inference method, the context likelihood relatedness algorithm (or CLR), of their own construction on a data set consisting of 445 *E. coli* Affymetrix microarray expression measurements coming from both published sources as well as new experiments (> 1/2 of the collated data was new). Once generated, they then validated their inferred network using a combination of *in vitro* and *in silico* methods. We will discuss the overall work and its findings shortly, but first let us describe the inference method they used.

3.9.9　The CLR algorithm

The CLR algorithm compares expression profiles of the genes by utilizing mutual information (MI), a commonly used information theoretic similarity measure. MI is defined as the *relative entropy* between the joint distribution and the product distribution of two random variables, X and Y, defined mathematically as:

$$I(X:Y) = \sum_x \sum_y p(x, y) \log \left(\frac{p(x, y)}{p(x)p(y)} \right)$$

where I(X:Y) is the MI for two variables (in this case the levels of two genes under a large number of conditions), $p(x)$ is the probability of seeing a value for x in the distribution, $p(y)$ is the probability of seeing a value for y in the distribution and $p(x, y)$ is the probability of seeing a given value for x and y in a single observation or sample.

Generally speaking, MI can be understood to be a measure of the coupling between the distributions of two random variables, or in the case of the CLR algorithm, the similarity between the distributions of two genes. So, to get an intuition of how this measure operates, consider an example of two genes that are completely independent of each other such that $p(gene_1, gene_2) = p(gene_1)$ $p(gene_2)$. As this situation would give us a fractional value (the fraction inside the log function, that is) equal to 1, we would get an MI of 0 as $\log(1) = 0$. Additionally, another key aspect of MI is that we are guaranteed that the MI between any two variables will be greater than or equal to 0. Thus, MI is a measure of the non-independence of two variables (or genes in our case). Importantly, the measure can detect relationships that would not be detected by a metric such as the Pearson correlation.

The CLR algorithm first calculates the background distribution of MI scores for each gene, estimated for each gene by determining the pairwise MI between it and the rest of the genes in the data set. Then, using this background distribution of pairwise MI scores, the CLR algorithm calculates the likelihood of their score. In doing so, this allows the CLR algorithm to filter out those genes that have spurious similarities with large numbers of other genes.

To improve the likelihood that high scoring gene pairs are causal and improve the stability and run time of their algorithm, Faith *et al.* (2007) selected a subset containing 328 known or putative transcription factors and used these as the centroids or mediods of their clustering scheme. In doing so, they correctly reduced both the overall search space of their algorithm, improved the stability of the result with respect to small changes in the data and reduced the cost of the requisite computation.

To validate the results from their CLR algorithm, Faith *et al.* (2007) used the RegulonDB database (Salgado *et al.*, 2006) for its set of known interactions for *E. coli*. Using these known interactions (culled from the literature) to calculate precision and recall (percent true positives and percent true positives found), Faith *et al.* (2007) found that at a 60% precision rate, CLR identified 1079 interactions, of which 338 were known and 741 putative. Additionally, Faith *et al.* (2007) further explored all the discovered putative regulons containing five or more genes using the *in silico* motif analysis tool MEME, discovering significant motifs in 28 of the 61 regulons examined, with 13 of these corresponding to known motifs. As yet another validation method, Faith *et al.* (2007) also performed in vivo validation using chromatin immunoprecipitation with quantitative PCR (ChIP-qPCR) for three of the transcription factors they considered significant, identifying 21 previously unknown interactions. Finally, the regulatory network identified a potential combinatorial transcriptional control of iron transport by the central metabolism of *E. coli*, which Faith *et al.* (2007) validated using real-time quantitative PCR.

While these are clearly impressive and interesting results, one should also note a few limitations of their approach; many of these limitations represent limitations for all methods given current data sets and thus future directions for the field of regulatory network inference. By limiting their search space to that of the known transcription factors, as many other techniques do, the CLR algorithm cannot detect autoregulated proteins such as the CtrA master control regulator in *C. caulobacter* (nor can any other method we are aware of). Potentially, this could be resolved if the upstream region of the transcription factor corresponding to a particular regulon was included in upstream sequences that were validated using either the *in silico* or in vivo methods they employed.

3.9.10 Dynamic models of regulation and metabolism

In contrast, dynamic cellular models, as their name would imply, attempt to simulate the internal physiology of a cell. A number of different approaches have been created to do this including thermodynamic (Loew and Schaff, 2001; Beard *et al.*, 2002; Edwards *et al.*, 2002; Moraru *et al.*, 2002), stochastic (Arkin *et al.*, 1998), cybernetic (Varner and Ramkrishna, 1998, 1999; Guardia *et al.*, 2000) and constraint-based models (Majewski and Domach, 1990; Edwards and Palsson, 2000; Covert *et al.*, 2001, 2004; Edwards *et al.*, 2001, 2002; Covert and Palsson, 2002; Reed *et al.*, 2003; Barrett *et al.*, 2005). However, of these, constraint-based models are the only approach that has been shown to be scalable to genome-wide models as the others depend upon highly specified parameterizations of attributes such as polymerase availability and quantity, as well as other environmental factors such as temperature. For this reason, they do not scale well to full genome-wide models, and we will focus on constraint-based methods below (Covert *et al.*, 2001; Price *et al.*, 2003; Reed and Palsson, 2003).

3.9.11 Constraint-based overview and basic stoichiometric matrix

The constraint-based approach described by Price *et al.* (2003) uses a matrix of pre-specified constraints as the central model. As such, rather than a single solution, a constraints-based model may have multiple valid solutions provided they do not violate these constraints. The earliest constraint-based models were designed to model the metabolism of a cell in steady state by using a matrix representation of the metabolic network for a given cell, denoted as S, that encodes the stoichiometry of each of the biochemical reactions within that cell. To find the allowable rates of each reaction (generally not known for more than a minority of reactions in any cell) we find the null space of S by setting $Sv = 0$, where v is the vector of the fluxes in the reactions described in S (and the unknown we are searching for). For readers less familiar with this type of modelling we expand this discussion below. For a more extensive discussion of this type of modelling we refer interested readers to Palsson's recent book, aptly named 'Systems Biology' (Palsson, 2006).

3.9.11.1 The S matrix explained
Let us first examine the matrix S. Each row of S represents a single metabolic compound or metabolite, while each column represents an individual

$$\begin{pmatrix} -1 & 0 & 0 & 0 & 0 & 0 \\ -1 & 0 & 0 & 0 & 1 & -1 \\ 1 & 0 & 1 & -1 & 0 & 0 \\ 1 & -1 & 0 & 0 & 0 & 0 \\ 0 & -1 & -1 & 1 & 1 & -1 \\ 0 & 0 & 0 & 0 & -1 & 1 \\ 0 & 0 & -1 & 1 & 0 & 0 \\ 0 & 1 & 1 & -1 & -1 & -1 \end{pmatrix} \begin{matrix} A \\ B \\ C \\ D \\ E \\ F \\ CD \\ DE \end{matrix}$$

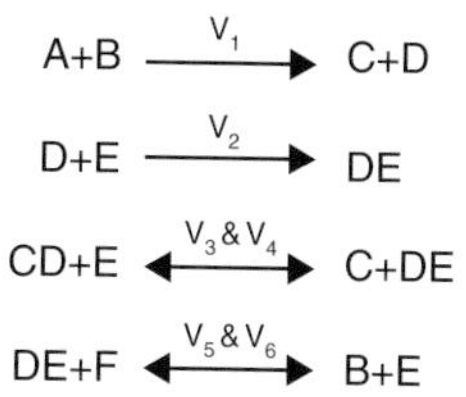

$$A+B \xrightarrow{\;v_1\;} C+D$$
$$D+E \xrightarrow{\;v_2\;} DE$$
$$CD+E \xleftrightarrow{\;v_3 \& v_4\;} C+DE$$
$$DE+F \xleftrightarrow{\;v_5 \& v_6\;} B+E$$

Figure 3.5 A stoiciometric matrix, S, for a system of four reactions involving eight reactants. Reaction 1 corresponds to column 1, reaction 2 to column 2, reaction 3 to columns 3 and 4 and reaction 4 corresponds to columns 5 and 6.

reaction that reflects the stoichiometry of that reaction. So, for example, the following hypothetical reaction involving four reactants

$$A + B \rightarrow C + D$$

would be represented by the vector $(-1, -1, 1, 1)^T$, where the symbol, ^{T}, is used to represent the transpose of the vector. As such, the first two values of the vector (the -1s) correspond to the compounds that are consumed in the reaction, namely, A and B, while the latter two values correspond to the compounds produced by the reaction, C and D. Continuing our example, a reversible reaction such as

$$CD + E \leftrightarrow C + DE$$

would require two vectors to represent the two possible reactions, that is $(-1, -1, 1, 1)^T$ and $(1, 1, -1, -1)^T$. A more complex system involving four reactions (two reversible and two irreversible) and eight reactants is illustrated in Fig. 3.5.

Given a stoichiometry matrix we still need to know the relative rate constants corresponding to each reaction, a vector of rates $\mathbf{v}$. One assumption is that the cell is at homeostasis (or will reach homeostasis following any perturbation). This assumption allows us to set $S\mathbf{v} = 0$, this equality combined with other assumptions about allowable rates (which impose only very broad constraints on rates, such as rates must be > 0) allows us to find sets of allowable rates, $\mathbf{v}$, which in turn allow us to predict the outcome of changes in metabolic flux following perturbations. This approach, forcing the solution to exist in the null space of S, centres on the simplifying assumption that the organism and/or cell operates in perfect homeostasis. Other uses of this encoding of metabolism do not require this assumption and are discussed briefly below, for example, Palsson's group has also performed analysis that do not require the assumption $S\mathbf{v} = 0$, and have also carried out analysis that couples the metabolic and regulatory networks to successfully predict systems-wide properties.

3.9.12 A general approach for using stoichiometric models in simulation

Shortly following the formulation of this framework, Varma and Palsson (1994) illustrated how these metabolic models could be applied in simulation. As an iterative approach, their algorithm divides the simulation into equal sized time slices or time points. Provided an initial condition for the first time point t_i, the solution v that produces optimal growth is chosen. Following this, any perturbations to the external media by this flux state are calculated, and then fed back into model to produce the state for the next time point t_{i+1}.

3.9.13 The first metabolic-only models

While, the first stoichiometric models of E. *coli* appeared as early as 1990 (Majewski and Domach, 1990), the first genome-wide model was the *i*JE660 *in silico* model that Edwards and Palsson developed in 2000 (Edwards and Palsson, 2000). To build their model, Edwards and Palsson relied on established databases such as EcoCyc, MPW and KEGG (Ogata *et al.*, 1998; Selkov *et al.*, 1998; Karp *et al.*, 2000) which contain massive collections of experimentally discovered enzymatic and metabolic reactions that have been manually culled from the available literature. Using these resources as the basis for their model, it contained 705 genes, as well as 436 metabolites involved in 720 reactions. Once completed, Edwards and Palsson used the model as a platform to perform a series of *in silico* gene knockout simulations, accomplished by removing the enzyme under consideration *in silico* by setting all relevant reaction rates (those involving that enzyme) to zero. To gauge the performance of their model, Edwards and Palsson next compared their *in silico* results with those from known experiments and found that their model had a predictive accuracy of 86%. In subsequent studies, their model was also used to predict optimal growth rates and evolutionary adaptation (Edwards *et al.*, 2001; Ibarra *et al.*, 2002).

3.9.14 Incorporating regulatory networks into constraints-based metabolic models

So, what was missing from these purely stoichiometric models? A careful reader will likely have noticed that regulatory interactions between transcription factors and enzymes were not part of the initial models. Originally, this stemmed from the assumption that an organism would regulate protein expression so as to optimize the metabolism of the compounds available to it; therefore, by focusing on metabolic rates, one could argue the model was implicitly taking these regulatory relationships into account (rolling regulatory influences on flux into the allowable rates found during the calculation of v). This initial lack of regulatory information was also a result of the fact that regulatory networks are less well determined than metabolic networks. However, to more realistically reflect the underlying biology, Covert *et al.* (2001) introduced into the model the use of Boolean logic to represent the various regulatory relationships between genes.

As an example of such Boolean logic, consider the hypothetical case of a microbe thriving happily on its preferred carbon source, *Carbon1*, while also having the capacity to utilize a secondary carbon source, *Carbon2*, when *Carbon1* is unavailable. Continuing the example, imagine that there exists a regulatory relationship such that the transcription of a protein to transport *Carbon2* into the cell is repressed when the microbe is in the presence of *Carbon1*. If we use *RPc1* to represent an external cell sensor protein for *Carbon1* and *tTc2* to represent a transcription factor that induces transcription of the transporter protein, this relationship can easily be encoded using the following Boolean logic rules:

$RPc1 = IF\ (Carbon1)$

$tTc2 = IF\ NOT\ (RPc1).$[2]

In their initial description of these Boolean rules, Covert *et al.* (2001) applied these to a simplified model that, as a proof of principle, covered only a few growth conditions. However, Covert and Palsson (2002) extended this approach to generate a model of the central metabolism of *E. coli*. Using a literature-based approach similar to that used to build the *i*JE660 model, Covert and Palsson generated a regulatory network consisting of 149 genes that regulated 73 enzymes and 16 other regulatory proteins. To produce their final model, this regulatory network was combined with the *i*JE660 metabolic model, with the final product containing 45 reactions whose availability was impacted by the regulatory relationships represented in the regulatory network. In a new set of *in silico* gene deletion simulations, using both the new regulatory network as well as the original metabolic network, Covert and Palsson discovered that the regulatory model improved the overall performance from 83% for the metabolic model to 91% correctly predicted growth responses.

Following their early success, Covert *et al.* (2004) reported that they had extended this approach to create *i*MC1010^{v1}, the first genome-wide metabolic and transcriptional model or *in silico* strain. The *i*MC1010^{v1} strain was actually an extension of an earlier metabolic model, *i*JR904, that had been reported the year before by Reed *et al.* (2003), who themselves had extended the earlier *i*JE660 model to include 904 genes following the release of the updated *E. coli* genome in 2001 (Serres *et al.*, 2001). This latest model was extended to

---

[2] Example taken from Covert *et al.* (2001).

include 1010 genes, 104 of which transcription factors that regulated 479 of the remaining 906 genes in the model. To validate their new model, Covert *et al.* (2004) compared the predicted growth responses with 13 750 known experimentally derived growth phenotypes available from the ASAP database (Glasner *et al.*, 2003), discovering that their model correctly predicted the growth response in 78.7% of the cases.

To improve on their model, Covert *et al.* (2004) analyzed those cases where the known response disagreed with those that the model predicted, in the process identifying several suspected cases of missing or unknown enzymes and transcriptional interactions. Furthermore, focusing on the organism's response to oxygen deprivation, Covert *et al.* (2004) also performed microarray expression profiling of several gene knockout strains they created to explore this response. From the results of the analysis of this expression data, a number of updates to the model's regulatory network were made, resulting in their next *in silico* model, iMC1010^{v2}, which too was tested using the same growth response cases that had been applied to iMC1010^{v1}. Unfortunately, the improvement in the number of correctly identified growth responses was negligible (+5 cases called correctly). Despite these disappointing results, Cover *et al.* (2004) observed, however, that the new model was far more successful in predicting the expression differences of genes that had been revealed to be differentially expressed by the microarray data.

Following this announcement of the iMC1010^{vX} *in silico* strains, in 2005 Barret *et al.* (2005) reported the result of an interesting experiment where they compared the simulations of the iMC1010^{v1} strain grown in various media. For their experiment, the media chosen was selected such that it would cover the full range of growth media that could be used by the iMC1010^{v1} strain. Enumerating all possible combinations of carbon, nitrogen, phosphate, sulfur and electron-acceptor sources resulted in 108 723 combinations, 15 580 of which induced sufficient predicted growth by the iMC1010^{v1} strain to be used in their comparison. Note, that rather than comparing the resulting growth phenotype, as was done by previous studies, they instead compared the predicted gene expression and activities *during* these simulations against one another. Using an agglomerative clustering algorithm in combination with principle-components analysis, Barret *et al.* (2005) discovered that most of the simulations were grouped together into a relatively small number of clusters – 36 or 13, depending upon whether gene expression or gene activity was compared. Moreover, for either type of comparison, Barret *et al.* (2005) discovered that these clusters were characterized by the terminal electron acceptor available in the *in silico* growth environment. These results led them to conclude that despite the multitude of possible environments *E. coli* could be subjected to, its genetic system is designed to function in a few dominant modes of response. Or, as they succinctly summarized it, their results were consistent with the hypothesis that 'system complexity is built in to robustly provide for simple behaviour.'

Though it would be interesting to see how these results would compare with a similar experiment using the MC1010^V (Handelsman, 2004) model, if we focus on just the technological aspects for the moment, the ability to perform simulations on nearly 110 000 different media is impressive, in and of itself. While acknowledging the current limitations of the existing models, it is clear that they still have the capacity to provide some important observations about the underlying nature of these organisms. Considering the fact that nearly 80% of their phenotype predictions (growth or not-growth) were accurate, this is clearly a milestone for global *in silico* modelling of global dynamics.

3.9.15   *E. coli* metabolomics

Metabolomics, briefly, is the study of all metabolites (small molecules), and their dynamics, for various conditions in an organism. The metabolome is crucial to our understanding of phenotype and fitness outcomes of different cell states (Fiehn, 2002) and the number of metabolites accessible is on the order of hundreds to thousands. There is evidence from comparing multiple complete genomes of a common core of enzymes that are fundamental for metabolism (Jardine *et al.*, 2002). Metabolism may be conserved to some degree at the enzyme level, but the processes and networks by which the various organisms convert metabolites vary significantly (Peregrin-Alvarez *et al.*, 2003). The field of metabolomics is advancing quickly. One example, important for industry and medicine, is the improvement of bacterial strains by metabolic engineering.

Nobeli *et al.* (Nobeli *et al.*, 2003) attempted to characterize the *E. coli* metabolome using 2D NMR to classify and identify metabolites systems wide from living cells. They compiled their data set of 745 metabolites, a subset of the complete metabolome, from publicly available, experimentally verified data from the EcoCyc (Keseler *et al.*, 2005) and KEGG (Kanehisa *et al.*, 2006) databases. Clustering of the metabolites revealed a continuum with significant overlap of clusters and no clearly defined classes of metabolites (with respect to the presence or absence under varying conditions). This early study demonstrated a novel systems-level perspective of the metabolome. Much 'omic' data are available and its integration is fundamental to understanding the complexities and robustness of a living system in its environment.

3.10   Halobacterium salinarium NRC-1

The archaeal *Halobacterium salinarum NRC-1* is a halophillic (salt loving) organism that can not only survive, but requires highly saline environments, flourishing in environments such as the Great Salt Lake in Utah with ~4.5 M salinity (or roughly 5–10 times the salinity of sea water). Halobacterium can also withstand a surprising variety of other stresses, such as oxidative stress, DNA-damaging chemicals, heavy metals, UV and gamma radiation, low oxygen and desiccation. To withstand high salt, it maintains an isoosmotic cytoplasm by eliminating some Na+ ions and maintaining a high intracellular K, Mg (and also Na) ion concentration. As such, its genome possesses multiple ion transporters such as active K+ transporters (KdpABC), Na+/ H+ antiporters (NhaC proteins), low affinity ion transporters driven by membrane potential (Trk proteins) and heavy metal (arsenic and cadmium) transporters. More importantly, *Halobacterium* flourishes in these environments by adjusting its physiology appropriately in response to numerous external stimuli. For example, it can relocate, in search of favourable environments, using sensors that can discriminate beneficial and detrimental spectra of light (Bogomolni and Spudich, 1982; Spudich and Bogomolni, 1984; Spudich *et al.*, 1989; Spudich, 1993), an aerotaxis transducer (HtrVIII) (Brooun *et al.*, 1998) and buoyant gas-filled vesicles (DasSarma, 1993). One of the hallmarks of *Halobacterium* is its ability to survive anaerobically using light and/or arginine as energy sources and aerobically as a chemoheterotroph. *Halobacterium* generates energy from light by its retinal-containing light-driven ion transporters, bacteriorhodopsin and halorhodopsin (Kolbe *et al.*, 2000; Luecke *et al.*, 2000). Additionally, *Halobacterium* can also ferment arginine via the arginine deiminase pathway with each mole of arginine fermented yielding one mole of ATP (Ruepp and Soppa, 1996). As such an extremeophile, it represents an interesting, yet still poorly understood class of organisms. Moreover, from a systems biology perspective, archaea present an interesting opportunity as while they are prokaryotic organisms, they share many attributes with eukaryotes such as eukaryotic-like transcription, translation and TATA boxes. Though they have been the subjects of study since the 1960s, the first *H. salinarum* genome was sequenced in 2000, opening the door for further systems-level study of the organism (Ng *et al.*, 2000; Dassarma *et al.*, 2006).

Below, we review how some of these efforts have been applied to *Halobacterium*. We will illustrate a systematic process consisting of the following steps:

1. Define all of the elements in the cell (or organism). Develop an initial model of the cell using existing knowledge, that is, literature review.
2. Perturb the system environmentally and/or genetically (knockouts, over expressions, etc.) and globally assay the relationships of the elements one to another (e.g. levels of mRNA and protein, protein/protein interactions, etc.). Integration of data from different sources is critical to a complete understanding.
3. Compare the model with the experimental results to formulate new hypotheses which explain the discrepancies.
4. Test these hypotheses with a new series of perturbations and update the model to more accurately reflect the experimental results.
5. Iterate steps 2–4.

3.10.1 Sequencing of Halobacterium

As mentioned above, sequencing of the *H. salinarium NRC-1* genome was completed by Ng *et al.* (2000), who used a whole-genome shotgun strategy to sequence the genome which consists of one large replicon and two, relatively smaller replicons. The larger of these contains ~2 Mbp (2 571 010 bp, exactly), while the two smaller replicons, pNRC100 and pNRC200 each contain roughly 200 and 350 Kpb, respectively (191 346 and 365 425 exactly). Using the *in silico* gene prediction programme, GLIMMER (Salzberg *et al.*, 1998; Delcher *et al.*, 1999), Ng *et al.* (2000) identified 2682 putative genes, of which 2111 were located on the large replicon, while 197 and 374 were found on the two smaller replicons, pNRC100 and pNRC200, respectively. To assign function to these, the putative genes were translated and then submitted to NETBLAST (Altschul *et al.*, 1997) to query for homologues in the non-redundant database of proteins hosted on the NCBI. The results from this search revealed that 1658 had significant matches, though of these matches, only 1067 had known function while the remainders were hypothetical proteins. Of these matches to genes with known function were genes involved in metabolism, cellular envelop maintenance, photobiology, DNA replication, transcription and translation. Interestingly, Ng *et al.* (2000) also identified 91 transposable insertion elements, with the majority of these (62) located on the two smaller replicons or minichromosomes, leading them to conclude that these play a significant role in *Halobacterium* evolution by allowing the organism to gain new genes.

3.10.2 Systems-wide exploration of energy production in differing environments

Following the sequencing of *Halobacterium*, the first system-level analysis was reported by Baliga *et al.* (2002) who explored the combined RNA and protein expression of *Halobaterium* during anaerobic energy production. As mentioned above, from earlier studies, it was already known that in anaerobic conditions, *Halobacterium* could generate energy from either arginine fermentation or photosynthesis. Additionally, from earlier studies, it was known that during phototrophic growth the organism generates numerous copies of a light-driven proton pump called bacteriorhodopsin (bR), which is a protein complex composed of the two proteins bacterioopsin (Bop), and retinal. During phototrophic growth these proton pumps are organized in a 2D lattice called the purple membrane. Furthermore, it was also known that another protein, Bat, regulated the expression of itself, as well as three others involved in bR synthesis, *bop*, *brp* and *crtB1*.

Thus, to explore the regulatory network driving phototrophic growth, Baliga *et al.* (2002) performed RNA and protein expression analyses of four different strains, including the NRC-1 wild type, a *bop* knockout strain (*bop-*), as well as both a *bat* overexpression (*bat+*) and knockout (*bat−*) strain. Using cDNA microarrays, they discovered that the *bop*− strain exhibited little expression difference from the *bat*+ strain. However, as would be expected of a transcription factor, the *bat*+ and *bat*− exhibited significant numbers of differentially expressed genes, with 151 and 157 differentially expressed genes, respectively. What was not expected, though, was that functionally, their expression profiles were inverted, as those genes involved in photosynthesis were induced in the *bat*+ strain, but repressed in the *bat*− strain, while the opposite was the case for those involved in arginine fermentation (repressed in the *bat*+ strain, but induced in the *bat*− strain). While the exact mechanism for this was unclear, Baliga *et al.* (2002) hypothesized that this inversion represents a strategy to maintain a steady level of ATP within the cell. Additionally, subsequent proteomics studies using the ICAT technique (Gygi *et al.*, 1999) found a number of differentially expressed proteins had no corresponding change in mRNA (33/50), indicating post-translational effects upon protein expression. Furthermore, *in silico* promoter analysis of the genes induced in the *bat*+ strain found only one additional gene containing the Bat-binding site, indicating that most of these were subject to indirect regulation by Bat. However, promoter analysis using MEME was able to identify a likely binding motif among five genes involved with arginine fermentation. In doing so, their study leads to new hypotheses later verified with future genetic modifications and later iterations of the group's systems-level analyses.

3.10.3 The functional annotation of Halobacterium proteome

In addition to these findings, Baliga *et al.* (2002) discovered that they had also been able to verify the existence (at the protein and transcript level) of 496 of the 971 hypothetical genes in the *Halobacterium* genome – those that had been predicted by gene finders, but had no homologues with other known genes. This annotation was further expanded by Bonneau *et al.* (2004) who in a paper from 2004 reported both a new functional, structure-based annotation of the *Halobacterium* genome, as well as a new contextual annotation of the genome that linked proteins by associations such as shared operon membership.

To update this proteome annotation, Bonneau *et al.* (2004) used a method which they had used previously in the critical assessment of structure prediction (CASP 3, 4 and 5) (Bonneau *et al.*, 2001a, b; Chivian *et al.*, 2003) which used two algorithms, Ginzu and Rosetta (Bonneau *et al.*, 2001b; Aloy *et al.*, 2003; Bradley *et al.*, 2003; Fischer *et al.*, 2003; Kinch *et al.*, 2003) to predict protein-domain boundaries and protein structure. The method is a hierarchical workflow that utilizes a protein domain-centric approach to identify function and structure starting only with the primary sequence of a predicted protein. As an initial, pre-processing step, each query sequence is filtered for regions that are likely to be either transmembrane, coiled coils, signal peptides or a disordered region. These regions are removed from further analysis, with the remainder submitted to their protein-domain parsing programme, Ginzu, which attempts to parse the primary sequence into likely domains and identify their functions by using a hierarchical workflow (with more accurate methods placed at the top of this hierarchy). The first step of this process is to use PSI-BLAST to search for sequence matches to the PDB, resulting in high-quality, high-likelihood domains of known function. For those regions of the protein not identified by this PSI-BLAST search, they are next queried using HMMER for matches in Pfam. If any regions still have not been identified by these previous searches, as a third step Ginzu next attempts to identify matches to protein structures using Fold Recognition. As the fourth and final step of Ginzu, any regions not recognized by the previous three methods are aligned to all known sequences using PSI-BLAST; multiple sequence alignments are parsed for block patterns indicative of domain structure. Finally, all domains not matched by a known structure using these methods are then passed to the Rosetta algorithm, a de novo structure prediction algorithm that uses information from the PDB to identify likely local structure confirmations.

With their functional annotation process, Bonneau et al. (2004) found 1077 of the 2596 protein coding genes in the *Halobacterium* genome had significant matches found by the initial PSI-BLAST search of the PDB. Additionally, 610 domains were identified by querying the Pfam database, with an additional 670 domains identified using the two de novo structure prediction methods (Rosetta). While 1234 protein domains could not be annotated by this method, this still translates into a nearly 30% improvement over the collection of sequence-based methods which had initially been used.

3.10.4 Protein associations and structure prediction to derive putative annotations for proteins

To generate their contextual annotation of associations, Bonneau et al. (2004) considered four possible association types, including protein–protein inter-actions, fusions of *Halobacterium* protein domains found in other genomes, proteins grouped into operons, and phylogenetic profile edges. To identify putative protein–protein interactions, they used the COG database (Uetz et al., 2000; Ito et al., 2001; Rain et al., 2001; Tatusov et al., 2001), along with other databases of known interactions to infer 1143 likely interactions. For the fusions of *Halobacerium* domains, a method described by Enright et al. (1999) was utilized to identify 2460 suspected associations. To identify oper-ons, two methods were used, one which considered clusters of genes with shared directionality, while the other considered nearby pairs of genes which had orthologues in other genomes that were similarly co-located (Mellor et al., 2002; Moreno-Hagelsieb and Collado-Vides, 2002). With these two methods, 1335 total putative operon associations were identified. Finally, 525 asso-ciation links were added using the phylogenetic profile method of Marcotte et al. to identify collections of genes which often co-occur in different genomes (Marcotte et al., 1999; Eisenberg et al., 2000). These associations and the prior proteome annotation effort provided a rich environment in which to explore protein function that was much greater than the sum of the individual parts.

3.10.5 *Halobacterium's* stress response following exposure to ultraviolet radiation

We now further review *Halobacterium's* stress response following exposure to UV radiation (Baliga *et al.*, 2004). Damage to DNA as a result of exposure to shortwave UV light (UV-C) falls into two categories, one being pyrimidine and pyrimidone phosphoproducts that are created between the C6 and C4 carbons of neighbouring pyrimidine nucleotides (i.e. T–C or C–C), while the other are cyclobutane pyrimidine dimers (CPD) that are created between the C4 and C5 positions of neighboring pyrimidines of the same type (i.e. C–C or T–T). Similarly, two repair mechanisms exist in most organisms, one of which is the nucleotide excision repair (NER) system that can occur at any time, but is better with repairing phosphoproducts. The second is a photolyase-catalyzed phosphoreaction that can only occur in the presence of light, and is more effective at repairing CPDs. Note, however, both repair pathways can repair both types of DNA lesions. Prior to the Baliga *et al.* (2002) exploration of the UV response, it had been known that *Halobacterium* had homologues for proteins in both systems, including homologues for both bacterial and eukaryotic NER proteins, though there were still questions regarding the exact machinery of these repair mechanism within the organism.

As an initial foray, Baliga *et al.* (2002) explored the UV-C resistance of the *Halobacterium*, by exposing *Halobacterium* in a thin liquid culture to UV-C radiation, finding that up to 110 J/m (Handelsman, 2004) there was no loss of viability and 37% survivability following 280 J/m (Handelsman, 2004). However, these initial tests also indicated that photoreactivation was a major UV repair mechanism (growth in light following exposure was 16 times more likely (16-fold) than growth in dark conditions). For this reason, they next focused on two photolyase homologues *phr1* and *phr2* within the *Halobacterium* genome. While it was already known that *phr2* was a photolyase, the role of *phr1* was still unknown. Using three strains, consisting of a *phr1* knockout (*phr1−*), a *phr2* knockout (*phr2−*), and a *phr1* and *phr2* double knockout (*phr1−/phr2−*), they found that their results clearly revealed that only *phr2* functioned as a CPD photolyase, as the *phr1−* strain exhibited no difference from the wild type following UV exposure. As they also found that both the *phr2−* and *phr1−/phr2-* strains exhibited ∼3.5-fold increased survivability when grown in the presence of light versus dark following UV exposure, they next explored the processes occurring during what they termed light versus dark repair following exposure to UV light.

To accomplish this, they used an experimental procedure where they examined the organism, grown in either light or dark conditions, at 30 and 60 min post-UV-exposure, as well as a control (no UV exposure) after 60 min growth in light. Thus, five separate assays were performed (L30, L60, D30, D60 and C60). Using new 70-mer oligonucleotide microarrays to assay the RNA expression at these time points, they found that a total of 420 genes whose mRNA was differentially expressed, with 273 of these only occurring during the repair tests, 40 of which occurred in both repair conditions and 61 that occurred in both the control and repair assays. One of the more interesting findings from these assays was the difference in number of genes that were repressed after 60 min repair growth in light (L60) assay versus those that were differentially expressed in the other repair assays. Specifically, while <2% of *Halobacterium's* genes were differentially expressed in any of the other repair assays, roughly 12% of the genome was found to be downregulated in the L60 assay, including nearly all the ribosomal and RNA polymerase genes. This massive downregulation has also been found to be a general stress response in other conditions, as well as other organisms.

Based on the structure-based reannotation of the genome, Baliga *et al.* (2002) were able to identify at least two transcription factors, genes VNG1318H and VNG0019H, whose function were unknown previously. In addition, using the association annotation that Bonneau *et al.* described, along with their own expression results and information from the Kyoto Encyclopedia of Genes and Genomes, Baliga *et al.* (2002) were able to identify and visualize the response of biomodules using Cytoscape (Shannon *et al.*, 2003), a genomic data visualization tool. We will discuss Cytoscape in greater detail below. However, all these combined tools and newly acquired information allowed Baliga *et al.* (2002) to formulate a number of new conclusions and hypotheses. Among these was the conclusion that *phr2*, and not *phr1*, was clearly a photolyase and the major mechanism of UV-C damage repair. Another conclusion, based on the number of genes downregulated in the L60 sample, was that the major response to UV-C damage is a halt in transcription and translation to allow the organism or cell to recover from the UV-C-induced damage before regular cell activity and division restarts (a result also seen in other organisms stress response). Furthermore, they identified three new putative transcriptional regulators involved in repair damage, including the VNG1218H gene that we mentioned above. Finally, the new experimental data and computational analyses techniques also allowed Baliga *et al.* (Baliga *et al.*, 2002) to speculate on two parallel mechanisms involving Cobalamin (B-12) biosynthesis.

3.10.6 Data Visualization: Cytoscape and the Gaggle

Cytoscape is a computer programme that Shannon *et al.* (2003) first reported in 2003, which displays the genes and associations of a given organism as a network where the genes represent nodes, and the associations represented as edges between the genes/nodes. Furthermore, attributes such as function and mRNA and protein expression data can then be assigned to each gene in the network. With this set-up, Boolean networks and active transcriptional paths calculated using mRNA expression data can then be explored in context of the other data types integrated into the network to gain systems-level insights and formulate hypotheses for further testing. See cytoscape.org for details, code and Cytoscape compatible tools (plug-ins).

3.10.7 The quest for the global Halobacterium regulatory network: Philosophy

Distilling regulatory networks from large genomic, proteomic and expression data sets is one of the most important mathematical problems in biology today (Yuh *et al.*, 1998; Friedman *et al.*, 2000; Wahde and Hertz, 2001; Ideker *et al.*, 2002; Lee *et al.*, 2002; Shmulevich *et al.*, 2003; Hashimoto *et al.*, 2004; Bonneau *et al.*, 2006;). The development of accurate models of global regulatory networks is key to the understanding of a cell's dynamic behaviour and its response to internal and external stimuli. A major goal of the Halobacterium project was thus to combine all data (including the data generated by the focused studies above) to generate a global regulatory network.

Methods for inferring and modelling regulatory networks must strike a balance between model complexity – a model must be sufficiently complex to describe the system accurately – and the limitations of the available data – in spite of dramatic advances in our ability to measure mRNA and protein levels in cells, nearly all biological systems are underdetermined with respect to the problem of regulatory network inference. We focus on further development of our algorithms for learning co-regulated modules and regulatory networks. Our aim is to learn models of regulation from data that include units of time, concentration (or at least relative concentration) and to explicitly model regulator-binding sites.

3.10.8 Halobacterium global regulatory network inference: methods, motivations, challenges and current progress

A major challenge is to distill, from large genome-wide data sets, a reduced set of factors describing the behaviour of the system. The number of potential regulators is often on the same order as the number of observations in current genome-wide expression and proteomics data sets. A further challenge in regulatory network modelling is the complexity of accounting for transcription factor interactions and the interactions of transcription factors with environmental factors (e.g. it is known that many transcription regulators form heterodimers, or are structurally altered by an environmental stimulus such as light, thereby altering their regulatory influence on certain genes). A third challenge and practical consideration in network inference is that biology data sets are often heterogeneous mixes of equilibrium and kinetic (time-series) measurements; both types of measurements can provide important supporting evidence for a given regulatory model if they are analyzed simultaneously. Last, but not least, is the challenge that data-derived network models be predictive, and not just descriptive: can one predict the system-wide response in differing genetic backgrounds, or when the system is confronted with novel stimulatory factors or novel combinations of perturbations?

We describe the methods we used to predict the global network from the *Halobacterium* data compendium as a two-part process (step 1, cMonkey, step

2, the Inferelator). We follow this discussion with a brief discussion of the tools that are used to explore this data, the resulting networks and associated annotation data (the Gaggle).

3.10.9 Step 1: cMonkey, the need for integrative biclustering

Learning and modelling of regulatory networks can be greatly aided by reducing the dimensionality of the search space prior to network inference. Two ways to approach this are (1) limiting the number of regulators under consideration, and (2) grouping genes that are co-regulated into clusters. In the first case, candidates can be prioritized based on their functional role, for example, limiting the set of potential predictors to include only transcription factors, and by grouping together regulators that are in some way similar. In the second case, gene-expression clustering, or unsupervised learning of gene-expression classes, is commonly applied. It is often incorrectly assumed that co-expressed genes correspond to co-regulated genes. However, for the purposes of learning regulatory networks it is desirable to classify genes on the basis of *co-regulation* (shared transcriptional control) as opposed to simple *co-expression*. Furthermore, many standard clustering procedures assume that co-regulated genes are co-expressed across all observed experimental conditions. Since genes are often regulated differently under different conditions, this assumption is likely to break down as the size and variety of data grows. *Biclustering* was developed to better address the full complexity of finding co-regulated genes under multi-factor control by grouping genes on the basis of coherence under *subsets* of observed conditions (Cheng and Church, 2000; Tanay *et al.*, 2002; Yang *et al.*, 2002, 2003; Kluger *et al.*, 2003; Segal *et al.*, 2003; Sheng *et al.*, 2003; Tanay *et al.*, 2004).

Co-regulated genes are often functionally (physically, spatially, genetically and/or evolutionarily) linked (Moreno-Hagelsieb and Collado-Vides, 2002; Harbison *et al.*, 2004). For example, genes whose products form a protein complex are likely to be co-regulated. Other types of associations among genes, or their protein products, that can imply functional couplings include (a) presence of common *cis*-regulatory motifs; (b) co-occurrence in the same metabolic pathway(s); (c) *cis*-binding to common regulator(s); (d) physical interaction; (e) common ontology; (f) paired evolutionary conservation among many organisms; (g) common synthetic phenotypes upon joint deletion with a third gene; (h) sub-cellular co-location and (i) proximity in the genome, or in bacteria and archaea, operon co-occurrence. These associations can be either derived experimentally or computationally (either pre-computed ahead-of-time, or on-the-fly during the clustering process); indeed it is common practice to use one or more of these associations as a post-facto measure of the biological quality of a gene cluster. However, it is important to note that these data types, to varying degrees, can contain a high rate of false positives, or may imply relationships that have no direct implication for co-regulation. Therefore in their consideration as evidence for co-regulation, these different sources of evidence should be treated as priors, with different amounts of influence on the overall procedure based upon prior knowledge of (or assumptions about) their quality and/or relevance.

Because a biological system's interaction with its environment is complex and gene regulation is multi-factorial, genes might not be co-regulated across all experimental conditions observed in any comprehensive set of transcript or protein levels. Also, genes can be involved in multiple different processes, depending upon the state of the organism during a given experiment. Therefore, a biologically motivated clustering method should be able to detect patterns of co-expression across subsets of the observed experiments, and to place genes into multiple clusters. So-called biclustering, clustering both genes and experimental conditions, is a widely studied problem and many different approaches to it have been published (Cheng and Church, 2000; D'Haeseleer et al., 2000; Tanay et al., 2002, 2004; Yang et al., 2002, 2003; Kluger et al., 2003; Segal et al., 2003; Sheng et al., 2003; Balasubramanian et al., 2004). Unlike standard clustering methods, most biclustering algorithms place genes into more than one cluster (genes can play more than one functional role in the cell). Because biclustering is an NP-hard problem (D'Haeseleer et al., 2000), no solution is guaranteed to find the optimal set of biclusters. However, many of these procedures have successfully demonstrated the value of biclustering when applied to real-world biological data (Balasubramanian et al., 2004; Reiss et al., 2006).

We compared the method to several other methods including but not limited to: Order Preserving Sub-matrix (Ben-Dor et al., 2003), Iterative Signature (Bergmann et al., 2003), xMotif (Murali and Kasif, 2003), Bimax (Prelic et al., 2006) and SAMBA (Shamir et al., 2005). We also compared our method to hierarchical clustering and k-means clustering. We used multiple parameterizations of each competing method. In addition, we performed these analyses on cMonkey runs with various model parameters up and down-weighted to demonstrate tolerance of the cMonkey method to different parameterizations of free parameters. Additional details on the analysis are provided previously (Reiss et al., 2006). All biclusters generated by the cMonkey as well as the other algorithms we tested are available for interactive exploration via Cytoscape and the Gaggle (Shannon et al., 2003, 2006) at ⟨http://labs.systemsbiology.net/baliga/cmonkey/⟩.

3.10.10 Comparison in the context of regulatory network inference

A major motivation of cMonkey is to provide a method for deriving co-regulated groups of genes for use in subsequent regulatory network inference procedures. Thus, we wish to find coherent groups of genes over those conditions with a large amount of variation. In other words, we are hoping to detect submatrices in the expression data matrix that are coherent and simultaneously have high information content or overall variance (and probability given the network and motif components). In addition, we need to find biclusters with many conditions/observations included, as this increases the significance of each bicluster and also of the subsequently inferred regulatory influences for that bicluster. In general, we see that cMonkey generates biclusters with a significantly greater number of experiments than the other methods (higher coverage). Even with this additional constraint (i.e. including a greater number of experiments in the clusters) and further constraints that cMonkey imposes with the association network and motif priors, the algorithm in general generates biclusters with a 'tighter' profile, as measured by mean bicluster residual. Thus, we find that biclusters generated by cMonkey are generally better suited for inference algorithms such as the Inferelator (and potentially other methods as well). We tested this by running the Inferelator on biclusters generated by SAMBA for Halobacterium and then comparing the predictive performance of the resultant regulatory network models on newly collected data, relative to those generated for cMonkey generated biclusters. We found that, largely due to the smaller number of experiments included in SAMBA biclusters, the inferred network was significantly less able to predict new experiments (an increase in the predictive error from 0.368 to 0.470; p-value of difference by t-test $< 1 \times 10^{-22}$) (Kanehisa *et al.*, 2004). We find that cMonkey performs well in comparison to all other methods when the trade-off between sensitivity, specificity and coverage is considered, particularly in context of the other bulk characteristics (cluster size, residual, etc.). Most importantly, cMonkey significantly improves the performance of downstream network inference procedures. cMonkey biclusters do a better job at regenerating the expression data than other methods, and a similar job at recapitulating the external (as well as internal) measures of bicluster quality.

3.10.11 Step 2: the inferelator

Given modules from a clustering/biclustering algorithm, for example cMonkey, we are then faced with the task of learning which genes and environmental conditions influence/control each module/cluster/bicluster/gene. We have described an algorithm for doing this, the Inferelator, which infers regulatory influences for genes and/or gene clusters from mRNA and/or protein expression levels. The method uses standard regression and model shrinkage (L1-shrinkage) techniques to select parsimonious, predictive models for the expression of a gene or cluster of genes as a function of the levels of transcription factors, environmental influences and interactions between these factors (Thorsson *et al.*, 2005). The procedure can simultaneously model equilibrium and time-course expression levels, such that both kinetic and equilibrium expression levels may be predicted by the resulting models. Through the explicit inclusion of time, and gene-knockout information, the method is capable of learning causal relationships. It also includes a novel solution to the problem of encoding interactions between predictors into the regression. We discuss the results from an initial run of this method on a set of microarray observations from the halophilic archaeon, *Halobacterium NRC-1*. We have found the network to be predictive of newly measured data and have also validated parts of the network using ChIP-chip.

3.10.12 Model formulation

We assume that the expression level of a gene, or the mean expression level of a group of co-regulated genes, y, is influenced by the level of N other factors in the system: $X = \{x_1, x_2, \ldots, x_N\}$. We consider factors for which we have measured levels under a wide range of conditions; in our work on *Halobacterium* we use transcription factor transcript levels and the levels of external/environmental conditions as predictors and gene and bicluster transcript levels as the response. The relation between y and X is given by the kinetic equation:

$$\tau \frac{dy}{dt} = -y + g(\beta \cdot Z) \tag{3.1}$$

Here, $Z = \{z_1(X), z_2(X), \ldots, z_P(X)\}$ represents a set of functions of the regulatory factors X. The coefficients *beta* describes the influence of each element of Z, with positive coefficients corresponding to inducers of transcription, and negative coefficients to transcriptional repressors (Wahde and Hertz, 2001). The constant *tau* is the time constant of the level y in the absence of external determinants. We use a novel encoding of interactions by allowing functions in Z to be either: (1) the identity function of a single variable or (2) the minimum of two variables (Richter-Gebert *et al.*, 2003). For example, the inner product of the design matrix and linear coefficients for two predictors that are participating in an interaction is:

$$\beta Z = \beta_1 x_1 + \beta_2 x_2 + \beta_3 \min(x_1, x_2). \tag{3.2}$$

Using this encoding, for example, if x_1 and x_2 represent the levels of components forming an obligate dimer that activates y (x_1 AND x_2 required for expression of y), we would expect to fit the model such that $\beta_1 = 0$, $\beta_2 = 0$, $\beta_3 = 1$. This encoding results in a linear interpolation of (linearly smoothed approximation to) the desired Boolean function. This and other interactions (OR, AND, XOR), as well as interactions involving more than two components, can be fit by this encoding. In regression terminology, the influencing factors, X, are referred to as regressors or predictors, while the functions Z specify what is often referred to as the 'design matrix'.

With this scheme for encoding interactions in the design matrix, we expect to capture many of the interactions between predictors necessary for modelling realistic regulatory networks, in a readily interpretable form. To date we have limited the procedure to binary interactions, as it is unlikely that the quantity of data used would support learning beyond these pairwise interactions. Many other methods for capturing transcription factor cooperatively exist as well (Das *et al.*, 2004). We have shown that removal of the capability to model interactions in this way reduces the predictive power of the Inferelator over the newly collected validation data set.

Various functional forms can be adopted for the function g, called the 'non-linearity' or 'activation' function for artificial neural networks, and the 'link' function in statistical modelling. The function g often takes the form of a sigmoidal, or logistic, activation function. This form has been used successfully in models of developmental biology (von Dassow *et al.*, 2000). The function is compatible with L1-shrinkage (the method for enforcing model parsimony) (van Someren *et al.*, 2000, 2002; Efron, 2003).

The simplified kinetic description of equation (1) encompasses essential elements to describe gene transcription, such as control by specific transcriptional activators (or repressors), activation kinetics, and transcript decay, while at the same time facilitating access to computationally efficient methods for searching among a combinatorially large number of possible regulators. To better understand specific details of regulation, it will almost certainly be required to follow up on specific regulatory hypotheses using more mechanistically detailed descriptions. Although this method (explicit time component) does not lessen the need for correct experimental design it does: (1) facilitate using data with reasonable variation in sampling structure and (2) allow for the simultaneous combination of data from equilibrium and time-series data.

3.10.13 Predictive power of the *Halobacterium* network over new data (performance on novel combinations of environmental and genetic perturbations)

Our initial application of the method to *Halobacterium* resulted in a statistically learned regulatory network that can predict, with reasonable accuracy, mRNA levels of ~1900 out of the total ~2400 genes found in the genome, using relative concentrations of transcription regulators and environmental factors as predictors. We find that applying cMonkey to our expression compendium, the metabolic network, comparative genomics edges and upstream sequences gives us a set of ~300 biclusters spanning ~2000 of the 2400 genes in this organism. This set of biclusters is also linked to a set of putative *cis*-acting regulatory motifs (some validated by prior experiments). The learned network controlling the 300 biclusters and 159 individual genes contained 1431 regulatory influences (network edges) of varying strength. Of these regulatory influences, 495 represent interactions between two TFs or between a TF and an environmental factor. We selected the null model for 21 biclusters (no influences or only weak regulatory influences found), indicating that we are stringently excluding under-determined genes and biclusters from our network model. The ratio of data points to estimated parameters is approximately 67 (one time constant plus three regulatory influences, on average, from 268 conditions). The explicit time component and interaction component (which distinguish this method from other such shrinkage methods) were essential for predictive performance over the validation data and the new data.

In order to test predictive performance we chose to test the network model (trained prior on the 268 conditions available at the time) over 130 additional new measurements, collected after model fitting. We found that the prediction error over the training set was essentially the same as that over the new data set. This is encouraging as the new data included environmental perturbations, new combination of environmental and genetic perturbations and time series measurements after novel entrainments of the cell. This predictive power is a pre-requisite to further interpretation of organization of key processes in the network. The ability of the same network to predict transcriptional control in novel environments (>130 new experiments) verifies that, irrespective of the nature of the environmental perturbation, Halobacterium utilizes a core set of regulatory mechanisms to maintain homeostasis under extreme conditions. The resultant network (as well as biclusters and supporting tools) for *Halobacterium NRC-1* in Cytoscape, available as a Cytoscape/Gaggle web start at: http://halo.systemsbiology.net/inferelator (Bonneau *et al.*, 2006; Shannon *et al.*, 2006).

3.11 Conclusion

We have shown four cases where the promise has been realized. We have listed four projects that we feel provide clear examples of prokaryotic systems biology. Each example illustrates a different field-wide approach to incorporating genomics into the mainstream effort for a given organism. These four examples barely scratch the surface, as new prokaryotic functional genomics projects are being reported each day. As new projects cover more and more of the prokaryotic tree of life we will begin to see studies showing the power of systems-wide analysis that span multiple organisms. Although it is not the focus of this work, several multiple-species works have recently been reported involving host-pathogen interactions, multiple closely related microbes, and others. These studies will simultaneously provide views of both the functioning of prokaryotic systems as well as the evolutionary paths/processes that led to the systems we observe today. New technologies, such as recent advances in ultra-high-throughput sequencing technologies, will make measurement of genome, transcriptome and proteome levels and activities/modifications directly from the environment for multiple species possible. In this way we will apply the techniques and tools described above, for single species, to consortia of organisms; only by approaching the true complexity microbial communities we find in the field or in the clinic can we hope to fully apply what we learn.

Although the future is difficult to predict we can safely say that the future is bright for prokaryotic systems biology.

References

Aggarwal, K. and Lee, K.H. (2003) Functional genomics and proteomics as a foundation for systems biology. *Brief Funct Genomic Proteomic* **2**, 175–184.

Allenby, N.E., O'Connor, N., Pragai, Z., Ward, A.C., Wipat, A. and Harwood, C.R. (2005) Genome-wide transcriptional analysis of the phosphate starvation stimulon of Bacillus subtilis. *J Bacteriol* **187**, 8063–8080.

Alon, U. (2007) *An Introduction to Systems Biology: Design Principles of Biological Circuits* (Boca Raton, FL: Chapman and Hall/CRC Press).

Aloy, P., Stark, A., Hadley, C. and Russell, R.B. (2003) Predictions without templates: new folds, secondary structure, and contacts in, CASP5. *Proteins* **53** (Suppl 6), 436–456.

Altschul, S.F., Madden, T.L., Schaffer, A.A., Zhang, J., Zhang, Z., Miller, W., *et al.* (1997) Gapped BLAST and PSI-BLAST: a new generation of protein database search programs. *Nucleic Acids Res* **25**, 3389–3402.

Arends, S.J. and Weiss, D.S. (2004) Inhibiting cell division in Escherichia coli has little if any effect on gene expression. *J Bacteriol* **186**, 880–884.

Arkin, A., Ross, J. and McAdams, H.H. (1998) Stochastic kinetic analysis of developmental pathway bifurcation in phage lambda-infected Escherichia coli cells. *Genetics* **149**, 1633–1648.

Ausmees, N. and Jacobs-Wagner, C. (2003) Spatial and temporal control of differentiation and cell cycle progression in Caulobacter crescentus. *Annu Rev Microbiol* **57**, 225–247.

Bailey, T.L. and Elkan, C. (1994) Fitting a mixture model by expectation maximization to discover motifs in biopolymers. *Proc Int Conf Intell Syst Mol Biol* **2**, 28–36.

Balasubramanian, R., LaFramboise, T., Scholtens, D. and Gentleman, R. (2004) A graph-theoretic approach to testing associations between disparate sources of functional genomics data. *Bioinformatics* **20**, 3353–3362.

Baliga, N.S., Bjork, S.J., Bonneau, R., Pan, M., Iloanusi, C., Kottemann, M.C., *et al.* (2004) Systems level insights into the stress response to UV radiation in the halophilic archaeon Halobacterium NRC-1. *Genome Res* **14**, 1025–1035.

Baliga, N.S., Pan, M., Goo, Y.A., Yi, E.C., Goodlett, D.R., Dimitrov, K., *et al.* (2002) Coordinate regulation of energy transduction modules in Halobacterium sp. analyzed by a global systems approach. *Proc Natl Acad Sci USA* **99**, 14913–14918.

Barbosa, T.M. and Levy, S.B. (2000) Differential expression of over 60 chromosomal genes in Escherichia coli by constitutive expression of MarA. *J Bacteriol* **182**, 3467–3474.

Bar-Joseph, Z., Gerber, G.K., Lee, T.I., Rinaldi, N.J., Yoo, J.Y., Robert, F., *et al.* (2003) Computational discovery of gene modules and regulatory networks. *Nat Biotechnol* **21**, 1337–1342.

Barrett, C.L., Herring, C.D., Reed, J.L. and Palsson, B.O. (2005) The global transcriptional regulatory network for metabolism in Escherichia coli exhibits few dominant functional states. *Proc Natl Acad Sci USA* **102**, 19103–19108.

Beard, D.A., Liang, S.D. and Qian, H. (2002) Energy balance for analysis of complex metabolic networks. *Biophys J* **83**, 79–86.

Ben-Dor, A., Chor, B., Karp, R. and Yakhini, Z. (2003) Discovering local structure in gene expression data: the order-preserving submatrix problem. *J Comput Biol* **10**, 373–384.

Ben-Yehuda, S., Fujita, M., Liu, X.S., Gorbatyuk, B., Skoko, D., Yan, J., *et al.* (2005) Defining a centromere-like element in Bacillus subtilis by identifying the binding sites for the chromosome-anchoring protein RacA. *Mol Cell* **17**, 773–782.

Bergmann, S., Ihmels, J. and Barkai, N. (2003) Iterative signature algorithm for the analysis of large-scale gene expression data. *Phys Rev E Stat Nonlin Soft Matter Phys* **67**, 031902.

Berka, R.M., Hahn, J., Albano, M., Draskovic, I., Persuh, M., Cui, X., *et al.* (2002) Microarray analysis of the Bacillus subtilis K-state: genome-wide expression changes dependent on ComK. *Mol Microbiol* **43**, 1331–1345.

Biondi, E.G., *et al.*, (2006) Regulation of the bacterial cell cycle by an integrated genetic circuit. *Nature*, **444**(7121), 899–904.

Biondi, E.G., Reisinger, S.J., Skerker, J.M., Arif, M., Perchuk, B.S., Ryan, K.R., *et al.* (2006a) Regulation of the bacterial cell cycle by an integrated genetic circuit. *Nature* **444**, 899–904.

Biondi, E.G., Skerker, J.M., Arif, M., Prasol, M.S., Perchuk, B.S. and Laub, M.T. (2006b) A phosphorelay system controls stalk biogenesis during cell cycle progression in Caulobacter crescentus. *Mol Microbiol* **59**, 386–401.

Blattner, F.R., Plunkett, G., III, Bloch, C.A., Perna, N.T., Burland, V., Riley, M., *et al.* (1997) The complete genome sequence of Escherichia coli K-12. *Science* **277**, 1453–1474.

Bogomolni, R.A. and Spudich, J.L. (1982) Identification of a third rhodopsin-like pigment in phototactic Halobacterium halobium. *Proc Natl Acad Sci USA* **79**, 6250–6254.

Bonneau, R., Baliga, N.S., Deutsch, E.W., Shannon, P. and Hood, L. (2004) Comprehensive de novo structure prediction in a systems-biology context for the archaea Halobacterium sp. NRC-1. *Genome Biol* **5**, R52.

Bonneau, R., Reiss, D.J., Shannon, P., Facciotti, M., Hood, L., Baliga, N.S., *et al.* (2006) The inferelator: an algorithm for learning parsimonious regulatory networks from systems-biology data sets de novo. *Genome Biol* **7**, R36.

Bonneau, R., Strauss, C.E. and Baker, D. (2001a) Improving the performance of Rosetta using multiple sequence alignment information and global measures of hydrophobic core formation. *Proteins* **43**, 1–11.

Bonneau, R., Tsai, J., Ruczinski, I., Chivian, D., Rohl, C., Strauss, C.E., *et al.* (2001b) Rosetta in CASP4: progress in ab initio protein structure prediction. *Proteins Suppl* **5**, 119–126.

Bradley, P., Chivian, D., Meiler, J., Misura, K.M., Rohl, C.A., Schief, W.R., *et al.* (2003) Rosetta predictions in CASP5: successes, failures, and prospects for complete automation. *Proteins* **53** (Suppl 6), 457–468.

Bray, D. (1995) Protein molecules as computational elements in living cells. *Nature* **376**, 307–312.

Britton, R.A., Eichenberger, P., Gonzalez-Pastor, J.E., Fawcett, P., Monson, R., Losick, R., *et al.* (2002) Genome-wide analysis of the stationary-phase sigma factor (sigma-H) regulon of Bacillus subtilis. *J Bacteriol* **184**, 4881–4890.

Brocklehurst, K.R. and Morby, A.P. (2000) Metal-ion tolerance in Escherichia coli: analysis of transcriptional profiles by gene-array technology. *Microbiology* **146** (Pt 9), 2277–2282.

Brooun, A., Bell, J., Freitas, T., Larsen, R.W. and Alam, M. (1998) An archaeal aerotaxis transducer combines subunit I core structures of eukaryotic cytochrome c oxidase and eubacterial methyl-accepting chemotaxis proteins. *J Bacteriol* **180**, 1642–1646.

Buttner, K., Bernhardt, J., Scharf, C., Schmid, R., Mader, U., Eymann, C., *et al.* (2001) A comprehensive two-dimensional map of cytosolic proteins of Bacillus subtilis. *Electrophoresis* **22**, 2908–2935.

Chada, V.G., Sanstad, E.A., Wang, R. and Driks, A. (2003) Morphogenesis of bacillus spore surfaces. *J Bacteriol* **185**, 6255–6261.

Cheng, Y. and Church, G.M. (2000) Biclustering of expression data. *Proc Int Conf Intell Syst Mol Biol* **8**, 93–103.

Chivian, D., Kim, D.E., Malmstrom, L., Bradley, P., Robertson, T., Murphy, P., *et al.* (2003) Automated prediction of, C. A.SP-5 *structures using the Robetta server.* Proteins **53** (Suppl 6), 524–533.

Courcelle, J., Khodursky, A., Peter, B., Brown, P.O. and Hanawalt, P.C. (2001) Comparative gene expression profiles following UV exposure in wild-type and SOS-deficient Escherichia coli. *Genetics* **158**, 41–64.

Covert, M.W., Knight, E.M., Reed, J.L., Herrgard, M.J. and Palsson, B.O. (2004) Integrating high-throughput and computational data elucidates bacterial networks. *Nature* **429**, 92–96.

Covert, M.W. and Palsson, B.O. (2002) Transcriptional regulation in constraints-based metabolic models of Escherichia coli. *J Biol Chem* **277**, 28058–28064.

Covert, M.W., Schilling, C.H. and Palsson, B. (2001) Regulation of gene expression in flux balance models of metabolism. *J Theor Biol* **213**, 73–88.

D'Haeseleer, P., Liang, S. and Somogyi, R. (2000) Genetic network inference: from co-expression clustering to reverse engineering. *Bioinformatics* **16**, 707–726.

D'Haeseleer, P., Wen, X., Fuhrman, S. and Somogyi, R. (1999) Linear modeling of mRNA expression levels during CNS development and injury. *Pac Symp Biocomput* **4**, 41–52.

Das, D., Banerjee, N. and Zhang, M.Q. (2004) Interacting models of cooperative gene regulation. *Proc Natl Acad Sci USA* **101**, 16234–16239.

Dassarma, S., Berquist, B.R., Coker, J.A., Dassarma, P. and Muller, J.A. (2006) Postgenomics of the model haloarchaeon Halobacterium sp. NRC-1. *Saline Syst* **2**, 3.

DasSarma, S. (1993) Identification and analysis of the gas vesicle gene cluster on an unstable plasmid of Halobacterium halobium. *Experientia* **49**, 482–486.

Delcher, A.L., Harmon, D., Kasif, S., White, O. and Salzberg, S.L. (1999) Improved microbial gene identification with GLIMMER. *Nucleic Acids Res* **27**, 4636–4641.

DeRisi, J.L., Iyer, V.R. and Brown, P.O. (1997) Exploring the metabolic and genetic control of gene expression on a genomic scale. *Science* **278**, 680–686.

Dittmar, K.A., Mobley, E.M., Radek, A.J. and Pan, T. (2004) Exploring the regulation of tRNA distribution on the genomic scale. *J Mol Biol* **337**, 31–47.

Dunn, W.B. and Ellis, D.I. (2005) Metabolomics: current analytical platforms and methodologies. *Trends Anal Chem* **24**, 285–294.

Earl, A.M., Losick, R. and Kolter, R. (2007) Bacillus subtilis genome diversity. *J Bacteriol* **189**, 1163–1170.

Edwards, J.S., Ibarra, R.U. and Palsson, B.O. (2001) In silico predictions of Escherichia coli metabolic capabilities are consistent with experimental data. *Nat Biotechnol* **19**, 125–130.

Edwards, J.S. and Palsson, B.O. (2000) The Escherichia coli MG1655 in silico metabolic genotype: its definition, characteristics, and capabilities. *Proc Natl Acad Sci USA* **97**, 5528–5533.

Edwards, J.S., Ramakrishna, R. and Palsson, B.O. (2002) Characterizing the metabolic phenotype: a phenotype phase plane analysis. *Biotechnol Bioeng* **77**, 27–36.

Efron, B. (2003) Robbins, empirical Bayes and microarrays. *Ann Statist* **31**, 366–378.

Eichenberger, P., Fujita, M., Jensen, S.T., Conlon, E.M., Rudner, D.Z., Wang, S.T., *et al.* (2004) The program of gene transcription for a single differentiating cell type during sporulation in Bacillus subtilis. *PLoS Biol* **2**, e328.

Eichenberger, P., Jensen, S.T., Conlon, E.M., van Ooij, C., Silvaggi, J., Gonzalez-Pastor, J.E., *et al.* (2003) The sigmaE regulon and the identification of additional sporulation genes in Bacillus subtilis. *J Mol Biol* **327**, 945–972.

Eisenberg, D., Marcotte, E.M., Xenarios, I. and Yeates, T.O. (2000) Protein function in the post-genomic era. *Nature* **405**, 823–826.

Enright, A.J., Iliopoulos, I., Kyrpides, N.C. and Ouzounis, C.A. (1999) Protein interaction maps for complete genomes based on gene fusion events. *Nature* **402**, 86–90.

Errington, J. (2003) Regulation of endospore formation in Bacillus subtilis. *Nat Rev Microbiol* **1**, 117–126.

Eymann, C., Dreisbach, A., Albrecht, D., Bernhardt, J., Becher, D., Gentner, S., *et al.* (2004) A comprehensive proteome map of growing Bacillus subtilis cells. *Proteomics* **4**, 2849–2876.

Fabret, C., Feher, V.A. and Hoch, J.A. (1999) Two-component signal transduction in Bacillus subtilis: how one organism sees its world. *J Bacteriol* **181**, 1975–1983.

Faith, J.J., Hayete, B., Thaden, J.T., Mogno, I., Wierzbowski, J., Cottarel, G., *et al.* (2007) Large-scale mapping and validation of Escherichia coli transcriptional regulation from a compendium of expression profiles. *PLoS Biol* **5**, e8.

Fawcett, P., Eichenberger, P., Losick, R. and Youngman, P. (2000) The transcriptional profile of early to middle sporulation in Bacillus subtilis. *Proc Natl Acad Sci USA* **97**, 8063–8068.

Ferreira, L.C., Ferreira, R.C. and Schumann, W. (2005) Bacillus subtilis as a tool for vaccine development: from antigen factories to delivery vectors. *An Acad Bras Cienc* **77**, 113–124.

Feucht, A., Evans, L. and Errington, J. (2003) Identification of sporulation genes by genome-wide analysis of the sigmaE regulon of Bacillus subtilis. *Microbiology* **149**, 3023–3034.

Fiehn, O. (2002) Metabolomics – the link between genotypes and phenotypes. *Plant Mol Biol* **48**, 155–171.

Fields, S. and Song, O.-K. (1989) A novel genetic system to detect protein-protein interactions. *Nature* **340**, 245–246.

Finn, R.D., Mistry, J., Schuster-Bockler, B., Griffiths-Jones, S., Hollich, V., Lassmann, T., *et al.* (2006) Pfam: clans, web tools and services. *Nucleic Acids Res* **34**, D247–D251.

Fischer, D., Rychlewski, L., Dunbrack, R.L., Jr., Ortiz, A.R. and Elofsson, A. (2003) CAFASP3: the third critical assessment of fully automated structure prediction methods. *Proteins* **53** (Suppl 6), 503–516.

Fleischmann, R.D., Adams, M.D., White, O., Clayton, R.A., Kirkness, E.F., Kerlavage, A.R., *et al.* (1995) Whole-genome random sequencing and assembly of Haemophilus influenzae Rd. *Science* **269**, 496–512.

Foster, J.W. (2004) Escherichia coli acid resistance: tales of an amateur acidophile. *Nat Rev Microbiol* **2**, 898–907.

Friedman, N., Linial, M., Nachman, I. and Pe'er, D. (2000) Using Bayesian networks to analyze expression data. *J Comput Biol* **7**, 601–620.

Ge, H., Walhout, A.J. and Vidal, M. (2003) Integrating 'omic' information: a bridge between genomics and systems biology. *Trends Genet* **19**, 551–560.

Glasner, J.D., Liss, P., Plunkett, G., III, Darling, A., Prasad, T., Rusch, M., *et al.* (2003) ASAP, a systematic annotation package for community analysis of genomes. *Nucleic Acids Res* **31**, 147–151.

Guardia, M.J., Gambhir, A., Europa, A.F., Ramkrishna, D. and Hu, W.S. (2000) Cybernetic modeling and regulation of metabolic pathways in multiple steady states of hybridoma cells. *Biotechnol Prog* **16**, 847–853.

Gygi, S.P., Rist, B., Gerber, S.A., Turecek, F., Gelb, M.H. and Aebersold, R. (1999) Quantitative analysis of complex protein mixtures using isotope-coded affinity tags. *Nat Biotechnol* **17**, 994–999.

Hamon, M.A., Stanley, N.R., Britton, R.A., Grossman, A.D. and Lazazzera, B.A. (2004) Identification of AbrB-regulated genes involved in biofilm formation by Bacillus subtilis. *Mol Microbiol* **52**, 847–860.

Handelsman, J. (2004) Metagenomics: application of genomics to uncultured microorganisms. *Microbiol Mol Biol Rev* **68**, 669–685.

Harbison, C.T., Gordon, D.B., Lee, T.I., Rinaldi, N.J., Macisaac, K.D., Danford, T.W., *et al.* (2004) Transcriptional regulatory code of a eukaryotic genome. *Nature* **431**, 99–104.

Hashimoto, R.F., Kim, S., Shmulevich, I., Zhang, W., Bittner, M.L. and Dougherty, E.R. (2004) Growing genetic regulatory networks from seed genes. *Bioinformatics* **20**, 1241–1247.

Hayashi, K., Ohsawa, T., Kobayashi, K., Ogasawara, N. and Ogura, M. (2005) The H2O2 stress-responsive regulator PerR positively regulates srfA expression in Bacillus subtilis. *J Bacteriol* **187**, 6659–6667.

Helmann, J.D., Wu, M.F., Kobel, P.A., Gamo, F.J., Wilson, M., Morshedi, M.M., *et al.* (2001) Global transcriptional response of Bacillus subtilis to heat shock. *J Bacteriol* **183**, 7318–7328.

Hilbert, D.W. and Piggot, P.J. (2004) Compartmentalization of gene expression during Bacillus subtilis spore formation. *Microbiol Mol Biol Rev* **68**, 234–262.

Holtzendorff, J., Hung, D., Brende, P., Reisenauer, A., Viollier, P.H., McAdams, H.H., *et al.* (2004) Oscillating global regulators control the genetic circuit driving a bacterial cell cycle. *Science* **304**, 983–987.

Holtzendorff, J., Reinhardt, J. and Viollier, P.H. (2006) Cell cycle control by oscillating regulatory proteins in Caulobacter crescentus. *Bioessays* **28**, 355–361.

Hommais, F., Krin, E., Laurent-Winter, C., Soutourina, O., Malpertuy, A., Le Caer, J.P., *et al.* (2001) Large-scale monitoring of pleiotropic regulation of gene expression by the prokaryotic nucleoid-associated protein, H-NS. *Mol Microbiol* **40**, 20–36.

Hong, R.W., Shchepetov, M., Weiser, J.N. and Axelsen, P.H. (2003) Transcriptional profile of the Escherichia coli response to the antimicrobial insect peptide cecropin A. *Antimicrob Agents Chemother* **47**, 1–6.

Hottes, A.K., Shapiro, L. and McAdams, H.H. (2005) DnaA coordinates replication initiation and cell cycle transcription in Caulobacter crescentus. *Mol Microbiol* **58**, 1340–1353.

Hung, D.Y. and Shapiro, L. (2002) A signal transduction protein cues proteolytic events critical to Caulobacter cell cycle progression. *Proc Natl Acad Sci USA* **99**, 13160–13165.

Ibarra, R.U., Edwards, J.S. and Palsson, B.O. (2002) Escherichia coli K-12 undergoes adaptive evolution to achieve in silico predicted optimal growth. *Nature* **420**, 186–189.

Ideker, T., Ozier, O., Schwikowski, B. and Siegel, A.F. (2002) Discovering regulatory and signalling circuits in molecular interaction networks. *Bioinformatics* **18** (Suppl 1), S233–S240.

Iniesta, A.A., McGrath, P.T., Reisenauer, A., McAdams, H.H. and Shapiro, L. (2006) A phospho-signaling pathway controls the localization and activity of a protease complex critical for bacterial cell cycle progression. *Proc Natl Acad Sci USA* **103**, 10935–10940.

Ito, T., Chiba, T., Ozawa, R., Yoshida, M., Hattori, M. and Sakaki, Y. (2001) A comprehensive two-hybrid analysis to explore the yeast protein interactome. *Proc Natl Acad Sci USA* **98** (8), 4569–4574.

Jacobs, C., Ausmees, N., Cordwell, S.J., Shapiro, L. and Laub, M.T. (2003) Functions of the CckA histidine kinase in Caulobacter cell cycle control. *Mol Microbiol* **47**, 1279–1290.

Jacobs-Wagner, C. (2004) Regulatory proteins with a sense of direction: cell cycle signalling network in Caulobacter. *Mol Microbiol* **51**, 7–13.

Jardine, O., Gough, J., Chothia, C. and Teichmann, S.A. (2002) Comparison of the small molecule metabolic enzymes of Escherichia coli and Saccharomyces cerevisiae. *Genome Res* **12**, 916–929.

Kanehisa, M., Goto, S., Hattori, M., Aoki-Kinoshita, K.F., Itoh, M., Kawashima, S., *et al.* (2006) From genomics to chemical genomics: new developments in KEGG. *Nucleic Acids Res* **34**, D354–D357.

Kanehisa, M., Goto, S., Kawashima, S., Okuno, Y. and Hattori, M. (2004) The KEGG resource for deciphering the genome. *Nucleic Acids Res* **32**, D277–D280.

Karp, P.D., Riley, M., Saier, M., Paulsen, I.T., Paley, S.M. and Pellegrini-Toole, A. (2000) The EcoCyc and MetaCyc databases. *Nucleic Acids Res* **28**, 56–59.

Keijser, B.J., Beek, A.T., Rauwerda, H., Schuren, F., Montijn, R., Van Der Spek, H., *et al.* (2007) Analysis of temporal gene expression during Bacillus subtilis spore germination and outgrowth. *J Bacteriol* **23**, 23.

Keseler, I.M., Collado-Vides, J., Gama-Castro, S., Ingraham, J., Paley, S., Paulsen, I.T., *et al.* (2005) EcoCyc: a comprehensive database resource for Escherichia coli. *Nucleic Acids Res* **33**, D334–D337.

Kinch, L.N., Wrabl, J.O., Krishna, S.S., Majumdar, I., Sadreyev, R.I., Qi, Y., *et al.* (2003) CASP5 assessment of fold recognition target predictions. *Proteins* **53** (Suppl 6), 395–409.

Kluger, Y., Basri, R., Chang, J.T. and Gerstein, M. (2003) Spectral biclustering of microarray data: coclustering genes and conditions. *Genome Res* **13**, 703–716.

Kobayashi, K., Ogura, M., Yamaguchi, H., Yoshida, K., Ogasawara, N., Tanaka, T., *et al.* (2001) Comprehensive DNA microarray analysis of Bacillus subtilis two-component regulatory systems. *J Bacteriol* **183**, 7365–7370.

Kolbe, M., Besir, H., Essen, L.O. and Oesterhelt, D. (2000) Structure of the light-driven chloride pump halorhodopsin at 1.8 A resolution. *Science* **288**, 1390–1396.

Kunkel, B., Kroos, L., Poth, H., Youngman, P. and Losick, R. (1989) Temporal and spatial control of the mother-cell regulatory gene spoIIID of Bacillus subtilis. *Genes Dev* **3**, 1735–1744.

Kunst, F., Ogasawara, N., Moszer, I., Albertini, A.M., Alloni, G., Azevedo, V., *et al.* (1997) The complete genome sequence of the Gram-positive bacterium Bacillus subtilis. *Nature* **390**, 249–256.

Laub, M.T., Chen, S.L., Shapiro, L. and McAdams, H.H. (2002) Genes directly controlled by CtrA, a master regulator of the Caulobacter cell cycle. *Proc Natl Acad Sci USA* **99**, 4632–4637.

Laub, M.T., McAdams, H.H., Feldblyum, T., Fraser, C.M. and Shapiro, L. (2000) Global analysis of the genetic network controlling a bacterial cell cycle. *Science* **290**, 2144–2148.

Lee, T.I., Rinaldi, N.J., Robert, F., Odom, D.T., Bar-Joseph, Z., Gerber, G.K., *et al.* (2002) Transcriptional regulatory networks in Saccharomyces cerevisiae. *Science* **298**, 799–804.

Lin, J., Lee, I.S., Frey, J., Slonczewski, J.L. and Foster, J.W. (1995) Comparative analysis of extreme acid survival in Salmonella typhimurium, Shigella flexneri, and Escherichia coli. *J Bacteriol* **177**, 4097–4104.

Lin, J.T., Connelly, M.B., Amolo, C., Otani, S. and Yaver, D.S. (2005) Global transcriptional response of Bacillus subtilis to treatment with subinhibitory concentrations of antibiotics that inhibit protein synthesis. *Antimicrob Agents Chemother* **49**, 1915–1926.

Loew, L.M. and Schaff, J.C. (2001) The virtual cell: a software environment for computational cell biology. *Trends Biotechnol* **19**, 401–406.

Luecke, H., Schobert, B., Cartailler, J.P., Richter, H.T., Rosengarth, A., Needleman, R., *et al.* (2000) Coupling photoisomerization of retinal to directional transport in bacteriorhodopsin. *J Mol Biol* **300**, 1237–1255.

Majewski, R.A. and Domach, M.M. (1990) Simple constrained-optimization view of acetate overflow in E. coli. *Biotechnol Bioeng* **35**, 732–738.

Malmstrom, L., Riffle, M., Strauss, C.E., Chivian, D., Davis, T.N., Bonneau, R., *et al.* (2007) Superfamily assignments for the yeast proteome through integration of structure prediction with the gene ontology. *PLoS Biol* **5**, e76.

Mangan, S. and Alon, U. (2003) Structure and function of the feed-forward loop network motif. *Proc Natl Acad Sci USA* **100**, 11980–11985.

Mangan, S., Zaslaver, A. and Alon, U. (2003) The coherent feedforward loop serves as a sign-sensitive delay element in transcription networks. *J Mol Biol* **334**, 197–204.

Marcotte, E.M., Pellegrini, M., Ng, H.L., Rice, D.W., Yeates, T.O. and Eisenberg, D. (1999) Detecting protein function and protein-protein interactions from genome sequences. *Science* **285**, 751–753.

Margulies, M., Egholm, M., Altman, W.E., Attiya, S., Bader, J.S., Bemben, L.A., *et al.* (2005) Genome sequencing in microfabricated high-density picolitre reactors. *Nature* **437**, 376–380.

Mascher, T., Helmann, J.D. and Unden, G. (2006) Stimulus perception in bacterial signal-transducing histidine kinases. *Microbiol Mol Biol Rev* **70**, 910–938.

Masuda, N. and Church, G.M. (2002) Escherichia coli gene expression responsive to levels of the response regulator EvgA. *J Bacteriol* **184**, 6225–6234.

Masuda, N. and Church, G.M. (2003) Regulatory network of acid resistance genes in Escherichia coli. *Mol Microbiol* **48**, 699–712.

Meile, J.C., Wu, L.J., Ehrlich, S.D., Errington, J. and Noirot, P. (2006) Systematic localisation of proteins fused to the green fluorescent protein in Bacillus subtilis: identification of new proteins at the DNA replication factory. *Proteomics* **6**, 2135–2146.

Mellor, J.C., Yanai, I., Clodfelter, K.H., Mintseris, J. and DeLisi, C. (2002) Predictome: a database of putative functional links between proteins. *Nucleic Acids Res* **30**, 306–309.

Miller, M.B. and Bassler, B.L. (2001) Quorum sensing in bacteria. *Annu Rev Microbiol* **55**, 165–199.

Milo, R., Shen-Orr, S., Itzkovitz, S., Kashtan, N., Chklovskii, D. and Alon, U. (2002) Network motifs: simple building blocks of complex networks. *Science* **298**, 824–827.

Molle, V., Fujita, M., Jensen, S.T., Eichenberger, P., Gonzalez-Pastor, J.E., Liu, J.S., *et al.* (2003a) The Spo0A regulon of Bacillus subtilis. *Mol Microbiol* **50**, 1683–1701.

Molle, V., Nakaura, Y., Shivers, R.P., Yamaguchi, H., Losick, R., Fujita, Y., *et al.* (2003b) Additional targets of the Bacillus subtilis global regulator CodY identified by chromatin immunoprecipitation and genome-wide transcript analysis. *J Bacteriol* **185**, 1911–1922.

Moraru, I.I., Schaff, J.C., Slepchenko, B.M. and Loew, L.M. (2002) The virtual cell: an integrated modeling environment for experimental and computational cell biology. *Ann N Y Acad Sci* **971**, 595–596.

Moreno-Hagelsieb, G. and Collado-Vides, J. (2002) A powerful non-homology method for the prediction of operons in prokaryotes. *Bioinformatics* **18** (Suppl 1), S329–S336.

Moszer, I. (1998) The complete genome of Bacillus subtilis: from sequence annotation to data management and analysis. *FEBS Lett* **430**, 28–36.

Moszer, I., Jones, L.M., Moreira, S., Fabry, C. and Danchin, A. (2002) SubtiList: the reference database for the Bacillus subtilis genome. *Nucleic Acids Res* **30**, 62–65.

Murali, T.M. and Kasif, S. (2003) Extracting conserved gene expression motifs from gene expression data. *Pac Symp Biocomput* **8**, 77–88.

Nakano, S., Kuster-Schock, E., Grossman, A.D. and Zuber, P. (2003) Spx-dependent global transcriptional control is induced by thiol-specific oxidative stress in Bacillus subtilis. *Proc Natl Acad Sci USA* **100**, 13603–13608.

Ng, W.V., Kennedy, S.P., Mahairas, G.G., Berquist, B., Pan, M., Shukla, H.D., *et al.* (2000) Genome sequence of Halobacterium species NRC-1. *Proc Natl Acad Sci USA* **97**, 12176–12181.

Nierman, W.C., Feldblyum, T.V., Laub, M.T., Paulsen, I.T., Nelson, K.E., Eisen, J.A., *et al.* (2001) Complete genome sequence of Caulobacter crescentus. *Proc Natl Acad Sci USA* **98**, 4136–4141.

Nishino, K., Inazumi, Y. and Yamaguchi, A. (2003) Global analysis of genes regulated by EvgA of the two-component regulatory system in Escherichia coli. *J Bacteriol* **185**, 2667–2672.

Nishino, K. and Yamaguchi, A. (2001) Analysis of a complete library of putative drug transporter genes in Escherichia coli. *J Bacteriol* **183**, 5803–5812.

Nobeli, I., Ponstingl, H., Krissinel, E.B. and Thornton, J.M. (2003) A structure-based anatomy of the E.coli metabolome. *J Mol Biol* **334**, 697–719.

Noirot-Gros, M.F., Dervyn, E., Wu, L.J., Mervelet, P., Errington, J., Ehrlich, S.D., *et al.* (2002) An expanded view of bacterial DNA replication. *Proc Natl Acad Sci USA* **99**, 8342–8347.

Ogata, H., Goto, S., Fujibuchi, W. and Kanehisa, M. (1998) Computation with the KEGG pathway database. *Biosystems* **47**, 119–128.

Ogura, M. and Fujita, Y. (2007) Bacillus subtilis rapD, a direct target of transcription repression by RghR, negatively regulates srfA expression. *FEMS Microbiol Lett* **268**, 73–80.

Ogura, M., Tsukahara, K., Hayashi, K. and Tanaka, T. (2007) The Bacillus subtilis NatK-NatR two-component system regulates expression of the natAB operon encoding an ABC transporter for sodium ion extrusion. *Microbiology* **153**, 667–675.

Ogura, M., Yamaguchi, H., Kobayashi, K., Ogasawara, N., Fujita, Y. and Tanaka, T. (2002) Whole-genome analysis of genes regulated by the Bacillus subtilis competence transcription factor ComK. *J Bacteriol* **184**, 2344–2351.

Ogura, M., Yamaguchi, H., Yoshida, K., Fujita, Y. and Tanaka, T. (2001) DNA microarray analysis of Bacillus subtilis DegU, ComA and PhoP regulons: an approach to comprehensive analysis of B.subtilis two-component regulatory systems. *Nucleic Acids Res* **29**, 3804–3813.

Ohashi, Y., Inaoka, T., Kasai, K., Ito, Y., Okamoto, S., Satsu, H., *et al.* (2003) Expression profiling of translation-associated genes in sporulating Bacillus subtilis and consequence of sporulation by gene inactivation. *Biosci Biotechnol Biochem* **67**, 2245–2253.

Palsson, B. (2006) *Systems Biology: Properties of Reconstructed Networks* (Cambridge: Cambridge University Press).

Peregrin-Alvarez, J.M., Tsoka, S. and Ouzounis, C.A. (2003) The phylogenetic extent of metabolic enzymes and pathways. *Genome Res* **13**, 422–427.

Piggot, P.J. and Hilbert, D.W. (2004) Sporulation of Bacillus subtilis. *Curr Opin Microbiol* **7**, 579–586.

Poetz, O., Schwenk, J.M., Kramer, S., Stoll, D., Templin, M.F. and Joos, T.O. (2005) Protein microarrays: catching the proteome. *Mech Ageing Dev* **126**, 161–170.

Polen, T., Rittmann, D., Wendisch, V.F. and Sahm, H. (2003) DNA microarray analyses of the long-term adaptive response of Escherichia coli to acetate and propionate. *Appl Environ Microbiol* **69**, 1759–1774.

Pomposiello, P.J., Bennik, M.H. and Demple, B. (2001) Genome-wide transcriptional profiling of the Escherichia coli responses to superoxide stress and sodium salicylate. *J Bacteriol* **183**, 3890–3902.

Prelic, A., Bleuler, S., Zimmermann, P., Wille, A., Buhlmann, P., Gruissem, W., *et al.* (2006) A systematic comparison and evaluation of biclustering methods for gene expression data. *Bioinformatics* **22**, 1122–1129.

Price, M.N., Arkin, A.P. and Alm, E.J. (2006) The life-cycle of operons. *PLoS Genet* **2**, e96.

Price, M.N., Dehal, P.S. and Arkin, A.P. (2008) Horizontal gene transfer and the evolution of transcriptional regulation in Escherichia coli. *Genome Biol* **9**, R4.

Price, N.D., Papin, J.A., Schilling, C.H. and Palsson, B.O. (2003) Genome-scale microbial in silico models: the constraints-based approach. *Trends Biotechnol* **21**, 162–169.

Qian, J., Kluger, Y., Yu, H. and Gerstein, M. (2003) Identification and correction of spurious spatial correlations in microarray data. *Biotechniques* **35**, 42–44; 46, 48.

Rain, J.C., Selig, L., De Reuse, H., Battaglia, V., Reverdy, C., Simon, S., *et al.* (2001) The protein–protein interaction map of Helicobacter pylori. *Nature* **409** (6817), 211–215.

Reed, J.L. and Palsson, B.O. (2003) Thirteen years of building constraint-based in silico models of Escherichia coli. *J Bacteriol* **185**, 2692–2699.

Reed, J.L., Vo, T.D., Schilling, C.H. and Palsson, B.O. (2003) An expanded genome-scale model of Escherichia coli K-12 (iJR904 GSM/GPR). *Genome Biol* **4**, R54.

Reiss, D.J., Baliga, N.S. and Bonneau, R. (2006) Integrated biclustering of heterogeneous genome-wide datasets for the inference of global regulatory networks. *BMC Bioinformatics* **7**, 280.

Ren, D., Bedzyk, L.A., Setlow, P., England, D.F., Kjelleberg, S., Thomas, S.M., *et al.* (2004) Differential gene expression to investigate the effect of (5Z)-4-bromo-5-(bromomethylene)-3-butyl-2(5H)-furanone on Bacillus subtilis. *Appl Environ Microbiol* **70**, 4941–4949.

Richmond, C.S., Glasner, J.D., Mau, R., Jin, H. and Blattner, F.R. (1999) Genome-wide expression profiling in Escherichia coli K-12. *Nucleic Acids Res* **27**, 3821–3835.

Richter-Gebert, J., Sturmfels, B. and Theobald, T. (2003) First steps in tropical geometry. In Proceedings of the Conference on Idempotent Mathematics and Mathematical

Physics, *2003*, G.L. Litvinov and V.P. Maslov, eds (American Mathematical Society), *Contemp Math* **377**, pp. 289–317.

Roth, F.P., Hughes, J.D., Estep, P.W. and Church, G.M. (1998) Finding DNA regulatory motifs within unaligned noncoding sequences clustered by whole-genome mRNA quantitation. *Nat Biotechnol* **16**, 939–945.

Ruepp, A. and Soppa, J. (1996) Fermentative arginine degradation in Halobacterium salinarium (formerly Halobacterium halobium): genes, gene products, and transcripts of the arcRACB gene cluster. *J Bacteriol* **178**, 4942–4947.

Rusch, D.B., Halpern, A.L., Sutton, G., Heidelberg, K.B., Williamson, S., Yooseph, S., *et al.* (2007) The sorcerer II global ocean sampling expedition: northwest Atlantic through eastern tropical Pacific. *PLoS Biol* **5**, e77.

Salgado, H., Gama-Castro, S., Peralta-Gil, M., Diaz-Peredo, E., Sanchez-Solano, F., Santos-Zavaleta, A., *et al.* (2006) RegulonDB (version 5.0): Escherichia coli K-12 transcriptional regulatory network, operon organization, and growth conditions. *Nucleic Acids Res* **34**, D394–D397.

Salzberg, S.L., Delcher, A.L., Kasif, S. and White, O. (1998) Microbial gene identification using interpolated Markov models. *Nucleic Acids Res* **26**, 544–548.

Sanger, F., Nicklen, S. and Coulson, A.R. (1977) DNA sequencing with chain-terminating inhibitors. *Proc Natl Acad Sci USA* **74**, 5463–5467.

Segal, E., Shapira, M., Regev, A., Pe'er, D., Botstein, D., Koller, D., *et al.* (2003) Module networks: identifying regulatory modules and their condition-specific regulators from gene expression data. *Nat Genet* **34**, 166–176.

Segal, E., Taskar, B., Gasch, A., Friedman, N. and Koller, D. (2001) Rich probabilistic models for gene expression. *Bioinformatics* **17** (Suppl 1), S243–S252.

Selinger, D.W., Cheung, K.J., Mei, R., Johansson, E.M., Richmond, C.S., Blattner, F.R., *et al.* (2000) RNA expression analysis using a 30 base pair resolution Escherichia coli genome array. *Nat Biotechnol* **18**, 1262–1268.

Selkov, E., Jr., Grechkin, Y., Mikhailova, N. and Selkov, E. (1998) MPW: the metabolic pathways database. *Nucleic Acids Res* **26**, 43–45.

Serizawa, M., Kodama, K., Yamamoto, H., Kobayashi, K., Ogasawara, N. and Sekiguchi, J. (2005) Functional analysis of the YvrGHb two-component system of Bacillus subtilis: identification of the regulated genes by DNA microarray and northern blot analyses. *Biosci Biotechnol Biochem* **69**, 2155–2169.

Serres, M.H., Gopal, S., Nahum, L.A., Liang, P., Gaasterland, T. and Riley, M. (2001) A functional update of the Escherichia coli K-12 genome. *Genome Biol* **2**, RE-SEARCH0035.

Setlow, P. (2003) Spore germination. *Curr Opin Microbiol* **6**, 550–556.

Shamir, R., Maron-Katz, A., Tanay, A., Linhart, C., Steinfeld, I., Sharan, R., *et al.* (2005) EXPANDER – an integrative program suite for microarray data analysis. *BMC Bioinformatics* **6**, 232.

Shannon, P., Markiel, A., Ozier, O., Baliga, N.S., Wang, J.T., Ramage, D., *et al.* (2003) Cytoscape: a software environment for integrated models of biomolecular interaction networks. *Genome Res* **13**, 2498–2504.

Shannon, P.T., Reiss, D.J., Bonneau, R. and Baliga, N.S. (2006) The Gaggle: an open-source software system for integrating bioinformatics software and data sources. *BMC Bioinformatics* **7**, 176.

Shapiro, B.J. and Alm, E.J. (2008) Comparing patterns of natural selection across species using selective signatures. *PLoS Genet* **4**, e23.

Sheng, Q., Moreau, Y. and De Moor, B. (2003) Biclustering microarray data by Gibbs sampling. *Bioinformatics* **19** (Suppl 2), ii196–ii205.

Shen-Orr, S.S., Milo, R., Mangan, S. and Alon, U. (2002) Network motifs in the transcriptional regulation network of Escherichia coli. *Nat Genet* **31**, 64–68.

Shmulevich, I., Lahdesmaki, H., Dougherty, E.R., Astola, J. and Zhang, W. (2003) The role of certain Post classes in Boolean network models of genetic networks. *Proc Natl Acad Sci USA* **100**, 10734–10739.

Silvaggi, J.M., Perkins, J.B. and Losick, R. (2006) Genes for small, noncoding RNAs under sporulation control in Bacillus subtilis. *J Bacteriol* **188**, 532–541.

Skerker, J.M. and Laub, M.T. (2004) Cell-cycle progression and the generation of asymmetry in Caulobacter crescentus. *Nat Rev Microbiol* **2**, 325–337.

Skerker, J.M., Prasol, M.S., Perchuk, B.S., Biondi, E.G. and Laub, M.T. (2005) Two-component signal transduction pathways regulating growth and cell cycle progression in a bacterium: a system-level analysis. *PLoS Biol* **3**, e334.

Slonim, N., Friedman, N. and Tishby, N. (2006) Multivariate information bottleneck. *Neural Comput* **18**, 1739–1789.

Soga, T., Ohashi, Y., Ueno, Y., Naraoka, H., Tomita, M. and Nishioka, T. (2003) Quantitative metabolome analysis using capillary electrophoresis mass spectrometry. *J Proteome Res* **2**, 488–494.

Sogin, M.L., Morrison, H.G., Huber, J.A., Welch, D.M., Huse, S.M., Neal, P.R., *et al.* (2006) Microbial diversity in the deep sea and the underexplored 'rare biosphere'. *Proc Natl Acad Sci USA* **103**, 12115–12120.

Spudich, E.N., Takahashi, T. and Spudich, J.L. (1989) Sensory rhodopsins I and II modulate a methylation/demethylation system in Halobacterium halobium phototaxis. *Proc Natl Acad Sci USA* **86**, 7746–7750.

Spudich, J.L. (1993) Color sensing in the Archaea: a eukaryotic-like receptor coupled to a prokaryotic transducer. *J Bacteriol* **175**, 7755–7761.

Spudich, J.L. and Bogomolni, R.A. (1984) Mechanism of colour discrimination by a bacterial sensory rhodopsin. *Nature* **312**, 509–513.

Steil, L., Serrano, M., Henriques, A.O. and Volker, U. (2005) Genome-wide analysis of temporally regulated and compartment-specific gene expression in sporulating cells of Bacillus subtilis. *Microbiology* **151**, 399–420.

Stevens, C.M. and Errington, J. (1990) Differential gene expression during sporulation in Bacillus subtilis: structure and regulation of the spoIIID gene. *Mol Microbiol* **4**, 543–551.

Stock, A.M., Robinson, V.L. and Goudreau, P.N. (2000) Two-component signal transduction. *Annu Rev Biochem* **69**, 183–215.

Streips, U.N. and Polio, F.W. (1985) Heat shock proteins in bacilli. *J Bacteriol* **162**, 434–437.

Stuart, J.M., Segal, E., Koller, D. and Kim, S.K. (2003) A gene-coexpression network for global discovery of conserved genetic modules. *Science* **302**, 249–255.

Tam Le, T., Antelmann, H., Eymann, C., Albrecht, D., Bernhardt, J. and Hecker, M. (2006) Proteome signatures for stress and starvation in Bacillus subtilis as revealed by a 2-D gel image color coding approach. *Proteomics* **6**, 4565–4585.

Tanay, A., Sharan, R., Kupiec, M. and Shamir, R. (2004) Revealing modularity and organization in the yeast molecular network by integrated analysis of highly heterogeneous genomewide data. *Proc Natl Acad Sci USA* **101**, 2981–2986.

Tanay, A., Sharan, R. and Shamir, R. (2002) Discovering statistically significant biclusters in gene expression data. *Bioinformatics* **18** (Suppl 1), S136–S144.

Tao, H., Bausch, C., Richmond, C., Blattner, F.R. and Conway, T. (1999) Functional genomics: expression analysis of Escherichia coli growing on minimal and rich media. *J Bacteriol* **181**, 6425–6440.

Tatti, K.M., Jones, C.H. and Moran, C.P., Jr. (1991) Genetic evidence for interaction of sigma E with the spoIIID promoter in Bacillus subtilis. *J Bacteriol* **173**, 7828–7833.

Tatusov, R.L., Koonin, E.V. and Lipman, D.J. (1997) A genomic perspective on protein families. *Science* **278**, 631–637.

Tatusov, R.L., Natale, D.A., Garkavtsev, I.V., Tatusova, T.A., Shankavaram, U.T., Rao, B.S., *et al.* (2001) The COG database: new developments in phylogenetic classification of proteins from complete genomes. *Nucleic Acids Res* **29** (1), 22–28.

Thorsson, V., Hornquist, M., Siegel, A.F. and Hood, L. (2005) Reverse engineering galactose regulation in yeast through model selection. *Stat Appl Genet Mol Biol* **4**, Article28.

Tomasinsig, L., Scocchi, M., Mettulio, R. and Zanetti, M. (2004) Genome-wide transcriptional profiling of the Escherichia coli response to a proline-rich antimicrobial peptide. *Antimicrob Agents Chemother* **48**, 3260–3267.

Tringe, S.G. and Rubin, E.M. (2005) Metagenomics: DNA sequencing of environmental samples. *Nat Rev Genet* **6**, 805–814.

Tucker, D.L., Tucker, N. and Conway, T. (2002) Gene expression profiling of the pH response in Escherichia coli. *J Bacteriol* **184**, 6551–6558.

Tucker, D.L., Tucker, N., Ma, Z., Foster, J.W., Miranda, R.L., Cohen, P.S., *et al.* (2003) Genes of the GadX-GadW regulon in Escherichia coli. *J Bacteriol* **185**, 3190–3201.

Tyson, G.W., Chapman, J., Hugenholtz, P., Allen, E.E., Ram, R.J., Richardson, P.M., *et al.* (2004) Community structure and metabolism through reconstruction of microbial genomes from the environment. *Nature* **428**, 37–43.

Uetz, P., Giot, L., Cagney, G., Mansfield, T.A., Judson, R.S., Knight, J.R., et al. (2000) A comprehensive analysis of protein–protein interactions in *Saccharomyces cerevisiae*. *Nature* **403** (6770), 623–627.

van Someren, E.P., Wessels, L.F., Backer, E. and Reinders, M.J. (2002) Genetic network modeling. *Pharmacogenomics* **3**, 507–525.

van Someren, E.P., Wessels, L.F. and Reinders, M.J. (2000) Linear modeling of genetic networks from experimental data. *Proc Int Conf Intell Syst Mol Biol* **8**, 355–366.

Vanet, A., Marsan, L., Labigne, A. and Sagot, M.F. (2000) Inferring regulatory elements from a whole genome. An analysis of Helicobacter pylori sigma(80) family of promoter signals. *J Mol Biol* **297**, 335–353.

Varma, A. and Palsson, B.O. (1994) Stoichiometric flux balance models quantitatively predict growth and metabolic by-product secretion in wild-type Escherichia coli W3110. *Appl Environ Microbiol* **60**, 3724–3731.

Varner, J. and Ramkrishna, D. (1998) Application of cybernetic models to metabolic engineering: investigation of storage pathways. *Biotechnol Bioeng* **58**, 282–291.

Varner, J. and Ramkrishna, D. (1999) Metabolic engineering from a cybernetic perspective. 1. Theoretical preliminaries. *Biotechnol Prog* **15**, 407–425.

Velculescu, V.E., Zhang, L., Vogelstein, B. and Kinzler, K.W. (1995) Serial analysis of gene expression. *Science* **270**, 484–487.

Vemuri, G.N. and Aristidou, A.A. (2005) Metabolic engineering in the -omics era: elucidating and modulating regulatory networks. *Microbiol Mol Biol Rev* **69**, 197–216.

Venter, J.C., Remington, K., Heidelberg, J.F., Halpern, A.L., Rusch, D., Eisen, J.A., *et al.* (2004) Environmental genome shotgun sequencing of the Sargasso Sea. *Science* **304**, 66–74.

von Dassow, G., Meir, E., Munro, E.M. and Odell, G.M. (2000) The segment polarity network is a robust developmental module. *Nature* **406**, 188–192.

Wahde, M. and Hertz, J. (2001) Modeling genetic regulatory dynamics in neural development. *J Comput Biol* **8**, 429–442.

Walhout, A.J. and Vidal, M. (2001) High-throughput yeast two-hybrid assays for large-scale protein interaction mapping. *Methods* **24**, 297–306.

Wang, A. and Crowley, D.E. (2005) Global gene expression responses to cadmium toxicity in Escherichia coli. *J Bacteriol* **187**, 3259–3266.

Wang, Q.Z., Wu, C.Y., Chen, T., Chen, X. and Zhao, X.M. (2006) Integrating metabolomics into a systems biology framework to exploit metabolic complexity: strategies and applications in microorganisms. *Appl Microbiol Biotechnol* **70**, 151–161.

Wang, S.T., Setlow, B., Conlon, E.M., Lyon, J.L., Imamura, D., Sato, T., *et al.* (2006) The forespore line of gene expression in Bacillus subtilis. *J Mol Biol* **358**, 16–37.

Weaver, D.C., Workman, C.T. and Stormo, G.D. (1999) Modeling regulatory networks with weight matrices. *Pac Symp Biocomput* **4**, 112–123.

Weber, A. and Jung, K. (2002) Profiling early osmostress-dependent gene expression in Escherichia coli using DNA macroarrays. *J Bacteriol* **184**, 5502–5507.

Wei, Y., Lee, J.M., Richmond, C., Blattner, F.R., Rafalski, J.A. and LaRossa, R.A. (2001) High-density microarray-mediated gene expression profiling of Escherichia coli. *J Bacteriol* **183**, 545–556.

Wodicka, L., Dong, H., Mittmann, M., Ho, M.H. and Lockhart, D.J. (1997) Genome-wide expression monitoring in Saccharomyces cerevisiae. *Nat Biotechnol* **15**, 1359–1367.

Wolff, S., Antelmann, H., Albrecht, D., Becher, D., Bernhardt, J., Bron, S., *et al.* (2007) Towards the entire proteome of the model bacterium Bacillus subtilis by gel-based and gel-free approaches. *J Chromatogr B Analyt Technol Biomed Life Sci* **849**, 129–140.

Wolff, S., Otto, A., Albrecht, D., Zeng, J.S., Buttner, K., Gluckmann, M., *et al.* (2006) Gel-free and gel-based proteomics in Bacillus subtilis: a comparative study. *Mol Cell Proteomics* **5**, 1183–1192.

Yamane, K., Bunai, K. and Kakeshita, H. (2004) Protein traffic for secretion and related machinery of Bacillus subtilis. *Biosci Biotechnol Biochem* **68**, 2007–2023.

Ye, R.W., Tao, W., Bedzyk, L., Young, T., Chen, M. and Li, L. (2000) Global gene expression profiles of Bacillus subtilis grown under anaerobic conditions. *J Bacteriol* **182**, 4458–4465.

Yooseph, S., Sutton, G., Rusch, D.B., Halpern, A.L., Williamson, S.J., Remington, K., *et al.* (2007) The sorcerer II global ocean sampling expedition: expanding the universe of protein families. *PLoS Biol* **5**, e16.

Yoshida, K., Kobayashi, K., Miwa, Y., Kang, C.M., Matsunaga, M., Yamaguchi, H., *et al.* (2001) Combined transcriptome and proteome analysis as a powerful approach to study genes under glucose repression in Bacillus subtilis. *Nucleic Acids Res* **29**, 683–692.

Yang, J., Wang, H., Wang, W., Yu, P. (2003) Enhanced biclustering on expression data. In *Proceedings of the 3rd IEEE Conference on BioInformatics and BioEngineering*, March 10–12, pp. 321–327.

Yang, J., Wang, W., Wang, H. and Yu, P. (2002) Delta-cluster: capturing subspace correlation in a large data set. In *Proceedings of the 18th IEEE International Conference on Data Engineering (ICDE)*, pp. 517–528..

Yuh, C.H., Bolouri, H. and Davidson, E.H. (1998) Genomic cis-regulatory logic: experimental and computational analysis of a sea urchin gene. *Science* **279**, 1896–1902.

Zhang, G., Spellman, D.S., Skolnik, E.Y. and Neubert, T.A. (2006) Quantitative phosphotyrosine proteomics of EphB2 signaling by stable isotope labeling with amino acids in cell culture (SILAC). *J Proteome Res* **5**, 581–588.

Zheng, L.B. and Losick, R. (1990) Cascade regulation of spore coat gene expression in Bacillus subtilis. *J Mol Biol* **212**, 645–660.

Zheng, M., Wang, X., Templeton, L.J., Smulski, D.R., LaRossa, R.A. and Storz, G. (2001) DNA microarray-mediated transcriptional profiling of the Escherichia coli response to hydrogen peroxide. *J Bacteriol* **183**, 4562–4570.

Annual Plant Reviews (2009) **35**, 137–166
doi: 10.1111/b.9781405175326.2009.00004.x

www.interscience.wiley.com

Chapter 4
ANIMAL SYSTEMS BIOLOGY: TOWARDS A SYSTEMS VIEW OF DEVELOPMENT IN C. *ELEGANS*

Anita Fernandez,[1,2] Fabio Piano[1] and Kristin C. Gunsalus[1]

[1] *Department of Biology and Center for Genomics and Systems Biology, New York University, New York, NY, USA*
[2] *Department of Biology, Fairfield University, Fairfield, CT, USA*

'The differences between the cell and even the most intricate artificial machine still remain too vast by far to be bridged by our present knowledge.'

E.B. Wilson

Abstract: The nematode worm *Caenorhabditis elegans* is the pre-eminent model for understanding animal development at a systems level. Embryonic development in particular has been studied intensively in *C. elegans*, and genes essential for early stages of embryogenesis and their specific phenotypes have been catalogued comprehensively. Combining these datasets with genome-scale studies of gene expression and protein–protein interaction leads to modular views of how genes and their products collaborate to control fundamental processes in early development. Studying groups of genes as functional modules allows the higher order relationships between different biological processes to be observed and suggests how different events during development are coordinated. Here, we review the systems-level approaches that have been used to study early development in *C. elegans* and how these are deepening our understanding of the complex molecular programmes underlying development.

Keywords: *C. elegans*; systems biology; networks; embryogenesis; phenome; development

4.1 Why *C. elegans* as a model for developmental systems biology?

The free-living roundworm *Caenorhabitis elegans* was selected by Sydney Brenner and colleagues in the 1960s as a reference animal species to study the mechanisms underlying what he considered the two most challenging uncharted frontiers in biology: development and behaviour (Brenner, 1974). The motivation was to identify an ideal experimental system in which the genetic and molecular strategies that had been so successful in simpler systems like phage and *Escherichia coli* could be applied to tackle these problems. The ideal species should be as simple as possible while maintaining the complexity of developmental programmes seen across multicellular organisms.

It is interesting to note that a systems approach towards understanding *C. elegans* biology was adopted from the outset. An early goal was to map the complete cell lineage of all adult somatic cells with an accompanying anatomical description reconstructed from serial sections imaged by electron microscopy (Sulston and Horvitz, 1977; Sulston *et al.*, 1983), which formed the basis for the first comprehensive description of the ontogeny and cellular architecture of any animal. This achievement laid the foundation for many subsequent seminal discoveries and led to the 2002 Nobel Prize in Physiology or Medicine for Brenner, John Sulston and Bob Horvitz for their work concerning 'genetic regulation of organ development and programmed cell death' (Check, 2002). Over the last 40 years, the vigorous and growing *C. elegans* community has worked at the forefront of molecular genetics to define the molecular basis for many developmental programmes in *C. elegans* and has made great strides in these efforts (Kemphues *et al.*, 1988; Kimble *et al.*, 1984; Avery, 1993; Seydoux *et al.*, 1993; Sundaram and Greenwald, 1993; Gönczy *et al.*, 1999; Hubbard and Greenstein, 2000). Many of the underlying molecular processes, such as signalling pathways, that were first characterized in *C. elegans* are now recognized to be broadly conserved across eukaryotes and have paved the way for progress in understanding related processes in other organisms.

Basic considerations qualifying *C. elegans* as a good reference species in the 1960s are still relevant today: for example a rapid generation time (<3 days), amenability to laboratory culture (either in liquid or on agar plates) and experimental manipulation and the potential for long-term storage as frozen stocks. Moreover, *C. elegans* is optically clear, facilitating the visualization and identification of most cells. The sexually mature adult, about 1 mm in length, is most commonly hermaphroditic, and can either self-fertilize or be crossed to rare males arising in the population. Like most nematodes, it is eutelic (with a constant number of adult cells), and – as was learned from the lineage studies – has a constant pattern of somatic cell divisions, allowing developmental events to be mapped with high precision. These features have made *C. elegans* an especially powerful reference species to develop and apply genetic, and now genomic, tools to the study of developmental mechanisms.

Today, *C. elegans* has become a pre-eminent reference species for systems-level approaches to development. What properties define a good reference species for developmental systems biology? A necessary but not sufficient condition is a fully sequenced and annotated genome from which a 'parts' list of components can be assembled. In 1998, *C. elegans* became the first metazoan to have its genome completely sequenced. The sequence is now known down to the last base pair, and to this day remains the only genome sequence from a multicellular species that contains no gaps. The genome annotation is of high quality, and rich community resources for genomic and functional data, including reannotation of the original anatomical studies, are available as online repositories that are updated regularly with up-to-date information (WormBase.org, WormGenes.org, WormAtlas.org, WormBook.org). Genomic sequence data for several related nematode species are now becoming available, providing comparative data for gene models, regulatory regions and evolutionary histories.

The availability of the *C. elegans* genome sequence has fostered the development of a rich toolbox of experimental and bioinformatic resources and techniques enabling comprehensive systems-level analyses. Well-established forward and reverse genetic methods enable bottom-up, top-down and hybrid strategies to address questions of interest. Collections of insertion and deletion mutants currently cover ~19% of the genes in the *C. elegans* genome (WormBase 180) (Kemphues, 2005), and clone libraries are available that enable depletion of ~98% of *C. elegans* gene products using RNA interference (RNAi) (Kamath *et al.*, 2003; Rual *et al.*, 2004). Large-scale 'omic' clone libraries are also being constructed for individual promoter regions (the Promoterome; Dupuy *et al.*, 2004), open reading frames (the ORFeome; Reboul *et al.*, 2003) and 3'UTRs (the UTRome; Mangone *et al.*, 2007). These collections, many of which have adopted the Gateway system (Walhout *et al.*, 2000), allow the rapid generation of recombinant constructs for a wide variety of downstream applications, including cell biology, biochemistry and molecular interaction studies.

Ultimately, the goal of developmental systems biology is to move beyond a parts list to obtain a holistic view of the complexity and dynamics of biological processes and their 'emergent properties'. Every event in development represents the collaboration of multiple processes that can incorporate many distinct complexes of proteins, signalling pathways and regulatory cascades. Thus, the vastly complex problem of understanding even just one cell and how it divides cannot be fully appreciated by studying single genes in isolation of all the others. A 'bottom-up' approach that synthesizes data from single-gene studies has been successfully applied to derive systems-level properties of gene regulatory cascades during sea urchin early development (Davidson *et al.*, 2002). This chapter instead focuses on a different strategy: the development of integrative approaches to analyze and model biological processes using genome-scale data (e.g. Gunsalus *et al.*, 2005).

In summary, the experimental tractability of *C. elegans*, combined with powerful molecular tools and bioinformatics resources, present a unique opportunity to determine, at a systems level, the nature and coordination of the many processes that underlie development from a single fertilized egg to an adult. The anatomy and lineage of C. elegans are well described, its genome is well annotated and many of its genetic parts have been conveniently sub-cloned and are available for use in genome-wide experimental analyses.

By highlighting examples of recent studies that have harnessed these tools to tackle global questions, the following discussion will illustrate how integrative systems approaches are working towards obtaining holistic views of the molecular mechanisms underlying development in this organism. To date, the largest body of systematic large-scale analysis has focused on developmental mechanisms in embryogenesis, and in particular the first few cell cycles that take place in the zygote after fertilization. This chapter will therefore concentrate mainly on embryogenesis and use it as a vehicle to explore larger issues regarding molecular networks.

4.2 Defining in vivo functions during development: towards a phenome map of C. elegans embryogenesis

In order to arrive at a comprehensive description of biological processes, it is important to define which genes are essential, what their contributions are to specific events and how they function together to coordinate cellular behaviour. In the context of development, this requires an understanding of the roles of genes within a living organism. This can be approached by analyzing organismal responses to perturbations of gene function in vivo: in other words, defining the phenotypic landscape, or 'phenome', of the organism during development.

In this section, we begin with an overview of C. elegans development, describing in some detail the earliest events during embryogenesis that will serve as a point of reference for much of the content to follow. We then discuss how global genetic requirements for these events have been uncovered through genome-scale in vivo analyses of gene function. Finally, we describe how systematic phenotypic analyses are leading us towards a view of how individual molecules are organized into functional modules within the molecular networks underlying developmental processes.

4.2.1 Overview of development in C. elegans and cellular phenomena during early embryogenesis

Development begins with fertilization of an oocyte by a single sperm to produce a zygote with maternal and paternal pronuclei, initiating embryogenesis (Fig. 4.1). Fertilization triggers completion of female meiosis and re-entry into the mitotic cell cycle, resulting in a series of very rapid cell divisions (~20 min per cell cycle), the establishment of blastomeres with distinct developmental potential, and key transitions that lead to the formation of a fully specified larval body plan: gastrulation and morphogenesis (Chisholm and Hardin, 2005; Nance et al., 2005). At the 26-cell stage, within 2 h of fertilization, gastrulation begins when the two endodermal precursors ingress from the exterior of the embryo to the interior blastocoel (reviewed in Nance *et al.*, 2005). Gastrulation organizes the major tissue types of the embryo and results in the internalization of endodermal, mesodermal and germline progenitors. By mid-gastrulation, at about 4 h post-fertilization, 85% of cells will have been specified as precursors that contribute to a single fate (Sulston *et al.*, 1983). At about this stage, the mother lays the fertilized egg. After the completion of gastrulation, morphogenesis ensues, in which cells intercalate and dramatically change shape, and the embryo begins to adopt a tubular shape. Soon after, the embryo starts to twitch and eventually move in a coordinated manner within the eggshell, as myoblasts and neuroblasts give rise to muscle and nerve cells that form functioning junctions. At 25°C, ~15 h after fertilization, the egg hatches and a larva emerges. The larva will molt four times, progressing through four larval stages and eventually (at ~3 days post-fertilization) becoming a sexually mature adult hermaphrodite or, rarely, a male (XO males arise at a rate of about 1 in 500 from self-fertilizing XX hermaphrodites due to occasional meiotic non-disjunction (Hodgkin, 1988)). Males and hermaphrodites can be distinguished by size and morphology based on differences in reproductive organs such as the gonads, hermaphrodite vulva and the fan-like structure of the male tail.

The major events of oogenesis and early embryogenesis are readily visualized at a subcellular level using a light microscope fitted with Nomarski optics, and recorded using time-lapse microscopy. Developing oocytes arrest at the pachytene stage and are aligned single file at the proximal end of the gonad. These are cuboidal in shape and contain a single 4 N pronucleus (or germinal vesicle) in the centre. Before ovulation, oocytes closest to the spermatheca, the sperm storage organ, undergo meiotic maturation. During this process, the nucleus moves towards the end of the oocyte furthest from the spermatheca, the nuclear envelope begins to disappear, and the shape of the cell changes from cuboid to ovoid (for a review, see Schneider and Bowerman, 2003). Upon ovulation the oocyte enters the spermatheca, where it is fertilized. The sperm delivers both paternal DNA and a pair of centrioles into the oocyte. As the zygote is extruded from the spermatheca into the uterus, dramatic cytoplasmic and cortical activities result in the movement of cytoplasmic granules towards the future posterior and of cortical granules towards the anterior (Hird and White, 1993). Actomyosin forces drive these cytoplasmic and cortical activities and are required for setting up the embryonic anterior–posterior axis (Strome and Wood, 1983; Munro *et al.*, 2004).

Further rapid changes take place in the zygote (Fig. 4.1). The paternal chromatin decondenses and forms a haploid pronucleus. This pronucleus initially remains confined to the posterior of the embryo. Meanwhile, at the future anterior end, the maternal DNA rapidly completes meiosis. Upon the completion of meiosis I, a 2 N polar body is extruded from the anterior and is anchored to the concomitantly forming eggshell. Meiosis II ensues and the second (1 N) polar body is extruded but remains associated with the cell membrane and is eventually reabsorbed, usually at the four-cell stage.

The process that leads to the maternal and paternal DNA to combine is rather elegant and complex. The mature maternal pronucleus travels to the

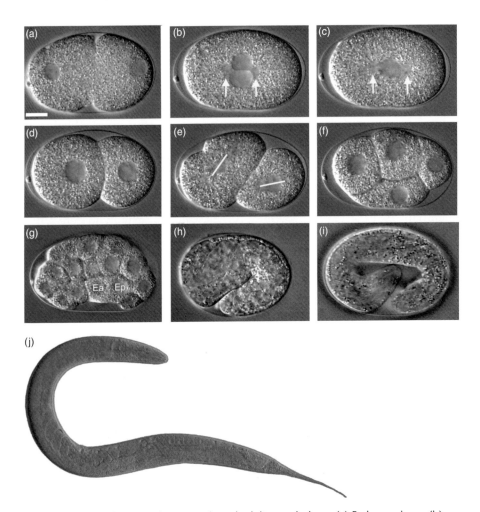

Figure 4.1 *C. elegans* embryogenesis and adult morphology. (a) Early prophase. (b) Late prophase. Arrows indicate centrosomes. (c) Metaphase. Arrows indicate centrosomes. (d) Two-cell stage. Anterior cell (AB) is larger than P1. (e) AB cell and P1 cells in preparation for next division. Spindle axes for each cell are delineated. (f) Four-cell stage. (g) Twenty-six-cell stage. The embryo begins gastrulation when E cells are pushed inward. (h) Comma stage. (i) Larva almost ready to hatch. (j) Adult hermaphrodite. Anterior is towards the left in all panels. Scale bars: 10 μm (a–i), 50 μm (j).

posterior to meet the paternal pronucleus, but fusion does not occur immediately. During this time the first mitotic cell cycle initiates, and chromosomes enter prophase and begin condensing before the maternal and paternal chromatin meet. After meeting in the posterior half, the pronuclei migrate towards the centre of the embryo without fusing. Once near the centre, the cell cycle has progressed to the point that the nuclear envelopes of both pronuclei break down and the chromosomes from the two parents align into a metaphase plate, which is visible as a clearing in the cytoplasm. During the metaphase to anaphase transition, sister chromatids are pulled apart by the mitotic spindle, and nuclear envelopes rapidly re-form around the decondensing DNA. A posteriorly displaced cytokinesis ensues, leading to a larger anterior cell (AB) and a smaller posterior cell (P1).

In addition to their difference in size, AB and P1 behave differently. Almost immediately upon reformation of the nuclear envelope in the AB cell, the DNA visibly condenses in preparation for the next cell division, and the centrosome appears round or circular. The P1 cell, which displays a disc-shaped centrosome, also enters prophase relatively quickly, but lags several minutes behind the AB cell while its spindle rotates about 90 degrees. As a result the spindles align orthogonally and AB and P1 divide perpendicularly to each other. Up until the four-cell stage of embryogenesis, there is no detectable zygotic transcription and all mRNAs are provided maternally (Hope, 1991; Edgar *et al.*, 1994; ·Seydoux and Fire, 1994). As discussed further below, the fact that subcellular events are easily observable and occur canonically in every individual in this species makes the early embryo an attractive subject for genome-wide investigation of the genetic mechanisms that orchestrate these phenomena.

4.2.2 What genes are required for embryogenesis in C. *elegans*?

In order to define genetic requirements for embryonic development and early embryogenesis in particular, multiple forward and reverse genetic screens have been performed to identify essential genes in C. *elegans* (for review see Jorgensen and Mango, 2002; Gunsalus and Piano, 2005). Genes whose mutant phenotypes affect embryonic development have usually been called *emb* or *zyg* for 'abnormal *emb*ryogenesis' or '*zyg*ote defective', respectively. These are also classified as *let* (*let*hal) or *mel* (*m*aternal *e*ffect *l*ethal), with subclasses defective in *fer*tilization (*fer*) or *ooc*yte development (*ooc*). Thus, any of these gene designations, as well as more specific ones such as *mom* (*mo*re of *M*S (muscle/skin lineage)) or *par* (abnormal embryonic *par*titioning of cytoplasm), have been used to refer to genes whose mutant phenotypes reflect their essential role in some aspect of embryonic development. Genes that are required in the embryo and have additional roles in oogenesis can also result in phenotypes such as '*ste*rile' (ste) or '*st*erile *p*rogeny' (stp).

4.2.2.1 Forward genetic approaches to embryogenesis

A recent review of essential genes (Kemphues, 2005) reports that genetic mutations in over 860 essential loci have been identified by forward genetic approaches by the C. *elegans* research community (including embryonic and post-embryonic lethals). This is a fraction of projected totals: previous estimates place the number of essential genes in the range of ~3000 (Johnsen and Baillie, 1991) to ~5000 (Hodgkin, 2001), and genes with potential to give rise to any visible phenotype at ~6000 (Jorgensen and Mango, 2002). Thus, forward genetic approaches have so far identified less than a third of genes estimated to be essential at any stage of development. Among genes with currently known roles in early development, less than 20% have been uncovered through forward genetics (see below).

Challenges in obtaining a global view of the phenome using classical forward genetics include issues of pleiotropy, partial penetrance and genetic redundancy or compensatory mechanisms. As discussed below, many genes required for early embryonic development are involved in basic cell biological functions, and thus are likely to affect multiple developmental processes and tissues, complicating genetic analysis. Coupled with the fact that early embryonic development is controlled by the maternal and not the zygotic genome, identifying genes with roles in early embryogenesis using classic forward genetic approaches necessitates screening for maternal effect (e.g. Kemphues *et al.*, 1988) or conditional alleles (e.g. temperature sensitive screens as in O'Connell *et al.*, 1998). Largely for practical reasons, such screens are also biased towards genes with at least one highly penetrant phenotype. All of these factors, along with the heterogeneity in approaches among genetic screens and a lack of consistency in reporting of results, make the task of combining all available forward genetic data together into a comprehensive view of the phenome practically impossible.

Moreover, identifying functions for those genes whose loss of function yields no visible phenotype on their own requires the co-occurrence of mutations in multiple genes in the same genome that uncover a new phenotypic effect. With the exception of yeast, where a full genome deletion set is available (Giaever *et al.*, 2002), systematic generation and analysis of double mutant combinations by forward genetics would be prohibitively time consuming, and thus comparatively little is currently known about global properties of genetic interactions (we return to this issue later in discussing how functional modules are coordinated within molecular networks). Since the vast majority of protein-coding genes currently have no known phenotype and the total number of these 'genetically obscure' genes is estimated at ~14 000, or almost 70% (Jorgensen and Mango, 2002), other methods to better address their function are needed.

4.2.2.2 Reverse genetic approaches

In species whose genome has been sequenced, reverse genetic approaches have allowed phenotypic studies to proceed in a systematic fashion, meaning that the effects of disrupting the function of each gene in the genome can be studied individually. In yeast and mouse, mitotic homologous recombination strategies have been successful in generating deletion alleles for many genes of interest. In yeast, this approach has been used to replace every open reading frame in the genome with a unique bar-coded selectable marker, generating a knockout for every gene (Giaever *et al.*, 2002). In *C. elegans*, although targeted homologous recombination has been reported (van Luenen and Plasterk, 1994), it has not yet become feasible to generate a genome-wide set of targeted deletions. However, two groups – the *C. elegans* Gene Knockout Consortium (celeganskoconsortium.omrf.org) and the National BioResource Project of Japan (www.shigen.nig.ac.jp/c.elegans) – are generating deletion libraries and characterizing specific deletion alleles using polymerase chain reaction-based assays, with the goal of generating a deletion allele for every gene in the genome.

By far the most advanced reverse genetic approach for genome-scale analysis is RNAi, a technique that depletes endogenous mRNA via the introduction of double-stranded RNA (dsRNA) corresponding to the targeted gene (Fire *et al.*, 1998). In *C. elegans*, RNAi can be applied to adult hermaphrodites and the progeny scored for defects, or young larvae can be treated to reveal post-embryonic phenotypes in larval or adult stages. RNAi may be performed by injecting animals with dsRNA (Fire *et al.*, 1998), feeding animals bacteria engineered to express dsRNA (Timmons *et al.*, 2001) or soaking animals in a solution containing dsRNA (Tabara *et al.*, 1998). RNAi is remarkably efficient in the *C. elegans* germline, and is thus a useful technique to study maternally controlled processes in early development. Because RNAi in *C. elegans* is systemic (Timmons *et al.*, 2003; Hunter *et al.*, 2006), dsRNA does not have to be delivered directly to germline to elicit effects in oocytes and embryos; injection into the gut (or elsewhere) is also effective, as are soaking and feeding (though feeding tends to produce less severe phenotypes in comparison with injection or soaking (Tabara *et al.*, 1998; Fernandez *et al.*, 2005)).

How reliable is RNAi as a tool to study gene function in *C. elegans*? RNAi is quite effective at phenocopying genetic null and hypomorphic mutant phenotypes. Comparing mutant and RNAi phenotypes from different large-scale studies shows that 'hit rates' (% of RNAi experiments that recapitulate the mutant phenotype) are estimated at between 60% and 80%, depending on the study (Kamath *et al.*, 2003; Fernandez *et al.*, 2005). When specific cases were quantitatively tested, between 80% and over 90% of the target protein was depleted after RNAi (e.g. Motegi and Sugimoto, 2006). Using a similar criterion, 'false positives', where RNAi gives rise to a phenotype not observed in a null mutant, are rare (no more than ~5% and probably lower) (Piano *et al.*, 2002; Fernandez *et al.*, 2005). Pairwise comparisons among RNAi studies reveal largely similar results for gene sets included in both studies, although typically some RNAi phenotypes are detected in one study and missed by another (Piano *et al.*, 2002; Fernandez *et al.*, 2005). As a result, RNAi phenotypes can reliably be considered to reflect the wild-type function of the gene, particularly when multiple studies give rise to a similar range of phenotypes. On the other hand, RNAi experiments that yield no detectable phenotypes are more difficult to interpret, since this could represent a technical failure, or perdurance of endogenous gene products may mask depletion of the targeted mRNA. Due to the ease of scaling up various RNAi protocols for high-throughput analyses, the vast majority of protein-coding genes in *C. elegans* have now been depleted by RNAi by multiple independent groups using different methodologies. All of these results are available in

WormBase (Rogers *et al.*, 2007) and RNAiDB (RNAi.org; Gunsalus *et al.*, 2004), such that all RNAi experiments and resulting phenotypes for a given gene can be viewed at once.

4.2.2.3 A bird's eye view of the embryonic phenome: global trends

What have we learned about the embryo from genome-wide RNAi studies? First, ~2500 genes, or ~13% of the genome, show a maternal sterile or embryonic lethal defect by RNAi (Fernandez *et al.*, 2005; RNAiDB v5.0). Given RNAi's false negative rates, this is likely an underestimate of the total number of genes required in these stages, but intriguing global trends have come out of initial genome-wide analyses (see Gönczy *et al.*, 1999; Fraser *et al.*, 2000; Piano *et al.*, 2000; Kamath *et al.*, 2003; Fernandez *et al.*, 2005). Genes required for basic cellular machinery, such as RNA or protein metabolism, are enriched among the set required for embryogenesis (Gönczy *et al.*, 1999; Fraser *et al.*, 2000; Piano *et al.*, 2000; Kamath *et al.*, 2003; Fernandez *et al.*, 2005). In addition, genes that are highly conserved across phyla are more likely to have lethal phenotypes in *C. elegans*, and the penetrance of lethality tends to parallel the degree of conservation (Piano *et al.*, 2002; Fernandez and Piano, 2006). In addition, the degree to which transcripts are enriched in the ovary is a great predictor of requirement in the embryo: genes with highly ovary-enriched transcripts are more likely to show sterile or embryonic lethal phenotypes by RNAi (Piano *et al.*, 2002; Fernandez and Piano, 2006). Another interesting trend observed was the dramatic paucity of genes required for embryonic viability on the X chromosome, which is likely related to transcriptional silencing of the X chromosome in the germline (Piano *et al.*, 2000, 2002; Kelly *et al.*, 2002; Kamath *et al.*, 2003; Fernandez and Piano, 2006).

Reverse genetics affords us the opportunity to study incompletely penetrant phenotypes in a way that would be impractical using forward genetic screens. Several genome-scale RNAi studies have scored and reported penetrance values for embryonic lethality as the percentage of progeny that fail to hatch into L1 larvae (Maeda *et al.*, 2001; Piano *et al.*, 2002; Kamath *et al.*, 2003; Fernandez *et al.*, 2005; Sönnichsen *et al.*, 2005). In one study, of 956 germline-enriched genes that gave rise to embryonic lethality by RNAi, 366 showed partially penetrant (5–79%) embryonic lethality (Fernandez *et al.*, 2005). It is striking that, as a set, those genes whose strongest reported lethality is partially penetrant exhibit different trends from genes that display either fully penetrant lethality or no RNAi phenotypes. In contrast to the high penetrance set, genes in the partial penetrance set are neither significantly under-represented on the X chromosome nor enriched for basic cellular functions, and their transcripts do not show strong germline enrichment (Fernandez *et al.*, 2005). In addition, a very interesting trend emerges regarding penetrance of embryonic lethality and sequence conservation across different phyla (Fernandez *et al.*, 2005): a high proportion (53%) of genes with the strongest phenotypes have identifiable homologs in both multicellular and single-celled eukaryotes (fly, human and yeast), and this proportion decreases steadily among genes with decreasing penetrance, down to 13% for genes with no detectable RNAi phenotypes. In contrast, the novel genes (genes specific to worms with no obvious homologs in other animals) show the opposite trend: the gene set with the strongest phenotypes contains the lowest proportion of novel genes (14%), and this proportion rises as penetrance decreases, such that 45% of genes with no detectable RNAi phenotype are worm-specific. Genes with homologs in fly and human but not in yeast (i.e. multicellular vs. unicellular organisms) show intermediate properties. One interpretation of this trend is that partially penetrant genes tend to be 'in transition', or en route to assuming increasingly essential roles over evolutionary time. Further study, including confirmation of partially penetrant RNAi phenotypes using the growing body of available genetic null alleles, will be needed to fully understand the roles of these interesting genes and their place in evolutionary history.

4.2.3 Digital phenotyping: the beginning of a modular view

Through large-scale RNAi analyses, ~2500 genes have been found to affect embryonic viability in some way when their function is depleted (RNAiDB 5.0). How can a systems approach be applied towards understanding how these genes specifically contribute to embryonic development? How can we better understand which groups of genes work together to affect similar processes?

The cornerstone of genetic logic is deduction of gene function via detailed study of phenotypes caused by gene perturbations. A corollary is that genes that cause similar phenotypes might be expected to function together in the same process. Phenotypic data are enormously informative about gene function; as greater phenotypic detail accumulates for a given gene, more specific inferences can be made about its wild-type function. Similarly, by analyzing and comparing phenotypes for different genes that impinge on the same process, it becomes increasingly possible to tease out specific roles for each gene and to understand their relationships with each other. Molecular geneticists have successfully used this powerful approach over the last several decades on a per-gene basis to map genetic requirements for numerous signalling pathways and cell biological events in several major model organisms.

The C. elegans early embryo is particularly amenable to high-content phenotypic analysis. Gravid hermaphrodites can be dissected and their embryos removed at an early stage for observation. The eggshell is optically clear, and the fertilized egg is a very large cell, about 60 μM long by 30 μM wide, allowing detailed visual inspection. Thus, many essential processes such as chromosome segregation, nuclear envelope breakdown and re-formation, cytokinesis and polarity establishment can be easily observed in real time using simple differential interference contrast (DIC) microscopy (Fig. 4.1). Remarkably, the cell divisions and subcellular activities of the early embryo proceed in a stereotypic and invariant manner, so deviations from canonical events in wild-type embryos caused by disrupting gene function are very easy to visualize and score. As a result, the *C. elegans* early embryo has been studied and described in great detail using both forward and reverse genetic approaches, and thus represents the best stage of development for which we can begin to achieve a systems-level understanding.

4.2.3.1 Digitizing phenotypes: generating comprehensive phenotypic signatures

How can detailed, comprehensive phenotypic data be accumulated and analyzed in large scale to obtain a global view of genetic requirements for early embryonic processes? The key to tackle this challenge has been to develop high-throughput RNAi methods coupled with time-lapse DIC recordings of early embryogenesis, captured every few seconds through the first 2–3 cell cycles (spanning ~30–50 min of development). For practical reasons, the first applications of this approach selectively examined genes with a high probability of being essential in the early stages, as identified from either an ovary cDNA library (Piano *et al.*, 2000) or germline-enriched transcripts (Piano *et al.*, 2002). This approach was scaled up to encompass systematic analysis of an entire chromosome (Gönczy *et al.*, 1999; Zipperlen *et al.*, 2001), and eventually to 98% of protein-coding genes in the *C. elegans* genome (Sönnichsen *et al.*, 2005).

To define specific roles for each gene required in the early embryo, a strategy was developed to systematically analyze DIC recordings for all detectable defects that could reproducibly be observed upon RNAi (Piano *et al.*, 2002). After compiling a comprehensive list of observable phenotypes, a set of characters was devised that explicitly describes discrete cell biological aberrations that can be detected in the early embryo as a result of depletion by RNAi (Piano *et al.*, 2002; Sönnichsen *et al.*, 2005). For each character, for example 'P0 cytokinesis defect', an RNAi-treated embryo is scored for the presence or absence of that phenotype, producing a vector of values for each RNAi experiment that includes a phenotypic score for every character. Data from multiple independent RNAi experiments for the same gene can then be combined to produce a consensus vector for each gene. As a result, complex temporal and spatial data are distilled into a digital 'phenotypic signature' that captures the complete phenotypic syndrome for each gene (Fig. 4.2).

4.2.3.2 Clustering phenotypes: identifying functional relationships

By converting analysis of a time-lapse recording into a phenotypic signature, therefore, complex phenotypic data are rendered amenable to computational analysis. The power of phenotypic signatures is that they provide a comprehensive systematic description that includes information about every observable trait that can be affected by RNAi and, importantly, an explicit record of whether each phenotype examined was distinguishable from wild type or not. This allows phenotypes to be directly and quantitatively compared using

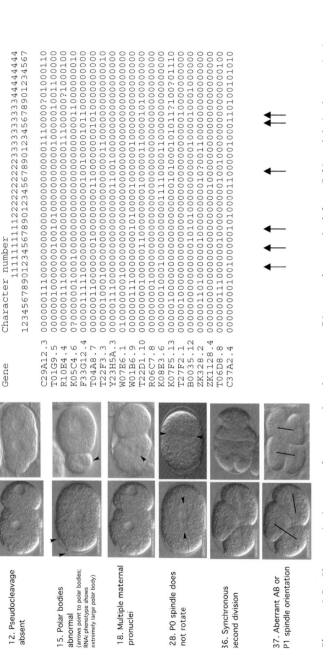

Figure 4.2 Phenotypic signatures from complex phenotypes. Discrete phenotypic defects (left) and their phenotypic signature representation (right), using early embryonic development in C. elegans as examples. Each of the phenotypic defects resulting from a genetic perturbation is systematically recorded into a digital representation. Arrows point to the characters shown on the left.

a mathematically defined measure of phenotypic similarity (or distance), for example a Pearson correlation coefficient, that reports on the degree to which two signatures are alike. This also allows genes to be grouped based on patterns of shared traits within their phenotypic signatures, either using hierarchical or other distance-based clustering methods, or using character-based phylogenetic methods (Piano *et al.*, 2002; Gunsalus *et al.*, 2005; Sönnichsen *et al.*, 2005). The resulting 'phenoclusters' are very often enriched for specific cellular functions, and can therefore be used to generate hypotheses about novel, uncharacterized genes that fall into a cluster containing genes of established function. Such groupings help define sets of genes, or modules, that are likely to function together in the same process.

What have we learned about *C. elegans* early embryogenesis from phenoclustering? First, this technique has pointed to putative roles for numerous genes of previously unknown function. For example the novel genes F35B12.5 and Y45F10D.9 were found in the same phenocluster as *zyg-1* and *sas-4*, two genes known to be required for proper duplication of the centrosome (the microtubule organizing centre) during mitosis (O'Connell *et al.*, 2001; Leidel and Gonczy, 2003; Gunsalus *et al.*, 2005). Significantly, detailed characterization of these novel genes (now called *sas-5* and *sas-6*) revealed that both are indeed components of the centrosome that are also required for centrosome duplication (Delattre *et al.*, 2004; Leidel *et al.*, 2005).

Second, phenoclustering has provided the first global view of functional modules, or groups of genes that work together to carry out specific cellular processes, that are required for early embryonic development. Using all 661 genes identified as having essential roles in the early embryo in a single study covering 98% of genes in the genome, around 20 clusters have been identified that are characterized by hallmark phenotypes spanning the range of essential cellular processes during this time (Gunsalus *et al.*, 2005; Sönnichsen *et al.*, 2005). These include such functions as nuclear membrane integrity, polarity and chromatin regulation. Many of these clusters are significantly enriched for genes encoding specific molecular functions, such as the ribosome, proteasome, nuclear pore complex, histones, etc. (Gunsalus *et al.*, 2005).

Third, phenoclusters can suggest relationships between different functional modules. For instance, two clusters diverging from the same branch in the hierarchical cluster dendrogram (Gunsalus *et al.*, 2005) are enriched for nucleoporins (components of the nuclear pore complex) or for genes regulating chromatin function (histones and factors involved in DNA replication licensing and chromosome segregation), suggesting that nuclear envelope function may be linked with DNA replication and chromosome segregation. The observations that nuclear envelope permeabilization states determine whether DNA replication can proceed in *Xenopus* egg extracts (Leno and Munshi, 1994; Lu *et al.*, 1999), and the identification of molecular links between nuclear membrane integrity and kinetochore function in *C. elegans* and vertebrates (Fernandez and Piano, 2006; Galy *et al.*, 2006; Rasala *et al.*, 2006; Franz *et al.*, 2007'), support the idea that these activities are coordinated in the cell. We return to this interesting topic later in discussing coordination between molecular modules.

A lesson distilled from these analyses is that digitally encoding phenotypic signatures followed by simple clustering analysis lead to biologically relevant predictions of functional gene groupings. However, representing gene relationships in the form of a dendrogram provides a limited and static view that may mask significant relationships between many genes. This is because each gene may occupy only one position in the tree, and will tend to cluster nearby genes with the most distinctive pattern of shared phenotypes. One way to circumvent this limitation is to use other tools such as PhenoBlast (Gunsalus *et al.*, 2004) that identify genes with similar signatures independent of the specific pattern of shared phenotypes. PhenoBlast is a data mining tool that 'blasts' a query gene or phenotypic signature against all other signatures in a database and returns a list of genes ranked by the number of shared phenotypes or the probability of finding the observed combination of phenotypes, based on the frequency of occurrence of different phenotypes in the population (PhenoBlast can also be adapted to use any similarity measure of interest). In this way, specific associations not revealed by a simple clustering approach can be identified. Additional relationships may be revealed using nested-effects models, which order genes based on subset relationships in shared phenotypes to identify potential genetic hierarchies based on patterns of perturbation effects (e.g. gene A may be inferred to act downstream of gene B if RNAi of gene A shows only a subset of phenotypes elicited by RNAi of gene B) (Markowetz *et al.*, 2007).

4.3 Data integration: towards a systems view of early embryogenesis

4.3.1 Colour EST E pluribus unus: putting it all together

Phenotypic analysis provides one important layer of information on gene function, but understanding how gene networks coordinate biological processes requires additional pieces of the puzzle, such as data on patterns of gene expression during development and physical interactions that mediate communication between cellular components. These layers of information are also being pursued using functional genomic approaches in *C. elegans*. For example expression levels of most of the genes in the genome have been measured across developmental time points and in many different conditions; many of these data have been combined in one compendium analysis representing the first group of ~500 microarray experiments performed in this organism (Kim *et al.*, 2001). Large-scale protein–protein interaction studies have also been performed, mainly using the yeast two hybrid method (Li *et al.*, 2004).

To obtain a systems view of early development, these different layers must be integrated into a unified framework for representing, conceptualizing, mining and eventually dynamic modelling of how developmental programmes operate (Fig. 4.3; reviewed in Piano *et al.*, 2006). A network paradigm is an intuitive and versatile way to represent relationships between genes and/or proteins that accommodate heterogeneous functional linkages, and effectively illustrates myriad connections between individual components and regional neighbourhoods. In a network graph, each component (e.g. gene, protein or metabolite) is represented as a node or vertex, and functional connections between pairs of components are represented as edges or links. For visualization, groups of functionally related nodes can be also collapsed into metanodes to facilitate interpretation of relationships between functional modules.

Many different types of functional relationships can be adapted to a network paradigm. Protein–protein interaction data easily lend themselves to this representation, particularly binary data such as generated from yeast two hybrid experiments. Interactions from co-immunoprecipitation (coIP) data are more challenging to represent, as not all associations in a molecular complex are direct; however these associations can be accommodated using either a 'spoke' model (containing only links between the assayed protein and each of the coIP partners in the complex) or a 'matrix' model (in which a link is included between every member of the complex). Other data reporting on single gene or protein attributes, such as phenotypic and expression data, must be transformed into an association matrix that captures shared similarities between genes. A simple way to do this is to generate a similarity score between two genes, using for example a correlation coefficient or a simple count of shared attributes. Associative data with quantitative values, such as confidence or similarity scores, can be incorporated either by creating weighted edges or by binarizing the data such that only edges meeting a selected threshold criterion are included in the model. Appropriate thresholds can be determined using various statistical models. For expression profiles in particular, however, global similarity measures may sometimes miss significant associations between genes that exist in only a subset of conditions; these may be recovered using alternative methods such as biclustering (e.g. Reiss *et al.*, 2006).

To integrate these heterogeneous data, a variety of approaches can be taken. The simplest is to overlay different dimensions of data. The resulting graphs can be filtered using a simple voting scheme to retain only edges with multiple types of evidence for functional linkage, for example defined as edges with a yeast two hybrid result *and* a highly correlated RNAi phenotype in the early embryo *as well as* a high expression correlation over the ~500 microarray experiments. Such 'multiply supported' networks are highly filtered, but retain biologically significant linkages that can highlight logical connections between cellular machinery, providing both global insights as well as specific hypotheses. More sophisticated computational approaches based on various machine learning methods can produce a single weighted score for each edge that captures the cumulative contribution of different data types, adjusted for the informative power of each. For example, Bayesian networks, support vector machines, decision trees, and logistic regression have all been applied with some success to the problem of predicting gene functions or interactions based on genome-scale data from different species (Troyanskaya *et al.*, 2003; Lee *et al.*, 2004; Wong *et al.*, 2004; Lewis *et al.*, 2006; Zhong and Sternberg, 2006. One drawback of these methods is that they generally require good positive and negative training datasets to tune performance, which can be a challenge to identify and will heavily influence output.

4.3.1.1 Integrated networks in *C. elegans* early embryogenesis
Incorporating different datasets into a network view using a simple overlay approach has revealed important insights into the organization of the molecular 'system' underlying early development in *C. elegans*. To draft a model of early embryogenesis, 661 genes that showed defects by RNAi within the first cell division in *C. elegans* (Sönnichsen *et al.*, 2005) were used to seed a network (Gunsalus *et al.*, 2005). Edges were generated between these nodes based on similarity of 45-digit phenotypic signature, protein–protein interaction and/or similarity of expression profile. The resulting early embryonic network had 31 173 individual edges. The network was then filtered so that only connections between genes with at least two types of evidence were retained. This created a 'multiple support' network composed of 305 nodes joined by 1036 merged edges representing the most strongly supported functional linkages between genes. In this multiply supported subnetwork, most nodes fell into highly interconnected groups that were enriched for a specific functional annotation, and these 'modules' were quite sparsely connected to each other. This modular network structure suggested that early embryogenesis is dominated by a handful of distinct modules composed of closely collaborating components.

Evidence of multiple connections between nodes can both strengthen the confidence of relationships between different genes and also be instructive as to the nature of the relationship between them. For example, genes whose protein products interact via yeast two hybrid are much more likely to have similar transcription profiles or phenotypes than those for which there is no evidence of protein–protein interaction (Walhout *et al.*, 2002; Li *et al.*, 2004; Gunsalus *et al.*, 2005). Interestingly, based on the dominant combination of evidence types within them, two types of molecular modules with fundamentally different functional characteristics could be distinguished in the integrated early embryonic network (Gunsalus *et al.*, 2005). Modules composed mainly of genes whose protein products interact and that give rise to the same phenotype when RNAi-depleted tended to represent 'molecular machines' whose members are subunits of well-known molecular complexes (such as the ribosome, proteasome, anaphase promoting complex and ATPases). However, modules composed mainly of edges supported by co-expression and phenotypic similarity, but with few protein–protein interaction edges, contained genes with diverse molecular functions that nevertheless affect the same biological processes (such as oocyte integrity or release from meiotic arrest). These can be interpreted instead to represent 'coordinated processes' whose members need to be present in the same time and place but not necessarily to interact physically, suggesting their components may be coordinated by other mechanisms such as signal transduction events.

4.3.1.2 Module biology: combined functional genomic approaches

One of the lessons of systems biology is that incorporating different kinds of data on a large scale can lead to a clearer view of how protein complexes function together as modules. Using the fusion of distinct types of information and a module-centric view, major progress has been made recently in understanding how specific subsystems in the early *C. elegans* embryo work.

One example is how the kinetochore is assembled and how it functions. The kinetochore is a multi-protein structure that binds to centromeric DNA during mitosis and serves as the site of spindle microtubule attachment, enabling chromatids to segregate properly to each daughter cell. Although the *C. elegans* kinetochore is holocentric, and thus dispersed along the entire chromosome rather than restricted to a single point, its molecular composition is similar to that of the other metazoans, including the presence of required components CENP-A (HCP-3) and CENP-C (HCP-4) (Maddox *et al.*, 2004). When *hcp-3* or *hcp-4* are depleted by RNAi, spindle microtubules fail to capture the chromosomes, leading to a set of specific defects including a failure of chromosome segregation and ectopic DNA remnants near the cytokinesis furrow. Using these 'kinetochore null' defects as a guide for examining RNAi data, two novel genes required for kinetochore function were identified, *knl-1* and *knl-3* (Desai *et al.*, 2003; Cheeseman *et al.*, 2004).

To identify every protein that complexes with proteins encoded by kinetochore null genes, immunoprecipitates of KNL-1 or KNL-3 were analyzed by mass spectrometry (Cheeseman *et al.*, 2004). Each of the new KNL-interacting proteins identified in this way was then tested by RNAi, and chromosome segregation defects were studied using a strain containing a green fluorescent protein (GFP)::histone marker. RNAi of most of the kinetochore complex genes gave rise to embryonic lethality, and chromosome segregation defects were observed (though not complete segregation failure, in contrast to kinetochore null genes). GFP fusions were also generated for proteins in the kinetochore complex, and all localized as expected to centromeric DNA during mitosis. The localization dependence of each kinetochore component on the others was determined in a set of experiments where each component was RNAi-depleted in the background of a GFP fusion strain representing one of the other components. These experiments revealed the assembly pathway for generating a fully functioning kinetochore. Subsequent biochemical studies showed that the kinetochore contains two distinct subcomplexes with microtubule-binding capabilities, and that these subcomplexes bind synergistically to the microtubules (Cheeseman *et al.*, 2006).

This example shows how a combination of phenotypic, protein–protein interaction and subcellular localization studies led to identification of the kinetochore module, and how further characterization of subcomplexes within this module led to an understanding of how the kinetochore binds to microtubules. Significantly, many of the kinetochore components identified in *C. elegans* are conserved in humans, illustrating the translational power of combined functional genomic approaches in a model system.

4.3.2 How are different modules coordinated?

There is higher order structure within biological networks: functions carried out by different individual modules are coordinated by connections between them. The nature of inter-module relationships specifies how different biological processes are orchestrated in space and time, which in turn will affect how robust or resistant the biological system is to perturbations. How do biological systems manage the many tasks that must be completed during development? How is this done reproducibly and with such exactitude that deviations from the goal developmental routine are rare? And how do these systems evolve? We are only beginning to be able to address these questions.

Developmental programmes are largely robust to genetic and environmental perturbation. Only ~13% of the genome gives rise to a detectable RNAi phenotype in the embryo. Are the rest of the genes in the genome dispensable for normal development? Unlikely. One possibility is that some of these are only required under specific environmental conditions not normally encountered in the laboratory. Another often-invoked explanation is that some modules may operate in parallel or redundant pathways, such that if one module is disabled the other module is able to provide a similar function. Perhaps more commonly, compensatory mechanisms in the cell may allow secondary pathways to take over missing functions when the primary mediators are compromised.

Many genetic diseases are now understood to be of complex origin, arising from the combined effects of allelic variations in multiple genes (Badano and Katsanis, 2002). Thus, moving beyond single-gene perturbations to learn how different cellular components synergize to maintain the integrity of cellular functions is an important frontier. This question is being actively pursued on a large scale in yeast, where systematic, high-throughput investigation of double mutant combinations is feasible using libraries of deletion or titratable promoter alleles combined with robotic manipulation and fast binary or quantitative readouts of growth and viability (e.g. Mnaimneh *et al.*, 2004; Tong *et al.*, 2004; ; Davierwala *et al.*, 2005; Schuldiner *et al.*, 2005; Pan *et al.*, 2007; reviewed in Boone *et al.*, 2007 and Komili and Roth, 2007). These studies are amassing a large body of data on genetic interactions within and between different cellular processes and are leading to a new view of how modules are coordinated in a single-celled organism.

4.3.2.1 *C. elegans* development: genetic interactions expose network logic

In *C. elegans*, similar efforts are underway to study the architecture of genetic networks involved in developmental programmes by systematic analysis of genetic interactions. Due to technical limitations, so far these require combining one genetic allele with RNAi of a second gene (since double RNAi in *C. elegans* is not uniformly effective or reproducible), and can only be performed on a limited scale (e.g. a genome-scale RNAi screen using a single query gene, or a few hundred RNAi assays for each of a handful of query genes).

In a recent analysis focusing on developmental signalling pathways, 37 mutant alleles of 31 query genes were tested for genetic interactions with a set of ~1750 genes whose function was depleted by RNAi (Lehner *et al.*, 2006). The query gene set was composed of genes representing conserved signalling pathways, and the target set was selected to include genes predicted to function in signal transduction, regulation of transcription or chromatin remodelling. Two genes were scored as genetic interactors when disrupting them simultaneously showed either a stronger phenotype or higher penetrance of phenotype than disrupting each single gene independently. In total, 349 genetic interactions were detected among 162 genes. One of the major findings from this study was that six genes, which encode known components of chromatin-modifying complexes, give rise to synthetic phenotypes with multiple partners representing several signalling pathways. Rather than dismiss these as 'non-specific' interactors, the authors interpret these as 'genetic interaction hubs' that may play important roles in genetic buffering. The observation that homologs of these genes also interact genetically with multiple signalling pathways in *Drosophila* suggests that their roles as putative genetic interaction hubs, and any buffering mechanism they may represent, are also conserved. This is potentially interesting since buffering mechanisms have been suggested to underlie the ability of life forms to accumulate masked genetic changes that allow the generation of morphological diversity in response to changing selection pressures (Hartman *et al.*, 2001; Queitsch *et al.*, 2002; Siegal and Bergman, 2002; Bergman and Siegal, 2003).

A more recent study probed 11 different query mutants (10 involved in signalling) against 858 target RNAi experiments, and recovered 1246 genetic interactions among 461 genes (Byrne *et al.*, 2007). The final network of genetic interactions, which ranged from weak to very strong, was determined by varying selection criteria to optimize precision and recall values against a 'gold standard' of shared gene ontology functional annotations (Ashburner *et al.*, 2000), based on the assumption that true interactions should exist between genes with similar functions. Gene pairs were scored as genetic interactors if doubly disrupted worms produced smaller broods than controls of mock-treated mutants or RNAi in a wild-type background; all interactions in the final network displayed stronger phenotypes than the expected product of the single disruption assays. Superimposing the genetic interaction network with other functional genomic datasets revealed an interesting trend: genetic 'synthetic interaction' edges tend not to overlap with protein–protein interaction, co-expression or phenotypic similarity edges, but instead appear to bridge different modules with each other. For example *sma-6* encodes an ortholog of a type I TGF-β receptor and is a component of the 'regulation of body size' module, which is highly interconnected by shared phenotype. This has five genetic interactions with components of the 'germline development' module, which is a subnetwork interconnected via shared expression profiles. *sma-6* is known to regulate body size, and TGF-β signalling is involved in regulation of germline proliferation in both *C. elegans* and *Drosophila melanogaster*. These *sma-6* genetic interactions may reflect overlapping functions between the two modules, and highlight the pleiotropic roles of common signalling pathways in different developmental contexts.

Large-scale genetic interaction studies in budding yeast have also found synthetic genetic interactions to be largely non-overlapping with other data types (Tong *et al.*, 2004; Boone *et al.*, 2007), potentially indicating a broadly conserved property of global network organization. However, specific genetic interactions do not seem to be conserved: homologs of gene pairs that give rise to synthetic phenotypes in *C. elegans* are not significantly more likely than randomly chosen pairs to interact in *S. cerevisiae*. This suggests that while the general theme of genetic interaction links between modules is preserved between yeast and worm, the specific connections between modules are not. One explanation for this could be that, as independent lineages diverge from a common ancestor, different molecular components acquire more prominent influence in balancing different cellular functions. Alternatively, differences observed between yeast and worm may arise largely from the fact that *C. elegans* is a multicellular organism, and must therefore co-opt subsets of the same building blocks to serve in different combinations in many different cellular contexts. The diversity of cellular identities in a multicellular organism could result in a different array of limiting factors in different cell types, so that pairwise depletion of factors does not always result in the same effect in each cell, leading to different outcomes in multicellular versus single cell systems.

4.3.2.2 Coordination of modules by shared components: a case study

Another means by which different modules might be coordinated is through direct functional interdependencies mediated by proteins that are shared components or that move from one complex to another. One such example is *C. elegans* MEL-28 (Fernandez and Piano, 2006), which plays a key role in the spatiotemporal coordination of nuclear processes during cell division in the early embryo. Consider the following events: DNA replication proceeds within the nucleus during interphase. By metaphase, the nuclear envelope has broken down, allowing the spindle microtubules to make contact with sister chromatids that have aligned at the metaphase plate. Soon after the metaphase to anaphase transition, the nuclear envelope reforms around decondensing chromosomes and DNA replicates in preparation for the next cell division. Clearly, the state of the nuclear envelope and the state of the chromatin are tightly coordinated throughout the cell cycle. The *C. elegans* early embryonic multiple support network (Gunsalus *et al.*, 2005) contains a highly interconnected 'chromatin maintenance' module that is enriched for proteins known to be required for DNA replication, chromosome congression and segregation and chromatin organization. There is also a nuclear envelope module that includes nucleoporins, nucleocytoplasmic exchange regulators and lamin. Most nodes in the multiple support network can easily be identified as members of individual modules. However, a novel gene now known as *mel-28* showed connections to both the chromatin maintenance and the nuclear envelope modules. Localization studies of the MEL-28 protein showed that it shuttles between the nuclear envelope and the chromatin throughout the cell cycle, as might be expected for a protein that coordinates nuclear envelope function with chromatin maintenance (Gunsalus *et al.*, 2005; Fernandez and Piano, 2006; Galy *et al.*, 2006). Further analysis revealed that MEL-28 is required for the proper localization of the nuclear lamina and of many of the nuclear pore complex components (Fernandez and Piano, 2006; Galy *et al.*, 2006). When MEL-28 is depleted by RNAi, the nuclear envelope is no longer capable of keeping nucleoplasm and cytoplasm separate (Fernandez and Piano, 2006; Galy *et al.*, 2006). Thus, MEL-28 is required for both the structural and functional integrity of the nuclear envelope. In addition, MEL-28 is necessary for proper chromosome congression, decondensation and segregation (Fernandez and Piano, 2006; Galy *et al.*, 2006). More recently, biochemical studies in *Xenopus* extracts demonstrated that the MEL-28 vertebrate homolog ELYS (*E*mbryonic *L*arge molecule derived from *Y*olk *S*ac (Kimura *et al.*, 2002)) coordinates nuclear pore complex assembly and DNA replication licensing via its association with chromatin (Gillespie *et al.*, 2007). Thus, the functions of both the chromatin maintenance module and the nuclear envelope module are dependent on MEL-28/ELYS, which has connections to both in an integrated functional genomic map of molecular networks in the early embryo. A simple model of how this might be achieved is through the formation of pre-complexes that are not active until a final 'key' is added to the multi-protein complex. The 'key', in this case MEL-28, shuttles between the kinetochore, where it helps complete assembly and function, and the nuclear envelope, where it functions to direct reformation and maintenance of nuclear membrane structures. When MEL-28 is recruited to the kinetochore, it leaves the nuclear envelope, signalling its disassembly. When the 'key' is released from the kinetochore and it becomes available to interact with nuclear envelope structural components, the effect is reversed: the kinetochore is disassembled and the nuclear envelope is reformed. This simple yet speculative model would ensure coordination of two critical processes during cell division by simply moving a key protein between two complexes.

4.4 Conclusion

An organism is far more than the sum of its parts. Myriad different genes, complexes, pathways and processes must collaborate seamlessly to allow development to proceed. The long-term goal of developmental systems biology is to comprehensively understand the trajectory whereby a single cell develops into a multicellular entity with many distinct cell types. Ideally a complete description of development would encompass every gene and chemical involved and their individual and synergistic contributions throughout.

Among multicellular organisms, *C. elegans* arguably has the most potential as a model for systems approaches to development, based on its fully characterized genome and cellular lineage combined with the availability of powerful molecular genetic and genomic tools. In the integrated approaches described here, combining different datasets, often on a global scale, led to novel insights about early development in *C. elegans*. However, much remains to be learned about even the very first cell division of worm development, and much more work will be needed to apply these approaches to study other aspects of development in a holistic way.

What sorts of information are now needed to move to the next level in our global understanding of developmental processes? The main limitation of the current integrated network map is that, while enormously useful, it provides only a static projection of dynamic processes; for example MEL-28 is seen as part of two submodules, whereas it alternates between them during the cell cycle. During development, rapid transitions in cellular composition, active processes and the relationships between different modules dictate a perpetual state of flux. Therefore, next-generation models will need to capture how various components work together dynamically: how components are regulated and localized spatiotemporally, how all the components fit together physically into complexes and supercomplexes, and which parts are required at what times and places for the activity of the others. We will also need to obtain a panoramic view of the extent to which different complexes are involved in multiple processes, and how different components may provide redundant or compensatory functions when others are impaired. Incorporating into network models information on how molecular interactions change through the cell cycle will revolutionize our understanding of module function and dynamics.

Some studies have begun to investigate the transient nature of interactions within global network models by integrating data on physical interactions and temporal expression patterns (Han *et al.*, 2004; Luscombe *et al.*, 2004). However, new levels of genome-scale information are still needed to obtain a high-resolution description of network dynamics: subcellular localization patterns of all proteins during each cell cycle throughout every stage of development, and a parallel dynamic map of transcriptional, post-transcriptional and post-translational regulation that captures signalling events. By integrating these new dimensions of function in a biologically relevant manner, we should be able to obtain a new perspective on how organisms faithfully accomplish cell division, determine cell identities and pattern an entire multicellular body plan.

Our progress to date suggests that integrative biology will yield more promising results in years to come: combined functional genomic and computational approaches are being developed and used with notable success both to elucidate global and local network properties and to further our mechanistic understanding of specific functional modules. It remains to be determined what level of experimental data and which types of computational approaches will eventually allow us to model developmental processes on a global scale with predictive power comparable to the best physical models of the universe. Achieving this goal will require the combined efforts of many individuals from different scientific disciplines and productive interdisciplinary, or cross-disciplinary, approaches. The creation of public consortia focused on global elucidation of functional elements encoded in the genomes of human, fly and worm (see Birney *et al.*, 2007 and modENCODE.org) will provide additional foundational data and may serve as a model on which to pattern future community efforts in developmental systems biology.

Acknowledgements

Our research has been supported by funds from NSF (BDI-0137617 to KCG and DBI-0408803 to AF), NIH (HD046236 to FP and HG004276 to FP and KCG) and NYU. KCG has also received support from the US Army Medical Research Acquisition (W23RYX-3275-N605) and NYSTAR (C040066).

References

Ashburner, M., Ball, C.A., Blake, J.A., Botstein, D., Butler, H., Cherry, J.M., *et al.* (2000) Gene ontology: tool for the unification of biology. The Gene Ontology Consortium. *Nat Genet* **25**, 25–29.

Avery, L. (1993) The genetics of feeding in *Caenorhabditis elegans*. *Genetics* **133**, 897–917.

Badano, J.L. and Katsanis, N. (2002) Beyond Mendel: an evolving view of human genetic disease transmission. *Nat Rev Genet* **3**, 779–789.

Bergman, A. and Siegal, M.L. (2003) Evolutionary capacitance as a general feature of complex gene networks. *Nature* **424**, 549–552.

Birney, E., Stamatoyannopoulos, J.A., Dutta, A., Guigo, R., Gingeras, T.R., Margulies, E.H., *et al.* (2007) Identification and analysis of functional elements in 1% of the human genome by the ENCODE pilot project. *Nature* **447**, 799–816.

Boone, C., Bussey, H. and Andrews, B.J. (2007) Exploring genetic interactions and networks with yeast. *Nat Rev Genet* **8**, 437–449.

Brenner, S. (1974) The genetics of *Caenorhabditis elegans*. *Genetics* **77**, 71–94.

Byrne, A.B., Weirauch, M.T., Wong, V., Koeva, M., Dixon, S.J., Stuart, J.M., *et al.* (2007) A global analysis of genetic interactions in *Caenorhabditis elegans*. *J Biol* **6**, 8.

Check, E. (2002) Worm cast in starring role for Nobel prize. *Nature* **419**, 548–549.

Cheeseman, I.M., Chappie, J.S., Wilson-Kubalek, E.M. and Desai, A. (2006) The conserved KMN network constitutes the core microtubule-binding site of the kinetochore. *Cell* **127**, 983–997.

Cheeseman, I.M., Niessen, S., Anderson, S., Hyndman, F., Yates, J.R., III, Oegema, K., *et al.* (2004) A conserved protein network controls assembly of the outer kinetochore and its ability to sustain tension. *Genes Dev* **18**, 2255–2268.

Chisholm, A.D. and Hardin, J. (2005) Epidermal morphogenesis. In *Worm-Book*, The *C. elegans* Research Community, ed., doi/10.1895/wormbook.1.35.1, http://www.wormbook.org.

Davidson, E.H., Rast, J.P., Oliveri, P., Ransick, A., Calestani, C., Yuh, C.H., *et al.* (2002) A genomic regulatory network for development. *Science* **295**, 1669–1678.

Davierwala, A.P., Haynes, J., Li, Z., Brost, R.L., Robinson, M.D., Yu, L., *et al.* (2005) The synthetic genetic interaction spectrum of essential genes. *Nat Genet* **37**, 1147–1152.

Delattre, M., Leidel, S., Wani, K., Baumer, K., Bamat, J., Schnabel, H., *et al.* (2004) Centriolar SAS-5 is required for centrosome duplication in *C. elegans*. *Nat Cell Biol* **6**, 656–664.

Desai, A., Rybina, S., Muller-Reichert, T., Shevchenko, A., Shevchenko, A., Hyman, A., *et al.* (2003) KNL-1 directs assembly of the microtubule-binding interface of the kinetochore in *C. elegans*. *Genes Dev* **17**, 2421–2435.

Dupuy, D., Li, Q.R., Deplancke, B., Boxem, M., Hao, T., Lamesch, P., *et al.* (2004) A first version of the *Caenorhabditis elegans* promoterome. *Genome Res* **14**, 2169–2175.

Edgar, L.G., Wolf, N. and Wood, W.B. (1994) Early transcription in *Caenorhabditis elegans* embryos. *Development* **120**, 443–451.

Fernandez, A.G. and Piano, F. (2006) MEL-28 is downstream of the Ran cycle and is required for nuclear-envelope function and chromatin maintenance. *Curr Biol* **16**, 1757–1763.

Fernandez, A.G., Gunsalus, K.C., Huang, J., Chuang, L.S., Ying, N., Liang, H.L., *et al.* (2005) New genes with roles in the *C. elegans* embryo revealed using RNAi of ovary-enriched ORFeome clones. *Genome Res* **15**, 250–259.

Fire, A., Xu, S., Montgomery, M.K., Kostas, S.A., Driver, S.E. and Mello, C.C. (1998) Potent and specific genetic interference by double-stranded RNA in *Caenorhabditis elegans*. *Nature* **391**, 806–811.

Franz, C., Walczak, R., Yavuz, S., Santarella, R., Gentzel, M., Askjaer, P., *et al.* (2007) MEL-28/ELYS is required for the recruitment of nucleoporins to chromatin and postmitotic nuclear pore complex assembly. *EMBO Rep* **8**, 165–172.

Fraser, A.G., Kamath, R.S., Zipperlen, P., Martinez-Campos, M., Sohrmann, M. and Ahringer, J. (2000) Functional genomic analysis of *C. elegans* chromosome I by systematic RNA interference. *Nature* **408**, 325–330.

Galy, V., Askjaer, P., Franz, C., Lopez-Iglesias, C. and Mattaj, I.W. (2006) MEL-28, a novel nuclear-envelope and kinetochore protein essential for zygotic nuclear-envelope assembly in *C. elegans*. *Curr Biol* **16**, 1748–1756.

Giaever, G., Chu, A.M., Ni, L., Connelly, C., Riles, L., Veronneau, S., *et al.* (2002) Functional profiling of the *Saccharomyces cerevisiae* genome. *Nature* **418**, 387–391.

Gillespie, P.J., Khoudoli, G.A., Stewart, G., Swedlow, J.R. and Blow, J.J. (2007) ELYS/MEL-28 chromatin association coordinates nuclear pore complex assembly and replication licensing. *Curr Biol* **17**, 1657–1662.

Gönczy, P., Schnabel, H., Kaletta, T., Amores, A.D., Hyman, T. and Schnabel, R. (1999) Dissection of cell division processes in the one cell stage *Caenorhabditis elegans* embryo by mutational analysis. *J Cell Biol* **144**, 927–946.

Gunsalus, K.C., Ge, H., Schetter, A.J., Goldberg, D.S., Han, J.D., Hao, T., *et al.* (2005) Predictive models of molecular machines involved in *Caenorhabditis elegans* early embryogenesis. *Nature* **436**, 861–865.

Gunsalus, K.C. and Piano, F. (2005) RNAi as a tool to study cell biology: building the genome-phenome bridge. *Curr Opin Cell Biol* **17**, 3–8.

Gunsalus, K.C., Yueh, W.C., MacMenamin, P. and Piano, F. (2004) RNAiDB and PhenoBlast: web tools for genome-wide phenotypic mapping projects. *Nucleic Acids Res* **32**, D406–D410.

Han, J.D., Bertin, N., Hao, T., Goldberg, D.S., Berriz, G.F., Zhang, L.V., *et al.* (2004) Evidence for dynamically organized modularity in the yeast protein-protein interaction network. *Nature* **430**, 88–93.

Hartman, J.L., Garvik, B. and Hartwell, L. (2001) Principles for the buffering of genetic variation. *Science* **291**, 1001–1004.

Hird, S.N. and White, J.G. (1993) Cortical and cytoplasmic flow polarity in early embryonic cells of *Caenorhabditis elegans*. *J Cell Biol* **121**, 1343–1355.

Hodgkin, J. (1988) Sexual dimorphism and sex determination. In The Nematode *Caenorhabditis elegans*, W.B. Wood, ed. (Cold Spring Harbor, NY: Cold Spring Harbor Laboratory), p. 256.

Hodgkin, J. (2001) What does a worm want with 20 000 genes? Genome Biol 2, COMMENT2008.

Hope, I.A. (1991) 'Promoter trapping' in *Caenorhabditis elegans*. *Development* **113**, 399–408.

Hubbard, E.J. and Greenstein, D. (2000) The *Caenorhabditis elegans* gonad: a test tube for cell and developmental biology. *Dev Dyn* **218**, 2–22.

Hunter, C.P., Winston, W.M., Molodowitch, C., Feinberg, E.H., Shih, J., Sutherlin, M., *et al.* (2006) Systemic RNAi in *Caenorhabditis elegans*. *Cold Spring Harb Symp Quant Biol* **71**, 95–100.

Johnsen, R.C. and Baillie, D.L. (1991) Genetic analysis of a major segment [LGV(left)] of the genome of *Caenorhabditis elegans*. *Genetics* **129**, 735–752.

Jorgensen, E.M. and Mango, S.E. (2002) The art and design of genetic screens: *Caenorhabditis elegans*. *Nat Rev Genet* **3**, 356–369.

Kamath, R.S., Fraser, A.G., Dong, Y., Poulin, G., Durbin, R., Gotta, M., *et al.* (2003) Systematic functional analysis of the *Caenorhabditis elegans* genome using RNAi. *Nature* **421**, 231–237.

Kelly, W.G., Schaner, C.E., Dernburg, A.F., Lee, M.H., Kim, S.K., Villeneuve, A.M., *et al.* (2002) X-chromosome silencing in the germline of *C. elegans*. *Development* **129**, 479–492.

Kemphues, K. (2005) Essential genes. In *WormBook*, The *C. elegans* Research Community, ed., doi/10.1895/wormbook.1.57.1, http://www.wormbook.org.

Kemphues, K.J., Kusch, M. and Wolf, N. (1988) Maternal-effect lethal mutations on linkage group II of *Caenorhabditis elegans*. *Genetics* **120**, 977–986.

Kim, S.K., Lund, J., Kiraly, M., Duke, K., Jiang, M., Stuart, J.M., *et al.* (2001) A gene expression map for *Caenorhabditis elegans*. *Science* **293**, 2087–2092.

Kimble, J., Edgar, L. and Hirsh, D. (1984) Specification of male development in *Caenorhabditis elegans*: the *fem* genes. *Dev Biol* **105**, 234–239.

Kimura, N., Takizawa, M., Okita, K., Natori, O., Igarashi, K., Ueno, M., *et al.* (2002)

Identification of a novel transcription factor, ELYS, expressed predominantly in mouse foetal haematopoietic tissues. *Genes Cells* **7**, 435–446.

Komili, S. and Roth, F.P. (2007) Genetic interaction screens advance in reverse. *Genes Dev* **21**, 137–142.

Lee, I., Date, S.V., Adai, A.T. and Marcotte, E.M. (2004) A probabilistic functional network of yeast genes. *Science* **306**, 1555–1558.

Lehner, B., Crombie, C., Tischler, J., Fortunato, A. and Fraser, A.G. (2006) Systematic mapping of genetic interactions in *Caenorhabditis elegans* identifies common modifiers of diverse signaling pathways. *Nat Genet* **38**, 896–903.

Leidel, S. and Gonczy, P. (2003) SAS-4 is essential for centrosome duplication in *C elegans* and is recruited to daughter centrioles once per cell cycle. *Dev Cell* **4**, 431–439.

Leidel, S., Delattre, M., Cerutti, L., Baumer, K. and Gonczy, P. (2005) SAS-6 defines a protein family required for centrosome duplication in *C. elegans* and in human cells. *Nat Cell Biol* **7**, 115–125.

Leno, G.H. and Munshi, R. (1994) Initiation of DNA replication in nuclei from quiescent cells requires permeabilization of the nuclear membrane. *J Cell Biol* **127**, 5–14.

Lewis, D.P., Jebara, T. and Noble, W.S. (2006) Support vector machine learning from heterogeneous data: an empirical analysis using protein sequence and structure. *Bioinformatics* **22**, 2753–2760.

Li, S., Armstrong, C.M., Bertin, N., Ge, H., Milstein, S., Boxem, M., *et al.* (2004) A map of the interactome network of the metazoan *C. elegans*. *Science* **303**, 540–543.

Lu, Z.H., Xu, H. and Leno, G.H. (1999) DNA replication in quiescent cell nuclei: regulation by the nuclear envelope and chromatin structure. *Mol Biol Cell* **10**, 4091–4106.

Luscombe, N.M., Babu, M.M., Yu, H., Snyder, M., Teichmann, S.A. and Gerstein, M. (2004) Genomic analysis of regulatory network dynamics reveals large topological changes. *Nature* **431**, 308–312.

Maddox, P.S., Oegema, K., Desai, A. and Cheeseman, I.M. (2004) 'Holo'er than thou: chromosome segregation and kinetochore function in *C. elegans*. *Chromosome Res* **12**, 641–653.

Maeda, I., Kohara, Y., Yamamoto, M. and Sugimoto, A. (2001) Large-scale analysis of gene function in *Caenorhabditis elegans* by high-throughput RNAi. *Curr Biol* **11**, 171–176.

Mangone, M., Macmenamin, P., Zegar, C., Piano, F. and Gunsalus, K.C. (2007) UTRome.org: a platform for 3'UTR biology in *C. elegans*. *Nucleic Acids Res* **36**, D57–D62.

Markowetz, F., Kostka, D., Troyanskaya, O.G. and Spang, R. (2007) Nested effects models for high-dimensional phenotyping screens. *Bioinformatics* **23**, i305–i312.

Mnaimneh, S., Davierwala, A.P., Haynes, J., Moffat, J., Peng, W.T., Zhang, W., *et al.* (2004) Exploration of essential gene functions via titratable promoter alleles. *Cell* **118**, 31–44.

Motegi, F. and Sugimoto, A. (2006) Sequential functioning of the ECT-2 RhoGEF, RHO-1 and CDC-42 establishes cell polarity in *Caenorhabditis elegans* embryos. *Nat Cell Biol* **8**, 978–985.

Munro, E., Nance, J. and Priess, J.R. (2004) Cortical flows powered by asymmetrical contraction transport PAR proteins to establish and maintain anterior-posterior polarity in the early *C. elegans* embryo. *Dev Cell* **7**, 413–424.

Nance, J., Lee, J.-Y. and Golstein, B. (2005) Gastrulation in *C. elegans*. In *Worm-Book*, The *C. elegans* Research Community, ed., doi/10.1895/wormbook.1.23.1, http://www.wormbook.org.

O'Connell, K.F., Caron, C., Kopish, K.R., Hurd, D.D., Kemphues, K.J., Li, Y., *et al.* (2001) The *C. elegans zyg-1* gene encodes a regulator of centrosome duplication with distinct maternal and paternal roles in the embryo. *Cell* **105**, 547–558.

O'Connell, K.F., Leys, C.M. and White, J.G. (1998) A genetic screen for temperature-sensitive cell-division mutants of *Caenorhabditis elegans*. *Genetics* **149**, 1303–1321.

Pan, X., Yuan, D.S., Ooi, S.L., Wang, X., Sookhai-Mahadeo, S., Meluh, P., *et al.* (2007) dSLAM analysis of genome-wide genetic interactions in Saccharomyces cerevisiae. *Methods* **41**, 206–221.

Piano, F., Gunsalus, K.C., Hill, D.E. and Vidal, M. (2006) *C. elegans* network biology: a beginning. In *WormBook*, The *C. elegans* Research Community Community, ed., doi/10.1895/wormbook.1.118.1, http://www.wormbook.org.

Piano, F., Schetter, A.J., Mangone, M., Stein, L. and Kemphues, K.J. (2000) RNAi analysis of genes expressed in the ovary of *Caenorhabditis elegans*. *Curr Biol* **10**, 1619–1622.

Piano, F., Schetter, A.J., Morton, D.G., Gunsalus, K.C., Reinke, V., Kim, S.K., *et al.* (2002) Gene clustering based on RNAi phenotypes of ovary-enriched genes in *C. elegans*. *Curr Biol* **12**, 1959–1964.

Queitsch, C., Sangster, T.A. and Lindquist, S. (2002) Hsp90 as a capacitor of phenotypic variation. *Nature* **417**, 618–624.

Rasala, B.A., Orjalo, A.V., Shen, Z., Briggs, S. and Forbes, D.J. (2006) ELYS is a dual nucleoporin/kinetochore protein required for nuclear pore assembly and proper cell division. *Proc Natl Acad Sci* **103**, 17801–17806.

Reboul, J., Vaglio, P., Rual, J.F., Lamesch, P., Martinez, M., Armstrong, C.M., *et al.* (2003) *C. elegans* ORFeome version 1.1: experimental verification of the genome annotation and resource for proteome-scale protein expression. *Nat Genet* **34**, 35–41.

Reiss, D.J., Baliga, N.S. and Bonneau, R. (2006) Integrated biclustering of heterogeneous genome-wide datasets for the inference of global regulatory networks. *BMC Bioinformatics* **7**, 280.

Rogers, A., Antoshechkin, I., Bieri, T., Blasiar, D., Bastiani, C., Canaran, P., *et al.* (2007) WormBase 2007. *Nucleic Acids Res* **36**, D612–D617.

Rual, J.F., Ceron, J., Koreth, J., Hao, T., Nicot, A.S., Hirozane-Kishikawa, T., *et al.* (2004) Toward improving *Caenorhabditis elegans* phenome mapping with an ORFeome-based RNAi library. *Genome Res* **14**, 2162–2168.

Schneider, S.Q. and Bowerman, B. (2003) Cell polarity and the cytoskeleton in the *Caenorhabditis elegans* zygote. *Annu Rev Genet* **37**, 221–249.

Schuldiner, M., Collins, S.R., Thompson, N.J., Denic, V., Bhamidipati, A., Punna, T., *et al.* (2005) Exploration of the function and organization of the yeast early secretory pathway through an epistatic miniarray profile. *Cell* **123**, 507–519.

Seydoux, G. and Fire, A. (1994) Soma-germline asymmetry in the distributions of embryonic RNAs in *Caenorhabditis elegans*. *Development* **120**, 2823–2834.

Seydoux, G., Savage, C. and Greenwald, I. (1993) Isolation and characterization of mutations causing abnormal eversion of the vulva in *Caenorhabditis elegans*. *Dev Biol* **157**, 423–436.

Siegal, M.L. and Bergman, A. (2002) Waddington's canalization revisited: developmental stability and evolution. *Proc Natl Acad Sci USA* **99**, 10528–10532.

Sönnichsen, B., Koski, L.B., Walsh, A., Marschall, P., Neumann, B., Brehm, M., *et al.* (2005) Full-genome RNAi profiling of early embryogenesis in *Caenorhabditis elegans*. *Nature* **434**, 462–469.

Strome, S. and Wood, W.B. (1983) Generation of asymmetry and segregation of germ-line granules in early *C. elegans* embryos. *Cell* **35**, 15–25.

Sulston, J.E. and Horvitz, H.R. (1977) Post-embryonic cell lineages of the nematode, *Caenorhabditis elegans*. *Dev Biol* **56**, 110–156.

Sulston, J.E., Schierenberg, E., White, J.G. and Thomson, J.N. (1983) The embryonic cell lineage of the nematode *Caenorhabditis elegans*. *Dev Biol* **100**, 64–119.

Sundaram, M. and Greenwald, I. (1993) Suppressors of a *lin-12* hypomorph define genes that interact with both *lin-12* and *glp-1* in *Caenorhabditis elegans*. *Genetics* **135**, 765–783.

Tabara, H., Grishok, A. and Mello, C.C. (1998) RNAi in *C. elegans*: soaking in the genome sequence. *Science* **282**, 430–431.

Timmons, L., Court, D.L. and Fire, A. (2001) Ingestion of bacterially expressed dsRNAs can produce specific and potent genetic interference in *Caenorhabditis elegans*. *Gene* **263**, 103–112.

Timmons, L., Tabara, H., Mello, C.C. and Fire, A.Z. (2003) Inducible systemic RNA silencing in *Caenorhabditis elegans*. *Mol Biol Cell* **14**, 2972–2983.

Tong, A.H., Lesage, G., Bader, G.D., Ding, H., Xu, H., Xin, X., *et al.* (2004) Global mapping of the yeast genetic interaction network. *Science* **303**, 808–813.

Troyanskaya, O.G., Dolinski, K., Owen, A.B., Altman, R.B. and Botstein, D. (2003) A Bayesian framework for combining heterogeneous data sources for gene function prediction (in *Saccharomyces cerevisiae*). *Proc Natl Acad Sci USA* **100**, 8348–8353.

van Luenen, H.G. and Plasterk, R.H. (1994) Target site choice of the related transposable elements Tc1 and Tc3 of *Caenorhabditis elegans*. *Nucleic Acids Res* **22**, 262–269.

Walhout, A.J., Reboul, J., Shtanko, O., Bertin, N., Vaglio, P., Ge, H., *et al.* (2002) Integrating interactome, phenome, and transcriptome mapping data for the *C. elegans* germline. *Curr Biol* **12**, 1952–1958.

Walhout, A.J., Temple, G.F., Brasch, M.A., Hartley, J.L., Lorson, M.A., Van Den Heuvel, S., *et al.* (2000) GATEWAY recombinational cloning: application to the cloning of large numbers of open reading frames or ORFeomes. *Methods Enzymol* **328**, 575–592.

Wong, S.L., Zhang, L.V., Tong, A.H., Li, Z., Goldberg, D.S., King, O.D., *et al.* (2004) Combining biological networks to predict genetic interactions. *Proc Natl Acad Sci USA* **101**, 15682–15687.

Zhong, W. and Sternberg, P.W. (2006) Genome-wide prediction of *C. elegans* genetic interactions. *Science* **311**, 1481–1484.

Zipperlen, P., Fraser, A.G., Kamath, R.S., Martinez-Campos, M. and Ahringer, J. (2001) Roles for 147 embryonic lethal genes on *C.elegans* chromosome I identified by RNA interference and video microscopy. *Embo J* **20**, 3984–3992.

Part II

Plant Systems Biology: Enumerating and Integrating the System Components

Annual Plant Reviews (2009) **35**, 169–195
doi: 10.1111/b.9781405175326.2009.00005.x

www.interscience.wiley.com

Chapter 5

SOFTWARE TOOLS FOR SYSTEMS BIOLOGY: VISUALIZING THE OUTCOMES OF N EXPERIMENTS ON M ENTITIES

Chris Poultney and Dennis Shasha

Courant Institute for Mathematical Sciences, Computer Science Department, New York University, New York, NY, USA

Abstract: Systems biology deals with genes, proteins, ions and other molecular entities. It involves the integrated analysis of these molecular entities from many experiments conducted under different conditions. This common feature – many treatment conditions and thousands or more entities – poses a challenge to finding the answers to natural questions such as: Which subsets of conditions yield similar outcomes? Which entities respond similarly in many conditions? and Which functional groupings of entities are most affected by which conditions? In this chapter, we describe a visualization tool to help answer these questions, complementary informatic tools to support alternate visualizations, and explore the lessons learned in constructing that tool both from the software designer's and biologist's perspective.

Keywords: visualization; genome; systems biology; experiments; interface

5.1 The worthwhile challenge of interdisciplinary work

When computer scientists and biologists start to work together, each person arrives with the goals of his or her discipline. The computer scientist wants to start with probably efficient algorithms, practice a disciplined software methodology, and end up with a very general tool. The biologist wants a tool to interpret data. The computer scientist may believe that designing the algorithm and publishing a paper is the end of his or her job. The biologist wants the tool. The computer scientist may talk about software design to the biologist. The biologist is uninterested – he or she still just wants the tool.

Many people's first reaction to this culture shock is to run back to their home disciplines which is why many such collaborations flounder. For those who continue to collaborate, a second reaction is to swallow the other discipline's goals entirely, forgetting the virtues of the home discipline. Computer scientists should be proud of the underlying virtues of their goals: efficient algorithms allow scalability, good software practices enhance maintainability, and a general tool will find new uses.

Systems biology presents particular challenges for data integration and visualization. For starters, omic data (genome, proteome, etc.) have no natural visualization because there is no natural physical layout for much of the data. On the other hand, an ome tends to have lots of instances (tens of thousands of genes or proteins); lists of responses are often unwieldy without visual support. Unfortunately, presenting a holistic picture, no matter how attractive, does not guide biologists to detailed study of individual molecules, which remains an important step in the biological process. In our visualization work developed in collaboration with the plant biology community, we have adopted the following goals in an attempt to unify the goals of the two disciplines:

1. The software should be easy enough to use that even biology lab directors can use it (biology-inspired goal).
2. The software should respond to a concrete present need as well as anticipated future needs (biology-inspired goal).
3. The software should give a holistic view but also the means to locate individual targets – individual genes, functional categories and experiments.
4. The software should be 'stupid', that is agnostic to its current application, so it can be extended to others (computer science-inspired goal).

Goal 1 is not meant to be impudent. To us, a litmus test of good visualization software is that our busy biological colleagues should be able to use the software easily even after having been away from it for some time or upon first being introduced to it. A second litmus test is that biologists, even those just starting to use the tool, quickly come to discuss the data rather than the interface. Goal 2 follows from our early observation that biologists, like any other scientists, have their own goals and inevitably think of people in other disciplines as purveyors of tools. We computer scientists have the same view of DNA chip makers. The utility of those tools is what counts. Goal 3 addresses a frustration we have felt with some visualization such as heat maps, in which a false colour picture appears to give substantial information, but fails to give any kind of cognitive zoom to individual genes or proteins. Goal 4 derives from the computer science aesthetic: the best software is software that can adapt to many situations. For biologists, this amortizes the cognitive cost of learning a new tool over its many uses.

This chapter presents a visualization tool called Sungear (Poultney et al., 2007) through increasingly functional examples and then derives lessons for future visualization efforts.

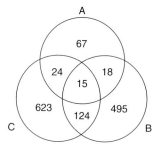

Figure 5.1 Venn diagrams are a primary source of inspiration for our software. They give much information at a glance. For example, the region in which A and B overlap corresponds to genes that are differentially expressed when A and B are at level 1 but C is at level 0. A value drawn outside of all circles (not shown) represents the case where A, B and C are at level 0.

5.2 Sungear design principles

The basic representation of much important system biological data is a list of molecules, commonly genes. Sometimes the genes have associated data, for example, expression values, rankings or connections to other genes. As useful as these lists are, it is difficult for people to gain insight from them, especially when the length of a list can grow into thousands and beyond. But sifting through vast quantities of data to gain insight is central to the biological enterprise, even if what is 'interesting' about the data can vary widely and depend on many things: organism, experimental design, data set, experimenter, etc. It is this relationship between M (thousands) of entities under N (several) conditions that we want to help visualize. The start of the visualization should be a picture that gives a holistic representation of what is going on, but should support a zoom to groups and then to individuals.

If the number of conditions were three or fewer, conventional Venn diagrams provide a good start. For example, consider a fictitious microarray experiment in which three input nutrients A, B and C are either at a level 0 or at a level 1, and we want to determine which genes are differentially expressed in which situation. One might represent the outcome as a Venn diagram (Fig. 5.1).

Given this Venn diagram, one can label each of the eight possible combinations resulting from the binary levels of the three inputs. There are however some problems with this visualization:

1. The Venn diagram gives no visual cue of which combinations of inputs gives the largest or smallest result – it is necessary to read the numbers.
2. The Venn diagram cannot be extended to more than three inputs, without using far more complicated shapes that make the image hard to read.
3. The Venn diagram does not by itself tell us which genes are in which intersection.

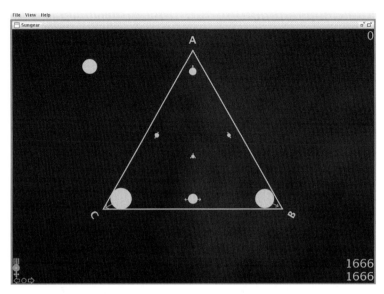

Figure 5.2 An equivalent Sungear representation of the Venn diagram. Each circle called a 'vessel' is pulled towards the 'anchors' from which it comes. For example, the vessel between anchors A and B corresponds to genes that are differentially expressed when input A is at level 1, input B is at level 1, but input C is at level 0. The case where A, B and C are all at level 0 is represented by the vessel located outside of the triangle.

Our software Sungear resolves the first objection in its basic window by representing A, B and C as the vertices of a polygon (a triangle when there are only three inputs) and representing the intersections as circles of different sizes depending on the number of genes (Fig. 5.2).

In contrast to Venn diagrams, the Sungear windows extend easily to more inputs. Imagine for example that we have seven inputs, and we are interested in all possible binary combinations of those inputs (Fig. 5.3). The data used here are from an analysis of hormone treatments of young Arabidopsis seedlings (Nemhauser *et al.*, 2006) (see the case study for more details).

The figure 5.3 gives a holistic view of the data, but it is crowded enough to make it difficult to understand which vessel corresponds to which collection of anchors. For that reason, mousing over a vessel paints the associated anchors pink and specifies the number of genes corresponding to that vessel (Fig. 5.4; arrows indicate highlighted anchors and vessels.).

Knowing how many genes are associated with each vessel does not tell us which genes they are, nor does it suggest what is special about those genes. To support that functionality, Sungear allows a click on a vessel to yield the genes and gene ontology (GO) annotations (Gene Ontology Consortium, 2001) that correspond to the set of conditions associated with the genes in that vessel (Fig. 5.5).

The genes and functional terms associated with a vessel still require the perusal of a long list. As an alternative, Sungear presents those functional

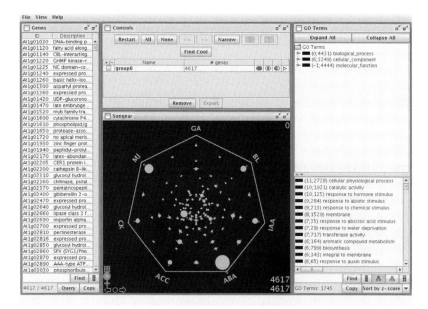

Figure 5.3 The 105 vessels in the figure reflect the relative numbers of genes corresponding to each combination (128 combinations are possible; the 23 null combinations are not drawn). GA, BL and IAA play the roles of A, B and C, respectively from Fig. 5.2.

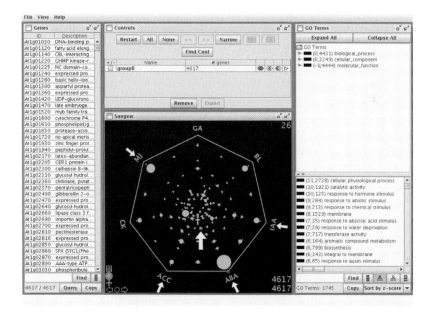

Figure 5.4 Moving the mouse over a vessel will highlight that vessel and its anchors in pink, and display the number of genes in the vessel at the upper right of the window. In this greyscale image, highlighted vessel and anchors have been marked with arrows.

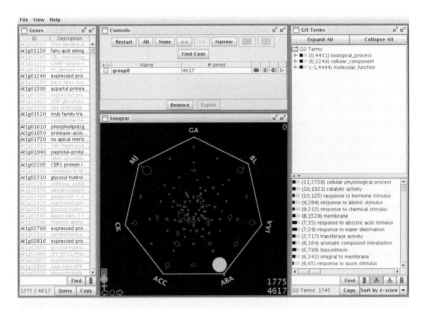

Figure 5.5 In the full Sungear display, clicking on a vessel yields the genes that correspond to that vessel as well as the GO terms corresponding to those genes.

terms that are significantly over-represented, and displays a z-score for each GO term at the top of the lower right hand window. Changing the selection will affect the z-scores, which are recalculated when Sungear 'narrows' its working set to a current selection. By narrowing to the vessel selected in Fig. 5.5, the z-scores are recalculated, and several of the top entries among the over-represented terms involve abscisic acid (Fig. 5.6). This is unsurprising in this case, since the ABA condition involves treatment with abscisic acid, and the vessel contains genes that responded only to that condition.

Besides clicking on a single vessel, one can also click on a single anchor (Fig. 5.7). This corresponds to the query 'Find all genes that are differentially expressed in this experiment, as well as the over-represented functional terms.' Similarly, one can choose a GO term and get a sense of the number of genes in each combination of conditions that correspond to that term (Fig. 5.8).

Having seen how the choice of a single vessel, anchor or GO term can influence the display, one can now ask how to combine such choices to answer more complex queries. For example, one may be curious about the transcription genes differentially expressed when both ABA and BL are at level 1. Clicking on the GO term 'transcription' and the anchors ABA and BL, and then clicking the 'set intersect' icon yields the 20 genes in this three-way intersection (Fig. 5.9).

More complicated operations are possible. To find genes with a z-score above 10 that are differentially expressed when both ABA and BL are at level 1,

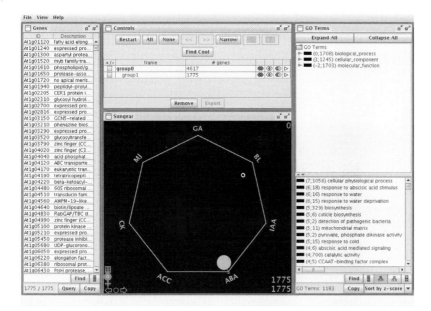

Figure 5.6 GO terms with recalculated z-scores after narrowing to the vessel selected in Fig. 5.5. The three highest z-scores are for 'cellular physiological process' (z-score 7; 1056 genes), 'response to abscisic acid stimulus' (6; 18) and 'response to water' (6; 16).

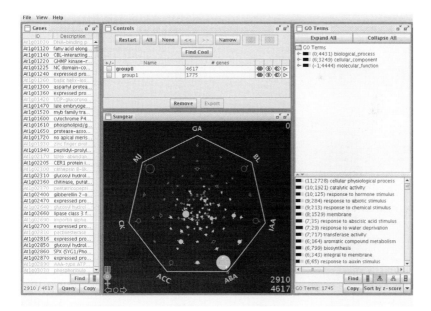

Figure 5.7 Clicking on anchor ABA leaves only those vessels that point to that anchor, indicating those vessels that require ABA to be at level 1 in order for any genes in those vessels to be differentially expressed.

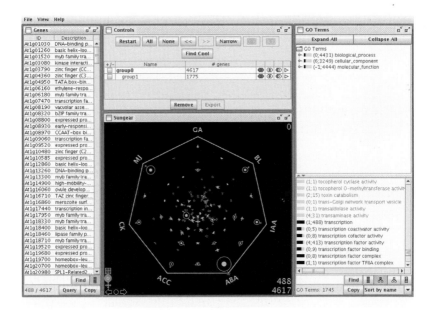

Figure 5.8 Clicking on the GO term 'transcription' makes some vessels hollow and leaves a hollow annulus in other vessels, corresponding to the genes that do not descend from the term transcription in the GO hierarchy.

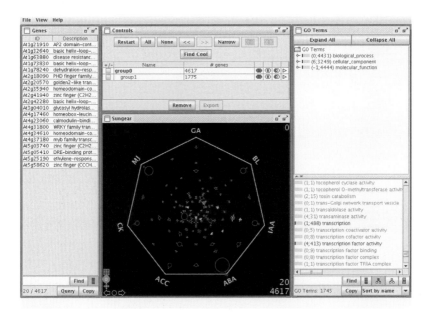

Figure 5.9 The result of using the set intersection operation to find genes associated to the GO term 'transcription' and differentially expressed when both ABA and BL are at level 1.

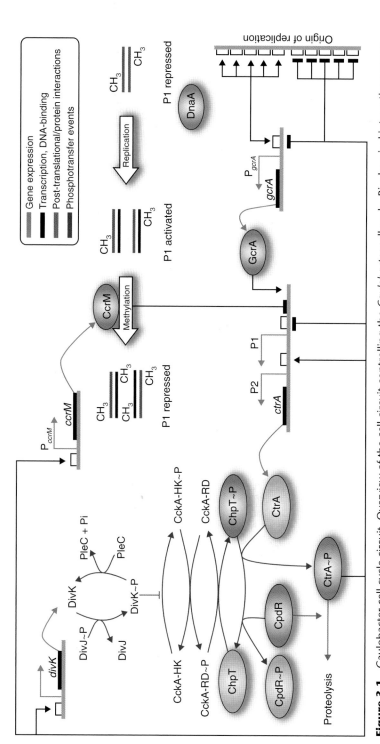

Figure 3.1 Caulobacter cell cycle circuit. Overview of the cell circuit controlling the *Caulobacter* cell cycle. Biochemical interactions are as indicated by the key in the figure. Proteins in their activated state are shaded in blue, while those in their deactivated state are shaded in grey. Adapted from Holtzendorff *et al.* (2004) and Biondi *et al.* (2006).

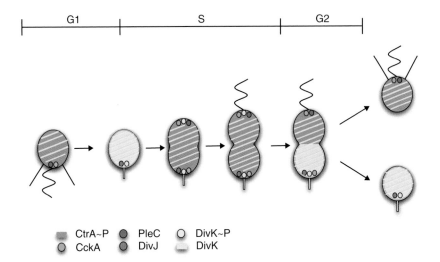

Figure 3.2 Caulobacter localization. Schematic of the proteomic localization during the cell cycle progression of CtrA~P, CckA, DivJ, PleC, DivK and DivK~P. Adapted from Holtzendorff *et al.* (2004).

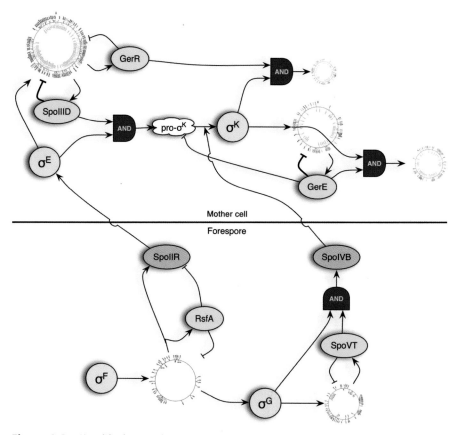

Figure 3.3 Simplified view of the mother cell and forespore regulatory circuits. Activation of transcription is represented by arrowheads (↓); repression is represented as a straight bar (⊥); AND symbols represent Boolean logic AND gates, meaning that the output requires the input from both transcription factors, that is FFLs (either coherent or incoherent). A signal sent from the vegetative cell initiates the sporulation regulatory cascade beginning with σ^F in the forespore and σ^E in the mother cell. Activation of the σ^F regulon, including RsfA, leads to RsfA repressing expression of some of the genes turned on by σ^F. The combined activation of σ^F AND repression of RsfA allows expression of SpoIIR (an incoherent FFL), which sends a signal to the mother cell to allow maturation of σ^E. Activation of the large regulon of σ^E, activates the genes GerR and SpoIIID, which then downregulate the expression of many genes in the σ^E regulon (resulting in a pulse of expression). Coherent FFLs are created by σ^G AND SpoVT, σ^E AND SpoIIID, σ^K AND GerR and σ^K AND GerE. Adapted from Eichenberger *et al.* (2004), Wang *et al.* (2006), Feucht *et al.* (2003) and Piggot and Hilbert (2004).

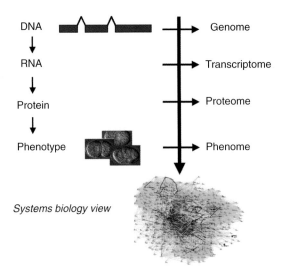

Figure 4.3 Achieving a systems view by integrating diverse 'omic' data. The transformation from single genes to 'omic' views has required developments in high-throughput technologies, as well as new data analysis and visualization tools. In the last step, diverse data types are brought together to build molecular pictures of biological processes that are based on combining all the different 'omic' views. The final systems view can be used to study 'local' parts of the network, revealing pathways and molecular complexes, and can also be used to study 'global' trends, such as topological patterns that may reveal biological properties such as buffering.

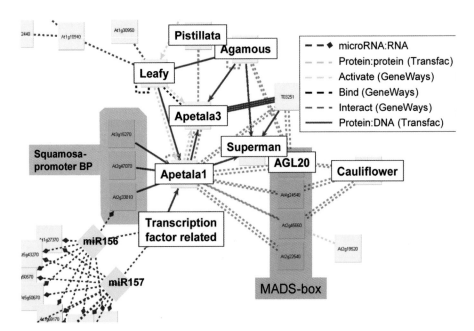

Figure 7.1 Sub-network of the multi-network model shows that miR156 and miR157 (large diamonds) may be associated with flower development. Note: This subnetwork was generated by R. Gutiérrez using the Arabidopsis multi-network described in Gutiérrez *et al.* (2007).

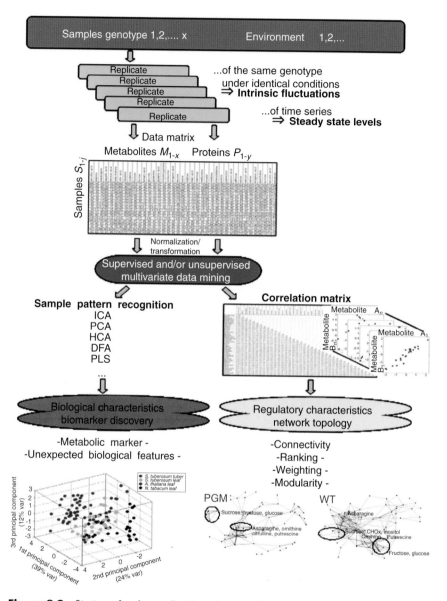

Figure 9.2 Strategy for the application of metabolite profiling as well as protein profiling (see Section 9.3.4) and multivariate data mining in plant physiology and biochemistry. The experiment is defined by using different genotypes (natural accessions, gene knockout mutants, RNAi lines or antisense plants) which are exposed to a specific environment depending on the biological question, for instance diurnal rhythm, abiotic or biotic stress, etc. (Adapted from Weckwerth and Morgenthal (2005)).

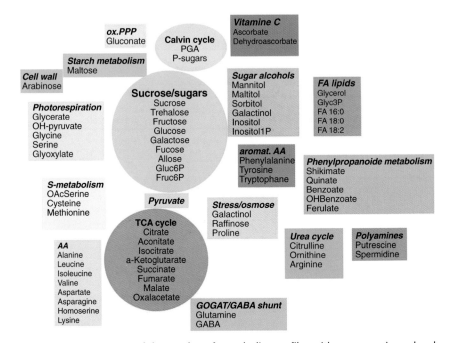

Figure 9.3 Overview of the overlap of metabolite profiles with processes in molecular plant physiology. Shown are metabolites which are typically identified and quantified in a GC-TOF-MS run. A lot of further chemical structures and unknown metabolites are identified as well but cannot yet be assigned to specific biological processes (Weckwerth, 2003; Weckwerth *et al.*, 2004a, b).

→

Figure 9.4 Metabolite profiling for the investigation of a starchless *A. thaliana* mutant lacking phospoglucomutase activity (PGM). (a) The PGM mutation and its location in the starch BS pathway is shown. Resulting changes in sugar levels are indicated by red arrows (Caspar *et al.*, 1985). (b) Principal components analysis of metabolite profiles of *A. thaliana* WT and PGM mutant during a diurnal rhythm. In the upper part, the scores plot for the separation of sample groups is seen. WT and PGM are separated on PC1; however, day–night sample discrimination is only visible for the mutant on PC2. In the lower part, specific metabolites are identified for the discrimination of either WT-PGM (PC1) or day–night samples (PC2). A clear overlap between the effects is visible and the assignment of metabolites to specific processes is not unambiguous (for further discussion see also Section 9.4.2). (c) Urea cycle intermediates and their response in the PGM mutant are shown. There is a clear link between the pathway and accumulation of the compounds in the PGM mutants. (d) Maltose concentrations are plotted during the diurnal rhythm. As no starch is synthesized in the PGM mutant during the day alterations in maltose metabolism are expected in these plants. This is seen in the plot. In WT plants maltose is peaking during the night according to Niittyla *et al.* (2004). This process is severely disturbed in the PGM mutants (Morgenthal and Weckwerth, unpublished data).

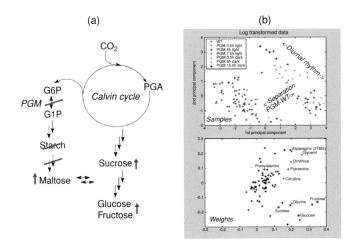

(a)

(b)

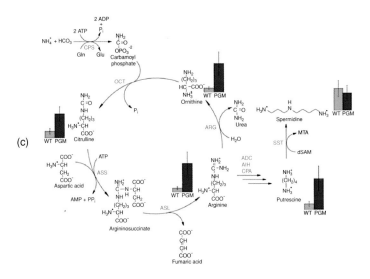

(c)

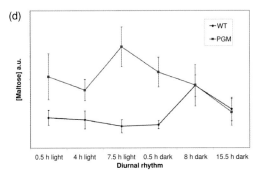

(d)

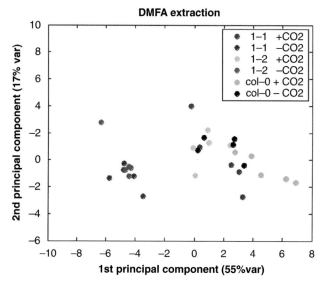

Figure 9.5 Photorespiratory HPR1-1 and HPR1-2 mutants show a viable phenotype under normal CO_2 concentrations. Under low CO_2 concentrations the mutants separate from the wild-type plants.

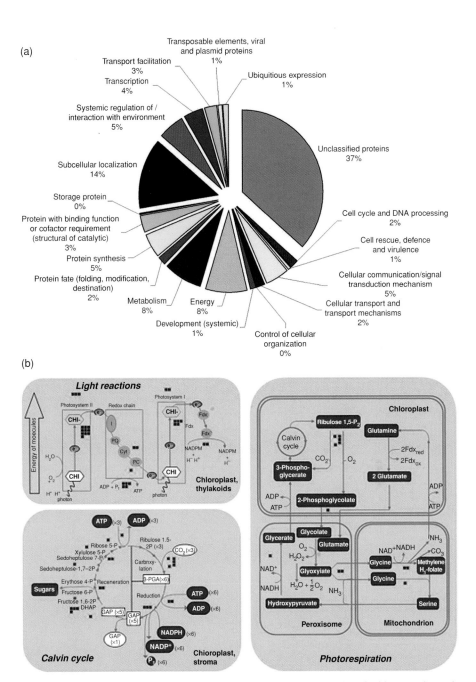

Figure 9.7 Functional classification of *A. thaliana* leaf proteins identified by coupling of protein and peptide multidimensional fractionation techniques and mass spectrometric analysis. (a) Pie chart diagram of the functional classification of 1032 *A. thaliana* leaf proteins. (b) Complete physiological processes in leaf metabolism are covered (red spots stand for identified proteins), for instance light reactions, calvin cycle and photorespiration (identified proteins loaded into 'Mapman'; Thimm *et al.*, 2004).

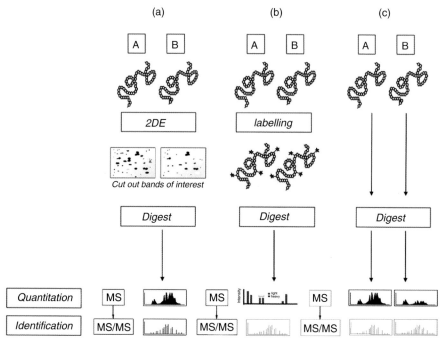

Figure 9.8 Overview of most commonly used protein quantification methods (adapted from Wienkoop and Weckwerth, 2006). (a) 2DE, (b) stable-isotope labelling and (c) direct analysis. For further details, see text.

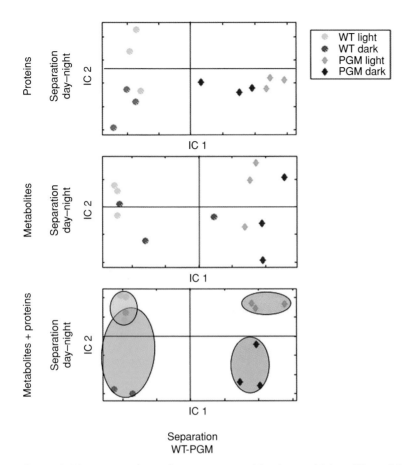

Figure 9.11 Improved sample pattern recognition by combining differential sample patterns based on the metabolites or proteins as separated data. Wild-type and *pgm* mutant data, as well as day and night samples, are well separated using the combined metabolite–protein profile and ICA (for details, see text and Morgenthal *et al.* (2005)).

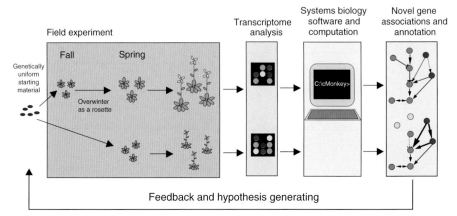

Figure 12.1 Depiction of an experiment that incorporates ecological experimental design with systems biology approaches. First, replicates of the same accession are grown in contrasting environments. Then, tissues of interest are collected at time intervals of interest and analyzed using genome-wide methods, such as microarrays. Bioinformatics can combine these genome-wide data with previously verified genetic interactions to learn putative gene interaction networks (GIN), suggesting novel associations. For example, in the figure, hypothetical novel genes are depicted as red in a known network of genes which are depicted as grey. These findings can then be used to generate new hypotheses and new experimental designs, thus completing one cycle of investigation. Iterating these procedures can lead to an extended and refined model of the GIN. (Adapted from Bossdorf *et al.*, 2008).

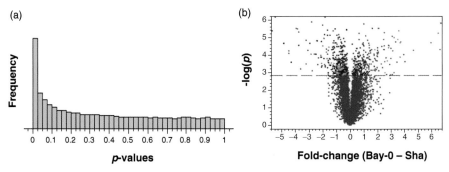

Figure 12.3 Microarray results for field-grown Bay-0 and Sha *A. thaliana* accessions after 3 weeks in the Cold Spring Harbor field site. (a) Distribution of *p*-values for the difference in expression between Bay-0 and Sha genes. (b) Volcano plot that indicates the expression fold-change of Bay-0 versus Sha (x-axis) and the corresponding log-transformed *p*-values for the expression difference (y-axis). The dashed line is the significance threshold with a 5% false discovery rate. The dots above the dashed line (in colour) are those genes that show significant difference in expression levels between Bay-0 and Sha in the field.

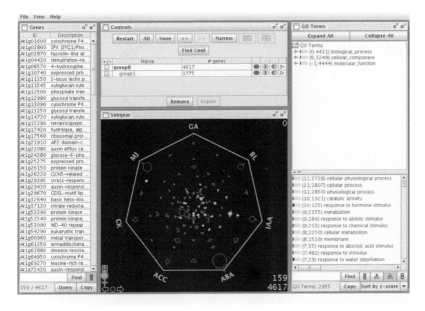

Figure 5.10 A two-stage operation, taking the union of the top six GO terms, then the intersection with ABA and BL.

first do a union operation over the six GO terms with z-scores over 10, and then perform an intersection between ABA and BL over the remaining genes (Fig. 5.10). As a colleague once told us after a similar demonstration, 'In 20 seconds, you have done what might take us 2 days to do.'

So far, we have imagined that each gene is either differentially expressed or not. Differential expression need not be the criterion for set membership – a researcher can choose any binary criterion – but membership is a binary quantity. However, expression levels can vary significantly (as can most other measures) and it is worthwhile to extend Sungear to analyze such continuous data. One of the design principles of Sungear is that all windows communicate with one another through their effect on a single master list of genes. This means that to add a new window W, all that is required is that (1) the window can adjust its display when an action in another window W changes the master list, and (2) the window W can send changes to the master list when there is a change to W. The module GeneLights, implemented by Delin Yang and Eric Leung, does this. The idea is simple. For every anchor, there exists a range of expression values. Each anchor corresponds to a line segment whose endpoints represent expression values. Superimposed on that line segment is a histogram composed of rectangles. The height of each rectangle corresponds to the number of genes in that range of expression values. Ranges of values can be chosen for one or more anchors and then linked with other windows in Sungear as illustrated below.

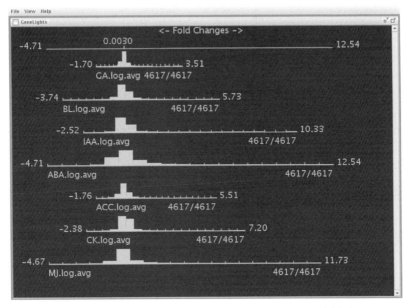

Figure 5.11 The basic GeneLights plot, showing log average fold change over the three time points that constitute each condition.

The basic GeneLights display (Fig. 5.11) shows the line segments, histograms and expression value ranges as described above. When a user selects two ranges on the ABA line, say, everything less than −0.5 and everything greater than 0.5, the result is a selection of all genes on that line whose absolute log fold expression is greater than 0.5 (Fig. 5.12). Since GeneLights is integrated into the Sungear framework, we can use it to perform more complex selection operations, such as finding all genes that have a GO annotation containing the word 'transcription' and that are twofold differentially expressed in one or more conditions (Fig. 5.13).

So far, we have focused on a genomic application. Sungear however is agnostic to the application, making almost no assumptions about the type of data upon which it operates. The assumptions are minimal: Sungear operates on several sets of items and these items can also be classified into categories, which, like GO terms, may be hierarchical. No assumptions are built into the software that the data pertain to a particular organism, or even that the data are of a biological nature. We have established a Sungear for baseball to demonstrate this point. Let us take a brief look at this application, just to underscore the point that any time a scientist has many entities and several conditions, regardless of the application, Sungear may be an appropriate tool.

Public repositories of baseball statistics such as The Baseball Archive (http://www.baseball1.com/statistics/) provide all the information necessary to adapt Sungear to display baseball information. In this baseball

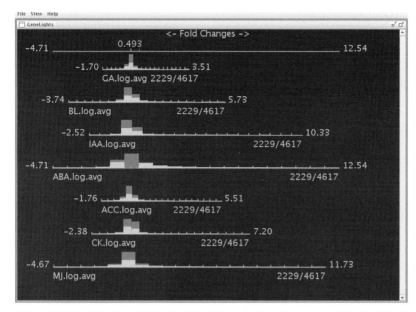

Figure 5.12 The GeneLights display after selecting all genes on the ABA line less than −0.5 or greater than 0.5 (log average differential expression greater than 0.5).

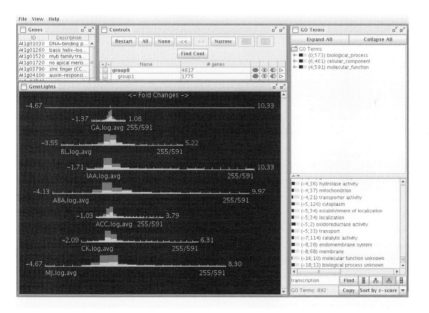

Figure 5.13 The intersection of all genes differentially expressed by a factor of two in any experiment (log base 2 ranges less than −1.0 or greater than 1.0) with and GO term containing the word 'transcription'. The active set was 'narrowed' to 591 genes before performing this selection.

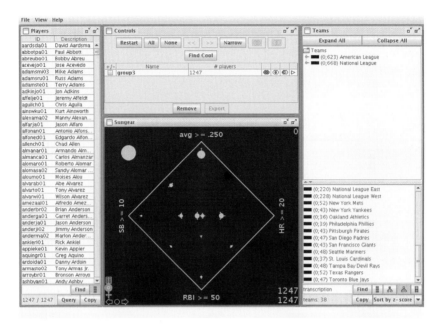

Figure 5.14 Sungear plot showing baseball data. The four anchors of the Sungear plot, clockwise from top, are: batting average of .250 or better (avg ≥ 0.250), 20 or more home runs (HR ≥ 20), 50 or more runs batted in (RBI ≥ 50) and 10 or more stolen bases (SB ≥ 10). The vessel outside the polygon represents the players who did not meet any of the above criteria in 2004.

example, individual players are the Sungear entities, and the various teams are the categories, structured hierarchically by division and league. The first step is to prepare the Sungear files that give player and team information, team hierarchy structure and player/team correspondence (for Sungear's default Arabidopsis set-up, these files would describe, respectively, TAIR and GO annotations, GO term parent–child relationships and TAIR-to-GO correspondence). Next, a simple 'experiment' data file is created based on the statistics from the 2004 season. Players are evaluated according to their performance in the four areas that constitute the Sungear plot (Fig. 5.14), with membership in a set being determined by a threshold. The categories are: batting average of 0.250 or above, 20 or more home runs, 50 or more runs batted in and 10 or more stolen bases. A quick perusal of the Sungear plot shows that (1) most players (835/1247) met none of the criteria during the 2004 season, and are represented by the vessel outside of the polygon, (2) only 23 players met all four criteria, with the American League claiming twice as many of those players as the National League (in fact, the American League champion Red Sox went on to win the World Series that season) and (3) 44 players played in both leagues during the course of the year, including

1 of the 23 four-criteria players, Carlos Beltran. This demonstrates that, after straightforward data preparation, Sungear is able to provide meaningful exploration of non-biological data.

5.2.1 A brief case study

We briefly show how Sungear can be used to help analyze experimental results. Nemhauser *et al.* analyze AtGenExpress data collected by Yoshida *et al.* under seven different hormone treatments at three fixed time points (descriptions available at http://arabidopsis.org/portals/expression/microarray/ ATGenExpress.jsp). They then classify genes as belonging to one of four categories: upregulated at one or more time points, downregulated at one or more time points, upregulated at some time points and downregulated at others, or none of the above. This is a perfect starting point for Sungear analysis. We coalesced the three types of changes (downregulated/upregulated/complex) for each hormone; the genes we consider to be at level 1 for an anchor were upregulated or downregulated during at least one time point. This illustrates how our binary set membership criterion can vary depending on the interests of the analyst. All of the Sungear plots in this chapter (except for the Venn diagram example in Fig. 5.2 and the baseball example in Fig. 5.14) use this data set.

Nemhauser *et al.* investigate the hypothesis that there is a core growth-regulatory module that collects inputs from multiple hormones. A quick glance at the Sungear plot (Fig. 5.15) shows that there are few target genes common to the hormones studied, and supports their conclusion that there is no core transcriptional growth-regulatory module. The largest vessels are around the periphery, where 1- and 2-anchor vessels are located, while the centre, which corresponds to more multi-anchor vessels, contains only small vessels. Using the vessel statistics window shows that, of the 4617 genes in the experiment, 3034 are associated with only one vessel (Fig. 5.15). In other words, the majority of genes in this experiment respond to only one hormone treatment.

Next, we examine the GO annotations of the genes. Using the GO term search feature, we find that of the 98 genes whose annotations contain the phrase 'cell wall', 63 are expressed under only one condition, providing finer grained support for the earlier statement (Nemhauser *et al.* report 60 of 101 genes; we suspect that our annotation file version was slightly different). We can also use the z-scores associated with the GO terms to examine over- and under-representation of groups of genes. The GO term 'metabolism' is strongly over-represented, with a z-score of 9 (2355 genes), while photosynthesis is slightly, if at all, over-represented with a z-score of 1 (24 genes). This is similar to the Nemhauser *et al.* survey of GO categories, which shows that genes annotated with 'metabolism' are over-represented in five of the experimental conditions, whereas the number of genes involved in photosynthesis was consistent with counts generated by a random distribution.

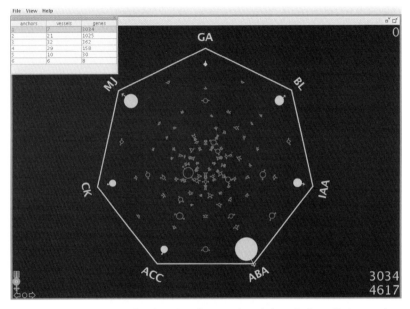

Figure 5.15 Sungear plot showing the vessel statistics window. Only vessels associated with exactly one anchor, and hence genes expressed in only one condition, are selected in the Sungear plot.

Sungear provides a shortcut to picking and examining GO terms that are expected to be interesting: the 'Find Cool' feature. This feature searches for vessels that are highly enriched for any GO terms. Each vessel is scored relative to the entire genome (as if a 'narrow' operation had been performed after selecting just that vessel), and any term associated with at least two genes in that vessel and with a z-score of 10 or more adds one to the 'cool' score for that vessel. The 'coolest' vessels are the ones with the highest scores. In this data set, the coolest vessel has a score of 11; its 14 genes responded to GA, ABA and MJ treatments. The highest scoring GO terms contributing to the score are lactose catabolism and lactose metabolism (both with z-score 19). The biological significance of this vessel, if any, is as yet undetermined; but the system's capability of finding a combination of treatments that correspond to an over-represented functionality can have many applications.

5.3 Combining visualization tools for plant systems biology

In this section, we describe complementary tools that are in wide use among our biological colleagues. The tools we discuss include features such as the in-depth analysis of individual genes, expression profiles in different networks

Table 5.1   Visualization tool websites discussed in this chapter

Software tool	URL
Sungear	http://virtualplant.bio.nyu.edu/cgi-bin/sungear/index.cgi
MapMan	http://www.gabipd.org/projects/MapMan/
Genevestigator	https://www.genevestigator.ethz.ch/
Cytoscape	http://www.cytoscape.org/
VirtualPlant	http://www.virtualplant.org/

and visualization of general networks (see Table 5.1 for websites of visualization tools discussed in this chapter).

As pointed out earlier, systems biology requires the integration of multiple experiments on multiple molecular entities. Sungear starts with a set of genes meeting some measure of interest (e.g. expression greater than or equal to a certain level) for each experiment. It then compares the experiments by displaying vessels containing those genes and relating them to functional categories. As we have seen, a paradigmatic use of Sungear is to see which combinations of experiments affect genes having similar functional categories.

Other visualization tools choose different conceptual groupings and contextualization of data. MapMan (Thimm *et al.*, 2004) shows gene expression values in the context of different pathways. Genevestigator (Zimmermann *et al.*, 2004) shows 'meta-profiles' of gene expression for several genes against a background chosen from a large database of experiments, and provides tools for working with these meta-profiles. Cytoscape (Shannon *et al.*, 2003) supports network visualization, with many options for adding, analyzing and visualizing node/edge annotations (including expression values); a plug-in framework allows for a wide variety of third-party extensions. VirtualPlant (Katari, in preparation) acts as a central data repository and launching point for a collection of browsing and analysis tools. Each of these tools encourages a different type of data exploration.

5.4   MapMan

MapMan displays gene expression values in the context of a particular pathway (Thimm *et al.*, 2004). That is, MapMan, at its core, associates expression values with a visual representation of a pathway or other process.

MapMan provides a list of pre-made pathways (currently around 45) for displaying expression data. Each pathway is arranged into a hierarchy of functional bins, which are assigned locations within the pathway image according to the pathway being represented. MapMan provides mappings from gene identifiers, such as Arabidopsis genome initiative (AGI) identifiers, to bins. When an expression data file is loaded, the value for each gene in the

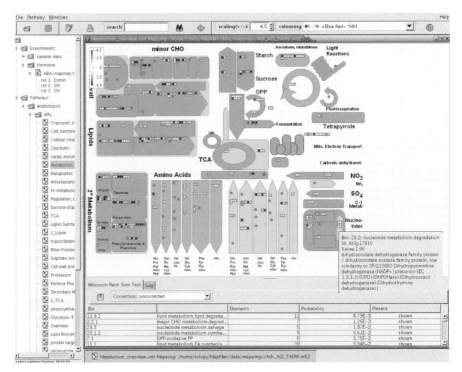

Figure 5.16 A MapMan pathway visualization. The main part of the screen shows gene expression data (log fold change) for one experiment over the Metabolism_overview pathway. In this greyscale plot, darker spots represent greater differential expression. The lower right of this window shows rollover information about one gene and its enclosing bin. The window below the visualization gives information on the hierarchical bins, and the window to the left allows a choice of experimental data and pathway visualizations.

file is displayed in its assigned bin(s) according to the mapping file; a layout component handles the organization of genes within a bin. Displaying the data is as simple as specifying a pathway file, an expression data file, and the appropriate mapping file. Only one expression data set can be shown on a pathway at a time, but multiple windows can mitigate this restriction.

Each gene in the data file is mapped to one or more bins (typically only one), and within each bin in the pathway, each gene's expression value is displayed using a continuous red-to-blue colouring (Fig. 5.16). Bins can also be displayed as histograms of expression values of genes assigned to that bin. The MapMan display makes it easy to spot bins with a large number of highly over or underexpressed genes; and the display as a whole gives a representation of where the genes in the data file are located in the overall pathway, and how strongly each gene is expressed. Gene names, expression values and other data are shown in a tooltip when the mouse is positioned over a gene, and displayed in the log window when the mouse is clicked.

Additional, gene information is provided via links out to several web-based data sources (e.g. TAIR; Rhee *et al.*, 2003). A separate window lists all the functional bins with descriptions and associated information. The tool also computes the degree to which each bin's expression profile differs from that of the other bins; this provides the default sort order for all bins represented in the network, so that the bins most likely to be interesting are shown at the head of the list.

By way of comparison, the MapMan display in Fig. 5.16 uses the same data as the 'ABA' line of the GeneLights displayed in Fig. 5.11. Whereas Sungear uses a histogram to represent the general pattern of expression in an experiment, MapMan represents the expression for each gene in the context of some larger pathway. Like Sungear, MapMan has a data-neutral architecture: although it is distributed with pathways and mapping for Arabidopsis, its text file-based architecture makes it a much more general tool for associating any type of continuous values with hierarchical locations in an image.

Here is a use scenario for MapMan: starting with a spreadsheet containing a list of genes and log fold change in expression for one or more experiments, we export that list as a tab-delimited file. We then open MapMan, choose a pathway (e.g. Metabolism_overview) and the mapping appropriate for our data (e.g. AGI) and take a look at the display. We look, either visually or using the bin list, for one or more bins with an unusual pattern of under or overexpressed genes. We can further examine these genes using one of the link-out data sources, or make a list of them for further analysis using tools such as Genevestigator.

5.5 Genevestigator

Genevestigator starts with a small number of genes specified by a user and gives information about those genes based on a set of microarray data also chosen by the user (Zimmermann *et al.*, 2004). The available microarrays are annotated according to four ontologies, which appear throughout Genevestigator: anatomy, development, stimulus and mutation. The conceptual core of Genevestigator is the 'meta-profile' of a gene, which is defined as its expression across the different categories within an ontology (the Genevestigator documentation refers to the categories within an ontology as 'conditions', but we will use the word 'categories' to avoid confusion with experimental conditions). For example, the 'anatomy' meta-profile of a gene is its expression, grouped by anatomical structure, across the microarrays chosen by the user; the 'stimulus' meta-profile of a gene is its expression across microarrays grouped by experimental stimulus.

Once an organism (currently either *Arabidopsis thaliana* or *Mus musculus*) and a set of microarrays have been chosen, Genevestigator provides four interrelated analysis tools: Meta-Profile Analysis, Biomarker Search, Clustering Analysis and Pathway Projector. Starting with a list of genes provided by the user, the Meta-Profile Analysis tool gives a number of different visualizations of the meta-profiles of the genes of interest. Two windows show expression values of selected genes over the entire set of chosen microarrays, or a subset thereof, in more detail. Four additional windows show meta-profiles of the selected genes across categories in each of the four ontologies, with different display options (e.g. heat map or scatter plot). The Biomarker Search tool is designed for the case where the starting point is not a known list of genes but a desired meta-profile, such as differential expression in one or more tissues or up/downregulation by different stimuli. Differential expression is determined by comparing a set of target categories to a set of base categories over all selected microarrays in the microarray set. For example, selecting the 'Development' tab, then selecting all stages as the base and seedling as the target, will specify a desired meta-profile of 'differential expression in the seedling relative to its expression across all developmental stages'. After specifying a meta-profile, the tool will search for matching genes, which can then be added as groups for the Meta-Profile Analysis tool.

Meta-profiles derived using either the Meta-Profile Analysis or Biomarker Search tools can be further analyzed using the Clustering Analysis tool (with a meta-profile on the screen, simply switch to the Clustering Analysis tool, select some options and click 'Run'). This tool can perform biclustering or hierarchical clustering (see Fig. 5.17) of genes and categories, with several different options (e.g. up/downregulation for biclustering, and distance measure for hierarchical clustering).

The fourth component, Pathway Projector, allows for visualization of expression data ('projection') on different pathways. It is somewhat separate from the other three tools: it starts with a 'comparison set' of expression according to base and target categories within an ontology (similar to the Biomarker Search tool) from one of the four main categories. Genes differentially expressed in this comparison set are shown on a predetermined pathway tree selected by the user; an interface is also provided for building a new network. While not as fully featured as Cytoscape, the Pathway Projector provides quick network visualization within a familiar interface.

A use scenario for Genevestigator might be as follows: a small number of Arabidopsis genes are identified as interesting (e.g. membership in a 'cool' vessel as determined by Sungear). In Genevestigator, we first specify all of the AtGenExpress microarrays as our microarray set, then enter the AGI IDs of the identified genes. We start with the 'Meta-Profile Analysis' tool, and see a scatter plot of the expression values of the specified genes across all arrays. We can then examine the 'meta-profile' of these genes according to different ontologies – in this case, we pick 'Stimulus' as the ontology, and examine the heat map of expression of these genes across different treatments. Switching to the 'Clustering Analysis' tool, we perform a hierarchical clustering of our genes of interest across the different stimuli. Noticing that several of our genes are strongly upregulated for some subset of treatments, we switch to the 'Biomarker Search' tool, and search for other genes that are strongly

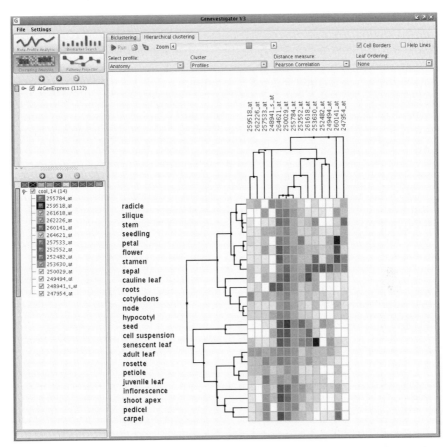

Figure 5.17 Genevestigator's Clustering Analysis tool. The main window shows 14 genes (from the 'coolest' Sungear vessel in the Sungear case study), with hierarchical clustering. Expression values are shown from light to dark, with dark representing greater differential expression. Genes (across the top of the main window) are clustered according to their expression profile across anatomical structures, and stimuli (along the left of the main window) are clustered by expression profile across genes. The microarray sets (all the AtGenExpress ATH1 22k arrays) are shown in the upper left window, and the query genes in the lower left window.

upregulated under the same treatments. This provides a new list of genes, which can be added to the original list or used to form a new group, and we can return to the 'Meta-Profile Analysis' tool to continue our analysis.

5.6 Cytoscape

Cytoscape helps to visualize networks where the nodes and edges can be defined and annotated by the user (Shannon *et al.*, 2003). A paradigmatic use of Cytoscape is to visualize regulatory networks among some subset of the genes of a genome. When used in concert with Sungear, one can imagine identifying a set of genes through Sungear and then visualizing the protein–protein interaction edges in that set through Cytoscape. Cytoscape is a comprehensive tool with many options and plug-ins; we provide here only an introduction to its full range of capabilities.

Cytoscape is built around its visualization of networks. The networks can be, for example protein–protein or protein–DNA interactions, with nodes representing either protein or DNA and an edge representing some type of interaction between two nodes. Networks can be created from scratch using Cytoscape's editing tools, but a more common route is to start with a network from a public source. Many such sources exist; for example, the Saccharomyces Genome Database and BIND (Bader *et al.*, 2001) will both export interaction data in Cytoscape-compatible formats. A third option is to use a Cytoscape plug-in like cPath (Cerami *et al.*, 2006) or Agilent Literature Search (Vailaya *et al.*, 2005) to create a network based on a literature search for user-specified terms (e.g. a protein of interest).

Once a network has been loaded, it usually needs to be organized on the screen in a human-readable format. Cytoscape provides a wide range of layouts for this purpose: layouts based on spring forces, layouts that try to detect certain types of graph structure, annotation-based layouts and many others. Which layout is best depends on the task and specific network topology. Once a layout is applied, networks are easily browsed: nodes and edges can be selected, and their attributes examined; and nodes can be searched for by ID.

As mentioned previously, nodes and edges may include additional information, referred to as attributes. As with the networks themselves, there are many available sources of node and edge attributes; one of the strengths of Cytoscape is its ability to incorporate external, publicly available data sources. Examples of node attributes are a common name, which can then be used for clarity in the network display; an alternate identifier, such as the AGI identifier where the node ID is the probe set identifier, to allow cross-referencing with data indexed by AGI identifier; a set of GO annotations; or expression values and p-values for a set of experiments. These attributes form the basis for Cytoscape's more powerful and interesting features. Once additional attributes have been loaded into the network, nodes and edges can be filtered based on their attributes. Filters can be used to select sets of nodes and edges comprising sub-networks, which can then be manipulated. For example, the sub-network comprises all the edges of a certain type, along with their corresponding nodes, can be used to create a new network; or all nodes with a p-value less than a threshold can be removed from the network.

Cytoscape can alter nearly every aspect of the network's appearance based on attribute values: node shape, node colour, node size, edge colour and edge line type are just a few of the appearance aspects that can be mapped to discrete or continuous attribute values. A set of such mappings of attributes to appearance is called a visual style. A canonical use of visual styles is to display expression data on a network, where nodes are coloured along a red-green continuum by expression value, and node shape is determined by a p-value cut-off. These separate appearance attributes can be combined to convey multiple channels of distinct information in one network, greatly increasing the amount of information communicated to the viewer.

The real power of Cytoscape lies in the combination of its annotation and visualization capabilities with sophisticated analysis, which is often provided by plug-ins. As usual, there is a wide range of plug-ins, providing functionality such as finding clusters (MCODE; Bader and Hogue, 2003) and active sub-networks (jActiveModules; Ideker et al., 2002) or performing literature searches (cPath, Agilent). The BiNGO (Maere et al., 2005) plug-in provides an excellent example of combining annotation, analysis and visualization to produce a meaningful display. Among the annotation types that Cytoscape can import are the GO annotation terms. Once these annotations have been loaded, the BiNGO plug-in analyzes the annotations associated with network nodes for significantly enriched terms; the result is a new graph where the nodes are GO terms, with directed connections reflecting the hierarchy of the terms, node colouring based on the calculated degree of significance, and node size based on the number of nodes in the original graph annotated with that term.

There are many use scenarios for Cytoscape, but one might be as follows: a set of interesting genes is identified. VirtualPlant's 'gene networks' function produces a network that associates our genes with enzymatic reactions (a simple example of such a network is shown in Fig. 5.18; Gutierrez et al., 2007). We add to this network the expression data whose analysis yielded these genes in the first place, and modify the network display to show up and downregulation. We may then be lucky enough to spot – either visually or by using a plug-in – that several of our genes are involved in a sub-network, which suggests further experiments to characterize the relationship among those genes. Network analysis of connectivity has been used to identify putative regulatory hubs that were experimentally validated in Arabidopsis (Gutierrez et al., 2008).

5.7 VirtualPlant

VirtualPlant is a web-based tool (currently annotated for Arabidopsis) that functions as both a data warehouse and a base of operations for a wide range of analysis tools. Items in the VirtualPlant 'warehouse' include many of the experimental and annotation data sets mentioned before: Affymetrix probes for Arabidopsis Affymetrix AG and ATH1 chips, TAIR and GO annotations and BIND protein interactions (Bader et al., 2001) among others. VirtualPlant uses an e-commerce metaphor: users can browse or search through these data sets, and place virtual 'orders' for sets of genes. These genes are then placed

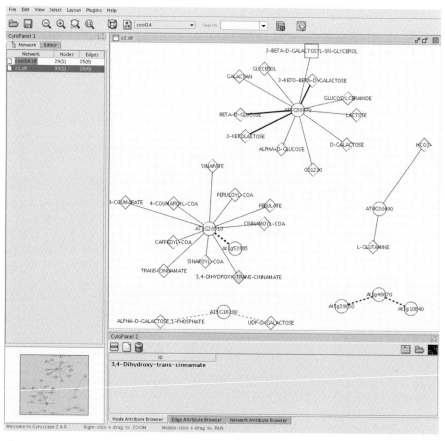

Figure 5.18　A Cytoscape network showing enzymatic reactions from KEGG and **AraCyc** for 5 of the 14 genes from Fig. 5.17. Genes are drawn as circles, and proteins as diamonds. Different interaction types are represented with different line styles. This network was generated from VirtualPlant using the 'gene networks' analysis tool.

in the 'gene cart' for further investigation and analysis (see Fig. 5.19 for an example of VirtualPlant in use).

Genes can be placed in the gene cart in one of three major ways. The first option is to upload one or more lists of genes of interest (e.g. favourite genes studied in your own lab, genes differentially expressed in a published microarray experiment). The second option is to browse through one of the VirtualPlant annotations (e.g. GO annotations) to find a category of interest (e.g. 'abscisic acid biosynthesis'), or to perform a text search within one or all of these annotations. Once the category is found, its associated genes can be added as a set of genes to the gene cart. The third option is to browse or search the available microarray experiments, each of which contains multiple slides. The experiment is first added to the gene cart as an experiment, but

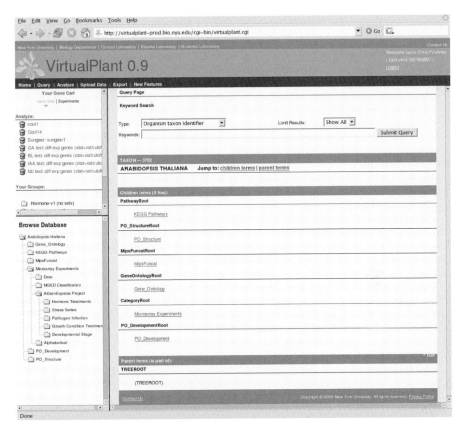

Figure 5.19 VirtualPlant. The main window shows the query interface and the top-level categories for browsing. The upper left window shows the gene cart, including the group 'Sungear: sungear1' which was exported from Sungear back to VirtualPlant. The lower left window shows the browse tree, expanded to show the folder containing the AtGenExpress hormone treatment experiments.

is converted to a gene set using one of several methods for comparing slides and thresholding (e.g. log base 2 ratio or RankProd; Hong *et al.*, 2006).

Once there are several sets of genes in the cart, VirtualPlant provides a wide range of analysis and visualization tools, including Sungear as well as on-the-fly network creation and display using Cytoscape. Most importantly, VirtualPlant not only brings together a wide range of tools, but is designed so that most of its tools can feed their output back to the gene cart, allowing for centralized, iterative analysis involving different tools, including new ones. The following extended use scenario illustrates the power of the centralized gene cart.

We start by navigating the microarray experiment hierarchy to find the AtGenExpress hormone time-course treatments. From this list, we choose the following experiments: Brassinolide time course in wild type and det2-1 mutant seedlings, GA3 time course in wild type and ga1-5 mutant seedlings and IAA time course in wild-type seedlings. We will refer to these as BL, GA and IAA from here on. For each of these three, we select the experiment, and choose the 'Create Experiment' option to add it to the gene cart. We select each experiment in turn from the gene cart, then select the mock treatments as the baseline and the hormone treatments as the treatment, being careful to avoid the mutant treatments in the GA and BL experiments; we then perform a log base 2 ratio test with a cut-off of 1.0. This will find genes in each experiment that are differentially expressed at one or more time points, albeit in a less sophisticated way than Nemhauser *et al.* (2006). When these three analyses are complete, we choose 'Analyze' from the top menu, select our three new groups and send them to Sungear for analysis.

Sungear now executes with our three groups as its anchor. Only one vessel corresponds to more than one condition (BL and IAA); we perform a union over it and several GO terms, then Narrow to create a new group. To send these genes back to VirtualPlant, we check the box next to the new group and click 'Export'. This group will now be listed among the sets of genes in the gene cart. We could now select that group for exploration by other VirtualPlant tools (such as viewing networks constructed from that set of genes, or searching for over-represented GO terms within that set); or save it as a file for analysis by a tool not directly callable from VirtualPlant.

5.8 Conclusion

Our experiences designing Sungear and observing other visualization tools have led us to the following core principles for designing tools for systems biologists:

1. Make the interface as transparent as possible. Good visualization software makes it straightforward for a skilled investigator to apply his or her experience to exploring the data. The data set-up should be straightforward and the first image insightful. After that, the investigator drives the exploration process.
2. Build a tool that does a small set of things well and with an intuitive interface. This reduces the length of the coding/feedback cycle, making it more likely that the tool will suit the investigators' needs.
3. Support different levels of abstraction ('conceptual zoom') to allow users to find additional data. It is usually impossible to show all the relevant details for a data set on-screen at one time without creating a display that is meaningless, overwhelming or both. Allowing the users to pan and zoom in a conceptual landscape effectively reduces the complexity of a visualization, making pertinent details available only on demand.

4. Make it easy for the tool to export and import data, so it can interact with complementary tools.
5. Make the tool as data generic as possible. Assumptions about the types of data that your first users care about may unnecessarily limit the usefulness of the software, so ensure your assumptions are minimal.

So much for the lessons for tool builders. What guidance can we give to users who must choose among tools? The obvious question to ask of any visualization is what it illustrates using its screen real estate. The less obvious question is the way in which the visualization reduces large numbers of items to a conceptually manageable small number. To appreciate this second point, consider a world without any visualization. You have just done 10 experiments, each on 20 000 genes. Do you stare at a table of text?

Using MapMan you might place gene expression on a particular pathway. Once you choose a pathway, each gene takes a very small amount of screen real estate, so MapMan can handle an experiment with thousands of genes. Using Cytoscape, you might want to find a small set of genes according to some criterion and then display some kind of interaction data, or find subnetworks with distinguishable biological meaning within a larger set. Unless the focus is on super-node analysis, you eventually need to reduce the set of genes to a small set to render the network comprehensible. Using Genevestigator, you again choose a small set of genes to do profile analysis. Using VirtualPlant, you can use multiple tools in a centralized iterative fashion to analyze lists of genes or microarray data to construct biological hypotheses.

Any tool that displays a small number of items from a larger set requires another tool to select that small number. A relational query language can serve this purpose but it requires too much knowledge (e.g. I want those genes that show up in the intersection of these three experiments, but how do you know which three experiments?). Sungear is an alternative in that it enables selection by experiment, functional category or commonality among experiments and/or functionalities. It does so in an interactive way in which most of the visual field is filled with circular objects whose size connotes quantity and thus suggests the queries to ask. As the amount of biological data explodes, the need for conceptually rapid filtering only increases.

Acknowledgements

We gratefully acknowledge our Sungear co-authors Rodrigo A. Gutiérrez, Manpreet S. Katari, Miriam L. Gifford, W. Bradford Paley and Gloria M. Coruzzi. Delin Yang and Eric Leung implemented GeneLights. This work was funded by grants from the National Science Foundation (IIS-9988345, IIS-0414763, DBI-0445666 and IBN-0115586) to D.E.S.; grants from the National Science Foundation – N2010 (IBN-0115586) and (DBI-0445666) to G.M.C.; a grant from the National Science Foundation (DBI-0445666) to R.A.G.; and

EMBO postdoctoral fellowship ALTF107–2005 to M.L.G. This support is greatly appreciated.

References

Bader, G.D., Donaldson, I., Wolting, C., Ouellette, B.F., Pawson, T. and Hogue, C.W. (2001) BIND–The biomolecular interaction network database. *Nucleic Acids Res* **29**, 242–245.

Bader, G.D. and Hogue, C.W. (2003) An automated method for finding molecular complexes in large protein interaction networks. *BMC Bioinformatics* **4**, 2.

Cerami, E.G., Bader, G.D., Gross, B.E. and Sander, C. (2006) cPath: open source software for collecting, storing, and querying biological pathways. *BMC Bioinformatics* **7**, 497.

Gene Ontology Consortium (2001) Creating the gene ontology resource: design and implementation. *Genome Res* **11**, 1425–1433.

Gutierrez, R.A., Lejay, L.V., Dean, A., Chiaromonte, F., Shasha, D.E. and Coruzzi, G.M. (2007) Qualitative network models and genome-wide expression data define carbon/nitrogen-responsive molecular machines in Arabidopsis. *Genome Biol* **8**, R7.

Gutierrez, R.A., Stokes, T.L., Thum, K., Xu, X., Obertello, M., Katari, M.S., *et al.* (2008) Systems approach identifies an organic nitrogen-responsive gene network that is regulated by the master clock control gene CCA1. *Proc Natl Acad Sci USA* **105**, 4939–4944.

Hong, F., Breitling, R., McEntee, C.W., Wittner, B.S., Nemhauser, J.L. and Chory, J. (2006) RankProd: a bioconductor package for detecting differentially expressed genes in meta-analysis. *Bioinformatics* **22**, 2825–2827.

Ideker, T., Ozier, O., Schwikowski, B. and Siegel, A.F. (2002) Discovering regulatory and signalling circuits in molecular interaction networks. *Bioinformatics* **18** (Suppl 1), S233–S240.

Katari, M., Nowicki, S.D., Aceituno, F.F., Nero, D.C., Kelfer, J., Thompson, L.P., *et al.* VirtualPlant: a software platform to support Systems Biology research in the post-genomic era (in preparation).

Maere, S., Heymans, K. and Kuiper, M. (2005) BiNGO: a Cytoscape plugin to assess overrepresentation of gene ontology categories in biological networks. *Bioinformatics* **21**, 3448–3449.

Nemhauser, J.L., Hong, F. and Chory, J. (2006) Different plant hormones regulate similar processes through largely nonoverlapping transcriptional responses. *Cell* **126**, 467–475.

Poultney, C.S., Gutierrez, R.A., Katari, M.S., Gifford, M.L., Paley, W.B., Coruzzi, G.M., *et al.* (2007) Sungear: interactive visualization and functional analysis of genomic datasets. *Bioinformatics* **23**, 259–261.

Rhee, S.Y., Beavis, W., Berardini, T.Z., Chen, G., Dixon, D., Doyle, A., *et al.* (2003) The Arabidopsis Information Resource (TAIR): a model organism database providing a centralized, curated gateway to Arabidopsis biology, research materials and community. *Nucleic Acids Res* **31**, 224–228.

Shannon, P., Markiel, A., Ozier, O., Baliga, N.S., Wang, J.T., Ramage, D., *et al.* (2003) Cytoscape: a software environment for integrated models of biomolecular interaction networks. *Genome Res* **13**, 2498–2504.

Thimm, O., Blasing, O., Gibon, Y., Nagel, A., Meyer, S., Kruger, P., *et al.* (2004) MapMan: a user-driven tool to display genomics data sets onto diagrams of metabolic pathways and other biological processes. *Plant J* **37**, 914–939.

Vailaya, A., Bluvas, P., Kincaid, R., Kuchinsky, A., Creech, M., Adler, A., *et al.* (2005) An architecture for biological information extraction and representation. *Bioinformatics* **21**, 430–438.

Zimmermann, P., Hirsch-Hoffmann, M., Hennig, L. and Gruissem, W. (2004) GEN-EVESTIGATOR. Arabidopsis microarray database and analysis toolbox. *Plant Physiol* **136**, 2621–2632.

Annual Plant Reviews (2009) **35**, 196–228
doi: 10.1111/b.9781405175326.2009.00006.x

www.interscience.wiley.com

Chapter 6

THE PLANT GENOME: DECODING THE TRANSCRIPTIONAL HARDWIRING

Erich Grotewold[1] and Nathan Springer[2]

[1] *Department of Plant Cellular and Molecular Biology and Plant Biotechnology Center, The Ohio State University, Columbus, OH, USA*
[2] *Department of Plant Biology, University of Minnesota, St. Paul, MN, USA*

Abstract: Transcription consists of the retrieval of information stored in the genome into mRNAs by DNA-dependent RNA polymerases (RNAPs). The regulation of transcription is controlled through *cis*-regulatory elements (CREs) that tether regulatory proteins, such as transcription factors (TFs), which act to direct the RNAP to transcribe specific genes with particular temporal and spatial patterns. Here, we provide a plant gene-centred perspective of how the regulatory code is believed to be hardwired and how TFs interpret this code to deliver the appropriate signals to the transcriptional machinery.

Keywords: *cis*-regulatory element; transcription factor; promoter; TATA; regulatory motif; chromatin

6.1 Introduction

A large fraction of the genome of each organism is dedicated to specify when, where and how much of each mRNA needs to be produced. Similar to the actual protein coding sequences, this regulatory information is hardwired into the genomic DNA and is essentially the same in every cell and is constant over time and generations. However, in contrast to coding sequences, which are translated identically in every cell of an organism, the same regulatory sequences are interpreted in very different ways, depending on the cell type or on the particular environmental circumstances. These regulatory sequences are often in close proximity to the genes they control, hence we will refer to them as the *cis*-regulatory apparatus. The *cis*-regulatory apparatus contains *cis*-regulatory elements (CREs). It is the function of a group of *trans*-acting proteins, the transcription factors (TFs), to interpret the sequence code hardwired in the *cis*-regulatory apparatus and execute it in the form of a signal to the basal transcription machinery that will result in RNA production by the corresponding DNA-dependent RNA polymerase (RNAP). Understanding the rules for deciphering the transcriptional regulatory code remains one of the most significant challenges in biology today.

Establishing which sequences in the genome contribute to the *cis*-regulatory apparatus and understanding the contribution of factors beyond DNA sequence, such as DNA structure and DNA accessibility, are significant hurdles in dissecting the transcriptional regulatory code. The large genome size of many plant species further enhances this problem. The increase in genome size is often accompanied by lower gene density, and thus larger stretches between genes, which often contain *cis*-regulatory information. CREs can also be present in introns, 5′ and 3′ UTRs (untranslated regions) and may even be embedded as part of protein coding sequences. Since the pioneering experiments by Jacob and Monod in the 1960s that established the basis for what is known today as the study of regulation of gene expression, few new rules have emerged to permit researchers to predict, from DNA sequence information, where the *cis*-regulatory code of a gene is located, less what that code actually is. Providing a new perspective of the current understanding of the *cis*-regulatory code of plant genomes is the main objective of this chapter.

6.2 The plant basal transcriptional apparatus

6.2.1 Plant RNA polymerase II core promoters

A logical place to begin to decipher the regulatory code of plants is the core promoter (a.k.a. basal promoter), which is the minimal region of contiguous DNA sequence sufficient to direct accurate transcriptional initiation by the RNAP II basal transcriptional apparatus (Butler and Kadonaga, 2002). By definition, the length of a core promoter needs to be experimentally determined, but they are operationally defined as the ~100 nucleotide region centred at the transcription start site (TSS) (Ohler *et al.*, 2002; Molina and Grotewold, 2005). Animal core promoters are generally characterized by the presence of one or several conserved DNA-sequence motifs (Fig. 6.1), including the TATA box (located around position −29 with respect to the TSS), the initiator element (Inr), positioned at the TSS, the transcription factor IIB (TFIIB) recognition element BRE (located around position −35) and the downstream core promoter element (DPE), located about 30 nucleotides after the TSS (position +30) (Smale and Kadonaga, 2003). The sequence motifs characteristic of the plant core promoter were largely unknown until very recently, largely due to the inability to position plant TSS. The generation and sequencing of large numbers of plant full-length cDNAs (Seki *et al.*, 2001, 2002; Alexandrov *et al.*,

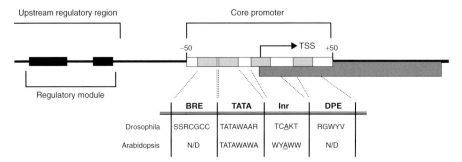

Figure 6.1 Structure of a canonical plant regulatory region. The core promoter is arbitrarily defined as the 50 bp region flanking each side of the transcription start site (TSS). The upstream regulatory region is formed by regulatory modules, each containing binding sites for transcription factors (indicated by shaded boxes). Several *cis*-regulatory elements (CREs) characterize core promoters, only some of them so far described in the plants. The nucleotide convention (IUPAC) is as follows: S = G or C; R = A or G; W = A or T; K = G or T; Y = C or T; V = A, C or G. The *Drosophila* core promoter consensus sequences were obtained from (Butler and Kadonaga, 2002; Ohler *et al.*, 2002), and information on the *Arabidopsis* core promoters from (Molina and Grotewold, 2005).

2006; Jia *et al.*, 2006) results in the precise mapping of TSS for many plant genes, with the subsequent possibility to analyze the sequence motifs that characterize plant core promoters.

6.2.1.1 Occurrence and properties of plant TATA motifs

In departure from the textbook version that TATA boxes are a general feature of plant promoters (Buchanan *et al.*, 2000), this motif is present in less than 30% of all the *Arabidopsis* genes (Molina and Grotewold, 2005). This is not a unique feature of plants, since TATA boxes are present in only 20–30% of *Drosophila* (Ohler *et al.*, 2002), mammalian (Shi and Zhou, 2006) and yeast promoters (Basehoar *et al.*, 2004). TATA-containing *Arabidopsis* genes generally have shorter 5′ UTRs and are expressed at higher levels (Molina and Grotewold, 2005), supporting a role for plant TATA motifs in accurate transcriptional initiation and an association with more robust mRNA production (Zhu *et al.*, 1995). Consistent with structural evidence showing that the distance from the TATA to the RNA Pol II active centre is 30 bp, the average distance of *Arabidopsis* TATA boxes to the TSS was found to be 31.7 bp (Molina and Grotewold, 2005), very similar to the position of mammalian TATA boxes (Carninci *et al.*, 2006). In mammals, TATA-less genes often show multiple TSS (Carninci *et al.*, 2006). We compared the preferred TSS between TATA-less and TATA-containing *Arabidopsis* genes, and no obvious difference in nucleotide distribution was observed (Table 6.1), suggesting that TATA-less promoters are regulated differently in animals and plants. In *Arabidopsis*, no over-represented core promoter element that obviously compensates for the absence of a TATA motif has yet been identified. Recent studies comparing the structural properties of mammalian and plant core promoters highlighted

Table 6.1 Nucleotide frequency matrices of TSS flanking sequences in ~13 000 *Arabidopsis* core promoters (Molina and Grotewold, 2005), comparing TATA-less and TATA-containing genes

All promoters

	–3	–2	–1	+1	+2	+3
A	0.302	0.244	0.152	0.542	0.346	0.344
C	0.223	0.231	0.377	0.145	0.187	0.245
G	0.147	0.139	0.065	0.218	0.128	0.142
T	0.328	0.386	0.406	0.095	0.340	0.269

TATA-less

	–3	–2	–1	+1	+2	+3
A	0.297	0.243	0.165	0.504	0.346	0.337
C	0.215	0.236	0.350	0.158	0.178	0.236
G	0.155	0.149	0.073	0.238	0.139	0.148
T	0.332	0.373	0.412	0.100	0.337	0.279

TATA-containing

	–3	–2	–1	+1	+2	+3
A	0.315	0.247	0.119	0.637	0.345	0.362
C	0.242	0.219	0.445	0.112	0.210	0.266
G	0.126	0.113	0.046	0.169	0.099	0.127
T	0.317	0.420	0.390	0.082	0.346	0.245

significant differences between these species (Florquin *et al.*, 2005). Interestingly, however, these studies did not detect any correlation between specific structural signals and the presence/absence of TATA motifs, suggesting that core promoter DNA-structure is unlikely to provide the difference between TATA-containing and TATA-less promoters. Nevertheless, results from these studies provide a reminder that, in addition to specific DNA-sequence motifs, DNA structure must be taken into account.

The picture that has emerged from studies in yeast and animals is that TATA boxes are enriched in stress-response genes and appear to be significantly under-represented in housekeeping or essential genes. While this correlation still needs strict verification in plants, it is already clear that specific classes of plant genes are devoid of TATA motifs, and nuclear genes participating in photosynthesis provide one good example (Nakamura *et al.*, 2002). When present, TATA motifs are essential or very important for promoter function (Grotewold *et al.*, 1994; Pan *et al.*, 2000; Zhu *et al.*, 2002; Grace *et al.*, 2004). TATA motifs may participate in making genes respond to particular environmental signals, as recently established for the effect on light responsiveness of promoters containing mutations in a prototypic plant TATA box (Kiran *et al.*, 2006). The presence/absence of a TATA motif in orthologous gene pairs across species also correlates with divergence in expression levels, a correlation that is independent of the specific function of the genes involved, holding true for plants, yeast and animals (Tirosh *et al.*, 2006). When considered together with the observed enrichment in non-essential genes, genes for which variations in gene expression are less likely to have immediate phenotypic consequences, these findings open the fundamental question of whether TATA boxes may play a central role in the evolution of gene expression (Tirosh *et al.*, 2006).

6.2.1.2 Other plant core promoter motifs

The large-scale analysis of *Arabidopsis* core promoters resulted in the identification of several other over-represented motifs. However, with the exception of the TATA box described above and microsatellite-like CT- and GA-rich regions, none of them were found in more than a few hundred promoters, from a total of 12 749 investigated (Molina and Grotewold, 2005). The high representation of microsatellites in core promoters and 5' UTRs may suggest an as yet unknown participation in the control of gene expression (Fujimori *et al.*, 2003), a role that is further supported by the conservation of many of these sequences in the regulatory region of orthologous genes pairs between *Arabidopsis thaliana* and *Brassica oleracea* (Zhang *et al.*, 2006). A sequence resembling the telobox (TAGGGTTT), which is CRE enriched in the promoters of co-expressed genes encoding proteins that participate in ribosome biogenesis and assembly (Vandepoele *et al.*, 2006), was found in ~450 core promoters (Molina and Grotewold, 2005). This motif, however, had the interesting property of being represented in ~200 core promoters within the first 50 bp after the TSS in the plus strand and in an additional ~250 core promoters, within 50 bp upstream of the TSS in the opposite orientation. The significance of this distribution is not known at present.

Another general characteristic of eukaryotic core promoters is the presence of an Inr element centred on the TSS. *Drosophila* Inr elements are characterized by $TC\underline{A}^G/_TT^T/_C$ and human elements by $PyPy\underline{A}N^T/_APyPy$ consensus sequences, in which the underlined residues indicate the TSS (Butler and Kadonaga, 2002). The recent large-scale profiling of cap analysis gene expression tags for mouse and human TSS, however, suggests that there is less of a requirement for adenine at position +1, and that the presence of Py$\underline{Pu}$ pairs (the +1 position underlined) is probably more important, with CG, TG and CA dinucleotides correlated with more active TSSs (Carninci *et al.*, 2006). The analysis of *Arabidopsis* TSS shows a similar preference for a Pu, preferentially adenine, at the TSS, with a strong over-representation of Py at position −1 (Table 6.1). Thus, Inr elements appear to be similar in plants and animals.

A BRE-like element, a sequence located immediately upstream of some TATA boxes and recognized by TFIIB (Lagrange *et al.*, 1998; Butler and Kadonaga, 2002), is necessary for the recruitment of rice TFIIB (OsTFIIB) to the phenylalanine ammonia-lyase (*pal*) promoter in *in vitro* transcription experiments. OsTFIIB physically interacts with and enhances the *in vitro* binding of recombinant TATA-binding protein 2 (OsTBP2) to the core *pal* promoter and stimulates its basal transcription (Zhu *et al.*, 2002). However, large-scale DNA motif search analyses did not result in the identification of BRE-like or other CREs characteristic of yeast or animals in plant core promoters (Fig. 6.1).

6.2.1.3 Known components of the plant basal transcription machinery

TBP is an important component of the RNAP I, RNAP II and RNAP III pre-initiation complexes, even in promoters that lack an obvious TATA motif. It is unclear whether TBP also participates in the activity of RNAP IV, an RNAP specific to the plant kingdom, which participates in the siRNA (small interfering RNA) pathway (Herr *et al.*, 2005; Onodera *et al.*, 2005). The structure of TBP, first elucidated for one of the two *Arabidopsis* proteins (TBP-2), has a saddle structure, which binds the minor groove of the DNA, and induces significant DNA bending. In contrast to yeast and animals, *Arabidopsis* and other plants contain two genes encoding TBPs (Vogel *et al.*, 1993; Baldwin and Gurley, 1996) and no recognizable TBP variant (Lagrange *et al.*, 2003), usually present in animal genomes (Berk, 2000). TBP associates with several TBP-associated factors (TAFs) (12–14) to form the TFIID, the major core promoter recognition complex (Buratowski *et al.*, 1989). Several loci encoding for potential *Arabidopsis* TAFs have been identified (Lago *et al.*, 2005).

In addition to TBP, TFIIB corresponds to another general TF that is evolutionarily conserved. *Arabidopsis* expresses several TFIIB proteins, three likely associated with RNAP II (TFIIB1, −2 and −3) and two (Brf1 and Brf2) with RNAP III (Lagrange *et al.*, 2003). Moreover, a TFIIB-related factor (pBrp), which appears to have molecular properties unique to the plants, was also identified from *Arabidopsis*. *In vitro* pBrp forms a stable complex with TBP2 on DNA (adenovirus major late promoter). However, in contrast to TBPs and TFIIBs, pBrp localizes to the outer membrane of the plastid envelope, and only upon a proteosome-dependent proteolytic cleavage, pBrp translocates to the nucleus (Lagrange *et al.*, 2003).

6.3 Plant transcription factors

6.3.1 Transcription factor definition and classification

Truly understanding the regulation of gene expression will require assembling a large puzzle in which half the pieces are CREs and the other half of the pieces represent TFs. By piecing together these parts we will understand the overall picture of regulation of gene expression. At this time, the majority of the TFs have been identified in several representative plant genomes, such as *Arabidopsis* and rice. However, we have a functional understanding for only a small fraction of the plant TFs. Current efforts are focused on the use of loss- or gain-of-function alleles and biochemical approaches to understand the cellular roles in which individual TFs participate. We will discuss the catalogue of plants TFs relative to animal TFs and the current efforts towards understanding the DNA-binding properties and regulatory activities of specific regulators.

TFs are commonly defined as modular proteins that contain a DNA-binding domain that interacts with *cis*-regulatory DNA-sequences and a protein–protein interaction domain that affects transcription, for example by interacting with components of the basal transcriptional machinery. This definition indeed omits the myriad of proteins that can affect gene expression without binding to specific DNA sequences. As these proteins often function by modulating the action of specific DNA-binding TFs, it will be necessary to determine how these co-regulators work to fully understand control of gene expression. However, the lack of conserved DNA-binding domains makes it quite difficult to accurately identify the full set of accessory proteins. The combination of biochemical (e.g. tandem affinity purification (Rubio *et al.*, 2005)) and genetic (e.g. two-hybrid (Ehlert *et al.*, 2006; Lawit *et al.*, 2007)) approaches aimed at identifying TF-interacting proteins will eventually result in a catalogue of TF accessory proteins.

In addition to the TFs and accessory factors, transcription is also regulated by groups of proteins that affect chromatin structure. These proteins include structural components of chromatin (such as histone variants), factors that covalently modify chromatin (such as DNA or histone methyltransferases) and factors that bind to specific forms of chromatin (such as chromodomain proteins). For this review, we will employ the term 'TF' to describe proteins that bind to DNA directly and participate in activating or repressing transcription and the term 'chromatin factor' to describe proteins that are structural parts of chromatin that covalently modify chromatin or that bind to chromatin.

The classification of transcription and chromatin factors is often accomplished through the use of a defining protein domain (Pabo and Sauer, 1992). For example, TFs can be placed in groups such as bZIP, MYB, WRKY or zinc finger-C2H2 that are based upon the presence of a conserved DNA-binding domain. This classification scheme can be quite useful for the identification and evolutionary analysis of a family of TFs; for comprehensive lists, see Riechmann *et al.* (2000), Riechmann and Ratcliffe (2000), and Shiu *et al.* (2005). Although there is some evidence that TFs from the same family will have similar functions, this is not necessarily the case (Itzkovitz *et al.*, 2006). Indeed, establishing how the divergence of recently duplicated regulatory genes results in novel TF-target gene interactions is largely unknown. An initial sub-functionalization resulting from the partial loss of function as a consequence of mutations that affect protein–protein interactions followed by neo-functionalization, has been proposed as one model to explain diversification among recently duplicated R2R3-MYB regulators (Grotewold, 2005).

Several databases including TRANSFAC, AGRIS, GRASSIUS, PlnTFDB and DATF (Table 6.2) maintain updated lists of TFs along with functional annotation for a number of different plant species (Davuluri *et al.*, 2003; Guo *et al.*, 2005; Matys *et al.*, 2006; Palaniswamy *et al.*, 2006; Riano-Pachon *et al.*,

Table 6.2 Websites with further information on plant TFs and CREs

Plant transcription factor databases

TRANSFAC	http://www.gene-regulation.com/pub/databases.html
AGRIS	http://arabidopsis.med.ohio-state.edu/
GRASSIUS	http://grassius.org/
PlnTFDB	http://plntfdb.bio.uni-potsdam.de/v1.0/
DATF	http://datf.cbi.pku.edu.cn/

Plant chromatin factor databases

ChromDB	www.chromdb.org

Protein structure database

PDB	http://www.pdb.org/

Plant *cis*-regulatory region databases

AGRIS	http://arabidopsis.med.ohio-state.edu/
TRANSFAC	http://www.gene-regulation.com/pub/databases.html
GRASSIUS	http://grassius.org/
PlantCARE	http://bioinformatics.psb.ugent.be/webtools/plantcare/html/
PLACE	http://www.dna.affrc.go.jp/PLACE/

2007). An up-to-date list of plant chromatin factors is maintained at ChromDB (Table 6.2). All these databases, with the exception of TRANSFAC, are freely available to the public.

TFs can either activate or repress gene expression. Transcriptional activation domains (TADs) can belong to one of several classes including acidic, proline or glutamine-rich domain (Roberts, 2000). Only a few TADs, primarily from the acidic type, have been dissected in detail from plant TFs (Sainz *et al.*, 1997a). A growing number of plant TFs have been shown to contain transcriptional repressors. Among them, the EAR motif, a short 12 amino acid domain sufficient to convert activators into repressors (Hiratsu *et al.*, 2003), is present in a number of stress and defence genes (Kazan, 2006). The recent identification of germline-restrictive silencing factor as a repressor that specifically targets germline-specific genes in somatic cells is likely to result in the characterization of novel mechanisms of transcriptional repression in plants (Haerizadeh *et al.*, 2006).

6.3.2 Expansion of transcription factor families in the plant kingdom

The availability of complete genome sequences for several plants (*Arabidopsis*, rice, poplar) and partial genomic sequences for numerous other plant species facilitates comparisons of TF families among different plants species, as well as between plant and animal species. Riechmann *et al.* (2000) and Shiu *et al.* (2005) have provided an analysis of the total number of TFs within the

Arabidopsis genome, and how these families have changed relative to several animal or plant species. In many cases, studies have been undertaken to describe phylogenetic relationships, genomic localization, copy number, expression patterns and duplication events for specific families of transcription or chromatin factors. For example, a detailed analysis of the WRKY family of TFs from *Arabidopsis* resulted in sub-classification of three major subgroups (Eulgem *et al.*, 2000). These genome-wide analyses can be quite useful in developing a framework for understanding the prevalence and potential redundancies within a transcription or chromatin factor family.

Plants have devoted a large capacity of their protein coding sequences to transcription and chromatin factors. In *Arabidopsis*, there are at least 1770 TFs (AGRIS; Table 6.2) and more than 390 chromatin factors (ChromDB; Table 6.2), which account for over 8% of the protein coding genes. Many of the animal model organisms contain fewer transcription and chromatin factors. While there is a general trend towards increasing numbers of transcription and chromatin factors in the plants, there is wide variation for the actual expansion in individual gene families. Shiu *et al.* (2005) found that 14/19 families of TFs that are present in plants and animals display greater expansion in plants than in the animals. Multi-cellular plants only encode ~150 zinc finger-C2H2 proteins while animal genomes encode 300–700 of these proteins. In contrast, plants express hundreds of proteins containing MYB domains, while there are less than 20 of them in metazoans (Lipsick, 1996). Indeed, MYB proteins provide one good example of how specific groups of TFs may have significantly expanded within one kingdom.

MYB domains were originally identified in the product of the *v-myb* oncogene, responsible for the transformation of myelomonocytic haematopoietic cells (Klempnauer *et al.*, 1982, 1983). MYB domains are formed by one or more imperfect MYB motifs, each characterized by the presence of three regularly spaced tryptophan residues, which in some instances can be replaced by hydrophobic amino acids (Kanei-Ishii *et al.*, 1990). Each MYB motif contains three α-helices, in which the second and third helices form a helix-turn-helix structure, with the third helix responsible for making contact with the major groove of DNA (Ogata *et al.*, 1992, 1994). Proteins containing MYB motifs have been typically characterized by the number of MYB motifs present in their MYB domain (Lipsick, 1996; Jin and Martin, 1999). Proteins with three MYB domain repeats, called 3RMYB (Kranz *et al.*, 2000), are characteristic of the animal c-MYB, B-MYB and A-MYB (Weston, 1998), and are also present in plants in the form of the small *pc-myb* gene family (Braun and Grotewold, 1999b). As is the case for *c-myb* and related genes in animals, the *pc-myb* genes appear to be required for cell cycle progression (Ito *et al.*, 2001; Ito, 2005; Haga *et al.*, 2007), and are likely essential. Two MYB repeat domains, most similar to the second and third motifs of animals 3RMYB proteins, are found in the expanded plant-specific R2R3-MYB family (Martin and Paz-Ares, 1997; Stracke *et al.*, 2001), with around 130 members in *Arabidopsis* (Stracke *et al.*, 2001) and more than 200 in maize and related grasses (Rabinowicz *et al.*, 1999;

Jiang *et al.*, 2004). *R2R3-MYB* genes derived from *pc-myb* genes by loss of the R1 MYB motif, likely before the invasion of land by plants (Dias *et al.*, 2003). The evolutionary steps that resulted in the plant-specific origin and dramatic amplification of these family of regulatory genes 250–400 million years ago have been established (Rabinowicz *et al.*, 1999; Dias *et al.*, 2003). The process of R2R3-MYB amplification and diversification appears to be ongoing, with evidence provided by the presence of particular groups of *R2R3-MYB* genes that have specifically amplified in maize and related grasses (Braun and Grotewold, 1999a). Among these recently expanded groups of *R2R3-MYB* genes is the Myb^{PtoA} clade, which is characterized by at least 10 genes in maize (Rabinowicz *et al.*, 1999), and that are likely to control individual branches of the general phenylpropanoid pathway (Grotewold, 2005; Heine *et al.*, 2007).

R2R3-MYB TFs may not be an exception with regards to amplifications unique to the plant kingdom, and a comparison with other types of proteins indicates that TF families are undergoing expansion at a much higher rate (Shiu *et al.*, 2005). There is evidence that TFs are more likely to be retained following a duplication event than many other types of proteins (Blanc and Wolfe, 2004; Seoighe and Gehring, 2004; De Bodt *et al.*, 2005; Maere *et al.*, 2005). A comparative analysis of *Arabidopsis* and rice reveals that in many orthologous groups there have been parallel expansions that have occurred independently in both genomes. However, in many cases there are examples of expansion of specific orthologous groups in certain genomes and not in others. For example, in the MBD family of chromatin factors there are some orthologous groups that have undergone expansion in monocots while other orthologous groups have expanded in dicots (Springer and Kaeppler, 2005).

The overall trend towards increasing sizes of transcription and chromatin factor families in plants leads to difficulties in molecular and genetic analyses. The presence of multiple genes with similar sequence can complicate expression analyses and the genetic redundancy can make it difficult to use reverse-genetics approaches to identify a function. However, it is worth noting that the uneven nature of expansion creates a situation in which different plant species can be more effective models for investigating function. Although it is often assumed that the 'small' *Arabidopsis* genome will contain the least redundancy, there are many examples in which there are more members of an orthologous group in *Arabidopsis* than in rice or maize (Springer *et al.*, 2003; Springer and Kaeppler, 2005).

6.3.3 DNA-binding properties of transcription factors

In putting together the puzzle of gene expression, we must learn how to connect the pieces of TFs and CREs. This problem can be approached from several different angles. Structural characteristics of specific DNA-binding domains can be used to make inferences about the DNA-binding properties of a TF. As of February 2007, 10% (4066) out of ~41 500 structures available at PDB (Table 6.2) correspond to DNA-binding proteins, but only 19 of them belong to plants, indicating a need to increase the structural analysis of plant TFs, particularly those that are unique to this kingdom. The specific sequences bound by a TF can be identified through *in vitro* protein–DNA interaction techniques such as electrophoretic mobility shift assays (EMSA) in combination with footprinting approaches, or by the systematic evolution of ligands by exponential enrichment (SELEX). However, it is often established that the *in vitro* DNA-binding activities of a particular TF do not explain the *in vivo* promoter occupancy (see Section 6.4.1.1).

It is often possible to make some inferences about the binding properties of a TF just based on sequence characteristics of the TF and structural characteristics of the conserved DNA-binding domain. There are a limited number of structural folds that are involved in sequence-specific DNA binding (Pabo and Sauer, 1992; Luscombe *et al.*, 2000; Stormo, 2000; Garvie *et al.*, 2001; Babu *et al.*, 2004; Aravind *et al.*, 2005; Itzkovitz *et al.*, 2006). Often, TFs within the same family exhibit very similar DNA-binding preferences, highlighting the necessity to identifying the factors that confer TFs regulatory specificity *in vivo*. The TFs can vary according to whether they make contacts within the major or minor groove of the DNA double helix, whether they act as monomers, homodimers or heterodimers and the number of base pairs that they contact. For example, basic helix-loop-helix (bHLH) proteins often bind DNA as dimers and make DNA contacts with a limited number of major groove hexamer sequences (CANNTG, where N represents any nucleotide) collectively known as E boxes (Murre *et al.*, 1989; Massari and Murre, 2000; Beltran *et al.*, 2005). Most of the base pair sequences are conserved and only two positions show variation. Through the formation of homo- or heterodimers, there is a potential of 185 total binding sequences that could be recognized by bHLH TFs (Itzkovitz *et al.*, 2006). This type of analysis was used to predict the theoretical limit for the number of TFs that could be encoded within each TF family (Itzkovitz *et al.*, 2006). In using the number of possible sequences to determine the theoretical limit for a number of TFs within a genome, it is useful to remember that related genes can exhibit divergence due to tissue-specific expression patterns in addition to functional divergence.

In a typical interaction, the TF and the DNA-binding site (D) are in a reversible equilibrium with the TF–DNA (TF–D) complex. The equilibrium can be represented by the equation TF + D $\leftrightarrow$ TF–D defined by an equilibrium dissociation constant K_d = (TF) (D)/(TF–D). In thermodynamic terms, K_d = $e^{\Delta G/RT}$, where ΔG represents the change in the Gibbs free energy upon the formation of the TF–DNA complex, which takes in account the enthalpic and entropic energies involved in the binding of the TF to DNA, as well as the ill-defined component involved in locating the specific site among all the others. Several interactions between amino acid side chains, backbone amide and carbonyl groups with both the sugar and phosphate backbones of base pairs as well as exposed chemical groups contribute to the enthalpy of the interaction. Among the main specific amino acid–base pair interactions are van der Waals contacts of aliphatic amino acids with the C5-methyl group of thyamine and

C5-hydrogen of cytosine, direct or water-mediated hydrogen bonds between exposed edges of base pairs and several amino acid side chains and polar groups, and amino acid intercalation between base pairs. Apparent K_d values for TF–DNA interactions can range from a low nM range (<0.1 nM) to about 500 nM for weaker interactions. K_d values higher than 500 nM are usually difficult to determine empirically because they get confounded with the non-specific DNA-binding activity that many basic proteins display *in vitro*. These DNA-binding kinetic considerations, together with the presence of multiple potential binding sites in the genome have resulted in calculations that estimate at 5–20 × 10^3 TF molecules per nucleus for efficient DNA binding in a typical eukaryotic cell (Wray *et al.*, 2003).

SELEX (Tuerk and Gold, 1990; Gold *et al.*, 1997; Levine and Nilsen-Hamilton, 2007) provides a powerful tool to determine the DNA-binding specificity of a TF. In SELEX experiments, a population of random double-stranded oligonucleotides, which can range in length from 8 to 26 base pairs and are flanked by conserved sequences, are challenged with the protein for which the DNA-binding specificity is to be determined. The protein–DNA complexes are separated by EMSA or affinity precipitation, the DNA amplified by PCR and subjected to additional rounds of selection. SELEX has been applied in the determination of the DNA-binding properties of several plant TFs (Huang *et al.*, 1993, 1996; Grotewold *et al.*, 1994; Sainz *et al.*, 1997b; Williams and Grotewold, 1997; Romero *et al.*, 1998). Information obtained from SELEX experiments permit the generation of position weight matrices (PWM), which can then be utilized to search for putative-binding sites in genome sequence information. The generation of an energy matrix, instead of a PWM, provides an even more accurate way to describe the binding of a TF to DNA (Djordjevic *et al.*, 2003). However, generation of energy matrices from SELEX data requires adjustments to the SELEX procedure (Djordjevic and Sengupta, 2006).

6.4 Hard wiring of regulatory sequences

6.4.1 *cis*-Regulatory elements

A gene's transcription pattern is largely specified by the identity and organization of specific CREs in the promoter and enhancers (Dynan, 1989). CREs thus provide the blueprints for the integration of cellular signals on the DNA, with the proper response furnished by the tethering of sets of TFs to specific DNA motifs and their interactions with the basal transcription machinery. Together with the modular nature of TFs (Keegan *et al.*, 1986), the modular organization of CREs into promoters and enhancers provides the basis for the combinatorial theory of transcriptional regulation (Britten and Davidson, 1969). The complex resulting from the interaction of specific sets of TFs with a particular group of CREs and which is responsible for the response to one particular signal has been termed the 'enhanceosome' (Maniatis *et al.*, 1998; Merika *et al.*, 2001). Establishing the identity of the enhanceosomes that participate in the spatial and temporal expression of all genes is a central thrust in biology today.

6.4.1.1 Identification and validation of *cis*-regulatory elements

CREs involved in the control of gene expression can be identified using experimental, computational or combinations of both approaches. Classical experimental methods to identify TF-binding sites involve investigating the formation of protein–DNA complexes, for example by EMSA or exploring the specific DNA sequence recognized by a TF on a given fragment of DNA using chemical or nuclease footprinting techniques. These techniques, however, involve *in vitro* protein–DNA interactions and their application depends on the availability of a DNA fragment containing the regulatory sequences. Chromatin immunoprecipitation (ChIP) provides an *in vivo* approach to detect protein–DNA interactions, and the careful analysis of probe signal intensities derived from ChIP-chip experiments using tiling arrays can provide a fairly accurate approximation of where binding sites for a TF might be located.

Using structural information or *in vitro* approaches to studying the DNA-binding properties of a particular TF can be quite informative about the sequence preferences for a TF. However, *in vivo* a TF often encounters highly dynamic chromatin as a substrate instead of naked DNA molecules, as usually utilized for *in vitro* DNA-binding assays. Two main approaches are currently available to identify the direct *in vivo* targets of TFs: (1) by expressing a fusion of the TF to GR (GR corresponds to the hormone-binding domain of the glucocorticoid receptor) and identifying the mRNAs induced/repressed in the presence of the GR ligand (dexamethasone, DEX), in the presence of an inhibitor of translation (e.g. cycloheximide, CHX); or (2) by identifying the DNA sequences that a TF binds *in vivo*, using ChIP assays. Both approaches have been successfully used to identify direct targets of a few plant TFs. The power of the analysis of GR fusions to identify and validate direct targets of TFs is exemplified in the identification of NAP (NAC-like activated by AP3/PI) as a direct target of the AP3/PI floral homeotic genes (analysis of 35S::AP3-GR plants) (Sablowski and Meyerowitz, 1998); in the discovery of genes involved in early flower development (Wellmer *et al.*, 2006); in the identification of PAL as a direct target of AtMYB21 (analysis of 35S::AtMyb21-GR plants) (Shin *et al.*, 2002); in the identification of several WRKY TFs as targets for the NPR1 regulator (Wang *et al.*, 2006), in establishing the regulatory networks associated with trichome initiation in *Arabidopsis* (Morohashi *et al.*, 2007) and in the recognition of DFR (dihydroflavonol 4-reductase) and the HLH factor JAF13 as direct targets of the Petunia AN1 HLH regulator (35S::AN1-GR) (Spelt *et al.*, 2000). ChIP has been widely used in animals and yeast to study chromatin-bound factors (Andrau *et al.*, 2002; Decary *et al.*, 2002). Indeed, ChIP-chip (or genome-wide location analysis) provided a first map of the transcription regulatory networks that govern budding yeast gene expression (Lee *et al.*, 2002). ChIP-chip analyses, however, have only recently started to be applied to plant TFs (Gao *et al.*, 2004; Thibaud-Nissen *et al.*, 2006; Lee *et al.*, 2007). ChIP-chip and TF–GR fusion methods each have strengths and limitations, and it is probably a combination of both, linked with clustering analyses of RNA profiling experiments, that will provide the most accurate image of the direct targets of a particular TF.

The computational discovery of TF-binding sites can be approached from various different perspectives, depending on the specific questions asked. For example, the availability of microarray expression data provides a dataset to inquire whether specific regulatory motifs are enriched in the promoters of co-regulated genes. Computational methods such as CARRIE (computational ascertainment of regulatory relationships inferred from expression) (Haverty *et al.*, 2004), *ModuleFinder* and *CoReg* (Holt *et al.*, 2006), and PhyloCon (Wang and Stormo, 2003) can provide robust information on sequence motifs over-represented in a specific group of promoters, compared with background models, a discovery that is further enhanced by combining it with conservation across species. Indeed, by combining the analysis of co-related genes with sequence conservation between *Arabidopsis* and *B. oleracea*, a large number of CREs with putative functions in gene expression were identified (Haberer *et al.*, 2006).

Although the de novo identification of CREs continues to be challenging, a large number of methods are currently available for motif prediction, and include probabilistic methods such as expectation-maximization (Cardon and Stormo, 1992), CONSENSUS (Stormo and Hartzell, 1989; Stormo, 1990; Ulyanov and Stormo, 1995; Hertz and Stormo, 1999), MEME (Bailey and Elkan, 1995) and Gibbs Sampling (e.g. AlignACE (Roth *et al.*, 1998)) approaches. The strengths and limitations of these and other methods have been extensively discussed and compared in recent years (Ohler and Niemann, 2001; Rombauts *et al.*, 2003; Tompa *et al.*, 2005).

A number of databases provide information on plant CREs (Table 6.2). PlantCARE (Lescot *et al.*, 2002) contains detailed information on a few hundred promoters from various plants, as well as information on plant TFs. PLACE (Higo *et al.*, 1998, 1999) provides a database of sequence motifs present in various plant promoters. AGRIS (Davuluri *et al.*, 2003; Palaniswamy *et al.*, 2006) in contrast, contains information only on *Arabidopsis* gene regulatory regions. A number of new databases are expected to provide similar resources for other plants in the next few years. Among them GRASSIUS contains information on TF, CRE and their interactions across the grasses.

Once identified, validating the participation of a CRE in gene expression is often performed by placing the promoter containing a mutant version of the CRE in front of a reporter gene such as β-glucuronidase (GUS), green fluorescent protein or luciferase and comparing the transcriptional activity furnished by the mutant promoter with that provided by the wild-type sequence. Transient expression experiments in plant tissues or plant cells in culture can provide information with regards to the quantitative contribution of the CRE on the overall promoter activity. The participation of the CRE on the spatial and temporal expression of the gene is often explored by integrating the promoter-reporter construct back into the genome through the generation of a transgenic plant. However, since gene replacement continues to be impossible in plants, transgenes lack any positional information, providing only partial information with respect to the function of CRE in the *in vivo* regulation of a gene. What is needed is a method to be able to analyze the effect of specific point mutations on CREs on the expression of a gene, without altering the spacing between adjacent CREs or the genomic location.

A back-of-the-envelope estimate of the total potential number of TF–CRE interactions in *Arabidopsis* can be obtained by multiplying the number of different TFs that bind to a given promoter (5–15 (Wray *et al.*, 2003)) by the number of binding sites for each TF in one promoter (we will assume a conservative 5 for each), by the number of TFs ($\sim$2000) by the number of genes ($\sim$30 000). This gives a rough estimate of $\sim 1.5 \times 10^9$ potential interactions. In contrast, the number of well characterized and experimentally demonstrated TF–CRE interactions remains in the two-digit range for *Arabidopsis*.

6.4.1.2 Establishing CRE-transcription factor relationships

Two important questions that remain to be established for plants are: (1) What is the density of CREs in a given promoter, and (2) How many target genes are regulated by any given TF? Partial answers to these questions might be obtained from looking at what happens in other eukaryotes. The sea urchin *Endo16* gene, controlled by one of the best-dissected promoters, contains at least 55 CREs recognized by 15 different TFs within $\sim$2.3 kb (Yuh *et al.*, 1998). The analyses of several others eukaryotic promoters suggest the presence of 15–50 CREs recognized by 5–15 different TFs (Arnone and Davidson, 1997; Wray *et al.*, 2003). These numbers are consistent with genome-wide location analyses in baker's yeast, showing a very large number of promoters recognized by four or more regulators (Lee *et al.*, 2002). Assuming regulatory regions that are 1–2 kb long and CREs that are 6–8 bp long, only 10–20% of a promoter region is occupied by TF-binding sites, prompting the question of whether the remaining promoter sequences are non-functional. The assumption that these sequences are non-functional, and therefore free to change, provides the rationale for phylogenetic footprinting (Wasserman and Sandelin, 2004).

The number of genes that a TF regulates follows a power law, consistent with the free-scale nature of regulatory networks (see below). In yeast, the number of target genes for TFs ranged from 0 to 181, with an average of about 38 promoters being recognized by each regulator (Lee *et al.*, 2002). The situation is likely much more complex in metazoans. For example, the MYC proto-oncogene recognizes more than 25 000 sites in the human genome, and likely participates in the regulation of 10% of all the human genes (Fernandez *et al.*, 2003; Adhikary and Eilers, 2005). A similar large number of target genes are regulated by the OCT4 (772 genes), NANOG (1687 genes) and SOX2 (1228 genes) regulators of human embryonic stem cell identity (Boyer *et al.*, 2005). It is currently unknown how many genes any plant TF can regulate. It is expected that data from the first genome-wide ChIP-chip experiments will become available soon, providing information on these issues.

6.4.1.3 Modular organization of CREs into promoters and enhancers

In contrast to animals, where the architecture and CRE organization of several promoters has been determined at a significant level of detail, similar information is sparsely available for plant promoters. In principle, there is no reason why the rules that apply for animal promoters should be different to those for the plants. Indeed, the few available examples show that plant promoters have also a modular structure with multiple binding sites for several TFs, often organized into regulatory motifs, each responsible for a subset of the overall expression output. It is common for CREs to be clustered in the promoters of the genes that they contribute to regulate (Davidson, 2001). This clustering presumably allows multiple TF molecules to act synergistically and in a combinatorial fashion for increased expression and specificity and is utilized as an additional feature in the identification of TF-binding sites (Berman *et al.*, 2002). A number of examples of plant gene regulatory regions highlighting the characteristics of the distribution of CREs are shown in Fig. 6.2.

Probably among the best-described promoters active in plants is that of the 35S transcript from the Cauliflower Mosaic Virus (*CaMV 35S*), a prototype of a constitutive plant modular promoter (Benfey *et al.*, 1990). Several CREs in the *35S* promoter have been identified, including the *activation sequence-1* (*as-1*) autonomous salicylic acid-responsive element recognized by bZIP factor of the TGA family (Lam *et al.*, 1989).

The maize *A1* gene, encoding the dihydroflavonol reductase enzyme, provides another example of a well-studied plant promoter (Fig. 6.3). *A1* is independently controlled by the assembly of at least two enhansosomes, each responsible for regulating the accumulation of distinct flavonoid pigments, the anthocyanins or the phlobaphenes (Grotewold, 2006). C1 (R2R3-MYB) and R (bHLH) proteins cooperate in the regulation of anthocyanin accumulation, and P1 (R2R3-MYB) controls the accumulation of phlobaphenes (Quattrocchio *et al.*, 2006). Supporting the activation by these two regulatory systems, the *A1* promoter has a compact and modular structure (Fig. 6.3) in which high- and low-affinity binding sites for the R2R3-MYB regulators are separated by the ARE, a motif conserved in the promoters of several other anthocyanin biosynthetic genes (Lesnick and Chandler, 1998). The *in vivo* dissection of the *A1* promoter using transposons showed that the ARE is uniquely necessary for the control of *A1* by C1 + R, but not by P1 (Pooma *et al.*, 2002). Indeed, while the ARE is not essential for the activation of *A1* by C1+R, it participates in the R-enhanced *A1* expression, probably by tethering the R to DNA directly through its bHLH region, or more likely, by the interaction of R with other DNA-binding proteins (Hernandez *et al.*, 2004). Thus, while

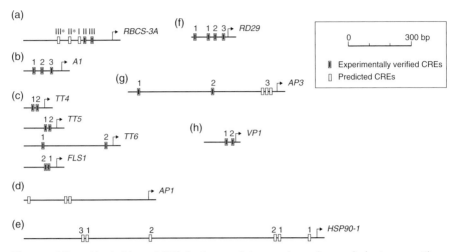

Figure 6.2 Hardwiring of CREs in the regulatory regions of several plant genes. The minimal promoter region shown to be sufficient for gene regulation is shown, when appropriate. (a) Pea *RBCS-3A* gene (Kuhlemeier *et al.*, 1988); II, III = GT-1-binding sites. (b) Maize *A1* gene (Sainz *et al.*, 1997b; Pooma *et al.*, 2002); 1, 3 = PBS; 2 = ARE. (c) *Arabidopsis TT5, TT6* and *FLS1* genes (Hartmann *et al.*, 2005); 1 = ACE; 2 = P box. (d) *Arabidopsis AP1* gene (William *et al.*, 2004); boxes indicate putative LFY-binding sites. (e) *Arabidopsis HSP90-1* gene (Haralampidis *et al.*, 2002); 1 = HSE; 2 = CCAAT box; 3 = MRE. (f) *Arabidopsis RD29* gene (Narusaka *et al.*, 2003); 1 = DRE; 2 = as-1; 3 = ABRE. (g) *Arabidopsis AP3* gene (Hill *et al.*, 1998; Lamb *et al.*, 2002); 1 = AG-binding sites; 2 = LFY-binding sites; 3 = conserved CArG elements. (h) Maize *VP1* gene (Cao *et al.*, 2007); 1 = ABRE; 2 = CE1. Abbreviations for CREs: ABRE, ABA-responsive element; ACE, ACGT-containing element; CE1, coupling element 1-like; DRE, dehydration-responsive element; HSE, heat shock element; MRE, metal regulatory element; P box, MYB recognition element; PBS, P1-binding site.

there is a certain level of redundancy among the CREs that participate in the expression of *A1*, robust expression requires the simultaneously occupancy of all these sites.

6.4.2 Regulatory activity of introns and 3′ UTR regions

In plants and animals alike, introns often have a positive influence on mRNA accumulation. This property has been exploited to increase plant gene expression. The first introns of the *Adh1* or *Bz1* genes are frequently included in many of the vectors utilized for high-level gene expression in monocots, such as maize (Goff *et al.*, 1990, 1991, 1992). The ability of introns to enhance plant gene expression decreases with the distance from the TSS, and the effect of the introns appears to be at several levels, including mRNA accumulation and translation (Rose, 2004). These findings are consistent with models that link transcription with key steps in the processing of the immature transcript including intron splicing, capping, polyadenylation and nuclear export. These activities might be associated with 'gene expression factories', meta-stable

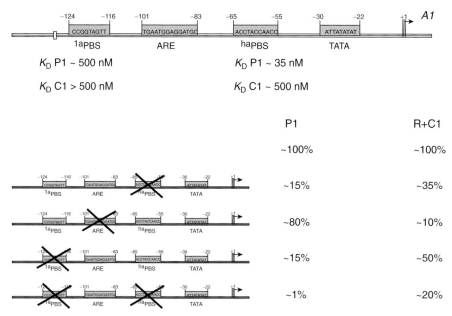

Figure 6.3 Regulatory module in the maize *A1* gene promoter sufficient for the regulation by the R2R3-MYB transcription factors P1 and C1. Mutations in each of the identified CREs affect the regulation by each activator in different ways suggesting complex synergistic and additive interactions are necessary for robust *A1* regulation. The high-affinity and low-affinity P1-binding sites ([ha]PBS and [la]PBS, respectively) are recognized by C1 and P1 with different affinities (illustrated by the K_D) (Grotewold *et al.*, 1994; Sainz *et al.*, 1997b). The ARE corresponds to the Anthocyanin Regulatory Element present in various flavonoid biosynthetic gene promoters (Lesnick, 1997) and necessary for the R-enhanced activation of *A1* (Pooma *et al.*, 2002; Hernandez *et al.*, 2004).

transcription hubs in which multiple steps of the gene expression pathway take place (Maniatis and Reed, 2002). According to these models, and consistent with the sequence-independent enhancement activity of some intervening sequences (Rose and Beliakoff, 2000), introns would not necessarily participate in the recruitment of gene-specific TFs, but rather contribute to linking splicing and transcription. Indeed, the enhancement of gene expression promoted by the maize *Sh1* first intron requires efficient splicing (Clancy and Hannah, 2002). However, in an increasing number of cases, introns appear to function in regulating gene expression at levels that are inconsistent with just a role as overall 'boosters' in mRNA or protein accumulation. For example, the expression pattern of the *Arabidopsis PRF2* gene, encoding one of five small actin-binding profilin proteins, is significantly affected by the PRF2 first intron. In the absence of it, the expression is largely restricted to the vasculature, but the presence of the intron expands the expression domain to most vegetative tissues. Transferring the intron from *PRF2* to a *PRF5* mini-gene-GUS fusion, affected the reproductive tissue-specific expression of

PRF5 (Jeong *et al.*, 2006). Similarly, DNA sequences within the 3 kb intron of the *Arabidopsis* floral organ specification gene *AGAMOUS* (*AG*), encoding a MADS box factor, are recognized by the BELLRINGER homeodomain protein, resulting in *AG* transcriptional repression (Bao *et al.*, 2004). Binding sites for TFs have also been identified in the first intron of the *Arabidopsis* elongation factor-1β gene (Gidekel *et al.*, 1996). The maize shoot apical meristem identity gene *Knotted1* is regulated by CREs located in intron sequences. The initial mutations that result in a knotted phenotype are dominant alleles that result from ectopic expression of the *Kn1* gene in developing leaf primordial (Hake *et al.*, 1989). A series of transposon insertions in the third intron of *Kn1* can produce this ectopic expression pattern and provide evidence for a CRE that that recruits negative regulators to the *Kn1* gene in a developmentally programmed manner (Greene *et al.*, 1994; Inada *et al.*, 2003). These studies increasingly suggest that introns play roles in gene regulation at multiple levels, including providing specific CREs that participate in transcriptional control.

Studies of the endosperm-specific cell wall invertase (INCW2) from maize provide an example of potential CREs in 3′ UTR regions (Cheng *et al.*, 1999). Two different transcripts, *Incw1*-S (small) and *Incw1*-L (large) with size differences in the 3′ UTR are produced. The two different transcripts are differentially regulated by metabolizable and non-metabolizable sugars. The *Incw1*-L form is induced by both metabolizable and non-metabolizable sugars but does not result in an increase in enzymatic activity. However, expression of the *Incw1*-S transcript is increased in response to metabolizable sugars and contributes to increases in enzymatic activity (Cheng *et al.*, 1999). It is likely that the difference in the two transcripts is caused by differences in transcription termination. This provides an example in which induction causes differences in transcript production. Other examples of regulation by 3′ UTRs include the *Me1* gene of the dicot *Flaveria bidentis* (Ali and Taylor, 2001) and the maize *Hrgp* gene encoding a cell wall protein (Menossi *et al.*, 2003).

6.4.3 Enhancers and long-range activity of CREs

Enhancers, classically defined as regulatory elements that function at a distance, often in an orientation-independent fashion (Serfling *et al.*, 1985), are also docking sites for TFs. In contrast to proximal locus control regions, which often function through the most proximal core promoter, enhancers frequently display a certain level of promiscuity, showing activity on multiple genes simultaneously. While many animal regulatory regions are positioned reasonably proximal to the TSS, enhancers have been detected at distances of up to a megabase from the gene that they control (West and Fraser, 2005). Plant gene regulatory regions have usually been assumed to be more compact than in metazoans. Indeed, in a dense genome like the one of *Arabidopsis*, with one gene approximately every 5 kb, most gene regulatory regions are concentrated within a few hundred base pairs upstream of the TSS, and when these regions are used in promoter-reporter constructs, they often recapitulate the corresponding gene expression patterns. In a less-dense genome such as the one from maize (one gene every 100–200 kb), many gene regulatory regions are also concentrated close to the TSS (Bodeau and Walbot, 1992; Grotewold *et al.*, 1994; Sainz *et al.*, 1997b; Lesnick and Chandler, 1998). However, *cis*-regulatory sequences in maize have also been found to function at significantly larger distances. For example, in the *P1* regulator of phlobaphene pigments, regulatory elements required for floral organ specific expression are located 1 kb and 5 kb upstream of the TSS (Sidorenko *et al.*, 2000). The *cis*-acting elements that enhance expression and contribute to paramutation of the *B-I* allele, a regulator of anthocyanin biosynthesis, are located 100 kb from the gene (Stam *et al.*, 2002a, b). The *Tb1* gene, which plays an important role in morphological changes associated with domestication in maize, is regulated by CREs that are ~60 kb upstream of the TSS (Clark *et al.*, 2006).

Chip-chip experiments being carried out on a number of *Arabidopsis* TFs using whole genome Affymetrix tiling arrays often detect binding sites for TFs in genomic regions distantly located far from genes affected by the gain- or loss of function of the corresponding TF (unpublished). These results suggest the possibility that *Arabidopsis* TFs can sometimes bind at significant distances from the genes that they regulate. Linking binding sites to distantly regulated genes poses significant technical challenges. Chromosome conformation capture (3C) (Dekker *et al.*, 2002; Dekker, 2006) provides an opportunity to explore such distal genomic interactions. After chromatin cross-linking, the solubilized DNA is digested with a convenient restriction enzyme followed by an intramolecular ligation reaction. If two distant chromosomal locations are in close physical proximity, then the ligation of the fragments will occur at a higher frequency than with a region that is distantly positioned. The abundance of the ligation products can then be evaluated by PCR using primers to the corresponding regions. A limitation of 3C is that the two regions coming in contact need to be known – this limitation is overcome in the Circular chromosome conformation capture (4C) (Zhao *et al.*, 2006) and 3C-carbon copy (5C) (Dostie *et al.*, 2006), in which the resulting ligation products can be used for a high-throughput search of distal chromosomal interaction through hybridization to tiling arrays or high-throughput sequencing. Chromosome conformation capture methods have not yet been applied to plants, but will likely provide a powerful tool for linking the activity of distant TF-binding sites to the genes that they control.

The growing availability of genomic sequences from related plant species allows the search for conserved non-coding sequences (CNSs). Many examples of CNSs have been identified in comparisons of *Arabidopsis* and other Brassicas (Koch *et al.*, 2001; Quiros *et al.*, 2001; Colinas *et al.*, 2002; Zhang *et al.*, 2006) and among the grasses (Morishige *et al.*, 2002; Guo and Moose, 2003; Inada *et al.*, 2003; Langham *et al.*, 2004). Given the relatively low conservation of general non-coding sequences it has been argued that conservation in these regions suggests selection for function as CREs, however there is relatively little data to support the actual functionality of these sequences in plant genomes. Analysis of some promoters, such as *adh1*, reveals little evidence for CNSs while the promoters of genes with regulatory roles, such as TFs, have increased numbers of CNSs (Inada *et al.*, 2003). The comparison of CNSs with functional data from EMSA, SELEX or ChIP experiments will provide a much better estimate of the utility and biological significance of CNS analyses. It is possible that many of these sequences will correspond to enhancers, as was determined for many human CNSs (Pennacchio *et al.*, 2006).

6.5 Plant transcriptional regulatory motifs, modules and networks

The integration of all the interactions linking TFs to the corresponding target genes provides the transcriptional regulatory network for that organism (e.g. (Barabasi and Oltvai, 2004) and references therein). As is the case for other biological networks, transcriptional regulatory networks are represented by nodes connected by edges, where the nodes represent the proteins and the edges correspond to direct interactions between a TF and the regulatory region of a target gene. Transcriptional regulatory networks are scale-free structures, where the number of nodes that make a large number of connections is significantly lower than the number of nodes with few connections (Bray, 2003; Barabasi and Oltvai, 2004). Nodes with a high connectivity are referred to as 'hubs', and the TFs that participate in these hubs can be viewed as 'global regulators', with a higher probability of being essential to the organism than those involved in 'fine tuning', represented in the network by a lower connectivity (Babu *et al.*, 2004; Blais and Dynlacht, 2005) and references therein). Transcriptional regulatory networks can be explored at four levels of detail. At the most basic level, the network is formed by the TFs, the CREs and their interactions. The second level consists of network motifs, which correspond to a discrete number of interconnection patterns that are repeated multiple times throughout the network (Milo *et al.*, 2002). Network motifs cluster into modules or sub-networks, which consists of semi-independent units. Connections between these modules provide the ultimate architecture to the regulatory network.

Regulatory networks are dynamic in space and time, reflecting the varying need for specific sets of genes to be deployed with particular temporal or spatial expression patterns. This creates a significant experimental challenge in multi-cellular organisms, where tissues are often formed by a plethora of cell types at different stages of differentiation. The problem is in part simplified when the network architecture of environmental responses is probed as the variable of environmental condition super-imposed on developmental networks. Indeed, significant progress has been made in identifying CREs and regulatory proteins that participate in osmotic and cold responses in *Arabidopsis*, permitting to move beyond the TF–CRE level to start building network motifs (Yamaguchi-Shinozaki and Shinozaki, 2005) and references therein). The identification of cell-specific regulatory networks involves the isolation of the specific cells to be investigated, either by using laser-capture microdissection (Nakazono *et al.*, 2003; Ohtsu *et al.*, 2006), or by driving fluorescent markers from cell-specific promoters, and then separating the cells using fluorescence-activated cell sorting. The latter provided a detailed gene expression map of the *Arabidopsis* root (Birnbaum *et al.*, 2003), which is significantly facilitating the identification of direct targets for TFs associated with root cell specification and development, and the elucidation of regulatory motifs that participate in these processes (Levesque *et al.*, 2006).

The experimental elucidation of network architecture is a time-consuming endeavour that increases in complexity with gene number and genome size. However, mathematical tools are becoming available to complement heuristic approaches. Such a combination is best exemplified by efforts aimed at elucidating the design principles that govern the circadian clock in *Arabidopsis*, and possibly in other plants as well (Rand *et al.*, 2004; Locke *et al.*, 2005, 2006).

6.6 Conclusion

The availability of complete genome sequences for several plants is permitting researchers to start decoding the hardwiring of plant regulatory sequences. Knowledge of the regulatory networks controlling gene expression levels and patterns in plants is critical to understanding how the genome specifies development and how plant genomes can respond to environmental cues and stress. Significant progress has been made towards understanding the TFs and CREs present in plant species. It is anticipated that in the coming years the catalogue of plant TFs and CREs will become much more detailed. However, we still have a relatively limited understanding of the factors that influence the interaction between TFs and CREs. Systems biology requires detailed knowledge of the interactions in a network as well as the pieces that comprise the network. There are many proteins that do not specifically bind DNA that act to affect how TFs and CREs interact by altering chromatin structure or modifying TFs. It will be important to develop technologies and approaches towards understanding how TFs and CREs interact and how this varies in different tissues or environments.

Acknowledgements

We thank Ramana Davuluri, Antje Feller and members of Yves Van de Peer's lab Eric Bonnet, Steven Maere, Tom Michoel, and Vanessa Vermeirssen for critical comments on the manuscript. We gratefully acknowledge support from the NSF (MCB-0418891 and MCB-0705415), DOE (DE-FG02–07ER15881) and USDA (NRICGP, 2006–03334) to EG, and from NSF (DBI-0421619) to NS.

References

Adhikary, S. and Eilers, M. (2005) Transcriptional regulation and transformation by Myc proteins. *Nat Rev Mol Cell Biol* **6**, 635–645.

Alexandrov, N.N., Troukhan, M.E., Brover, V.V., Tatarinova, T., Flavell, R.B. and Feldmann, K.A. (2006) Features of Arabidopsis genes and genome discovered using full-length cDNAs. *Plant Mol Biol* **60**, 69–85.

Ali, S. and Taylor, W.C. (2001) The 3′ non-coding region of a C4 photosynthesis gene increases transgene expression when combined with heterologous promoters. *Plant Mol Biol* **46**, 325–333.

Andrau, J.C., Van Oevelen, C.J., Van Teeffelen, H.A., Weil, P.A., Holstege, F.C. and Timmers, H.T. (2002) Mot1p is essential for TBP recruitment to selected promoters during *in vivo* gene activation. *EMBO J* **21**, 5173–5183.

Aravind, L., Anantharaman, V., Balaji, S., Babu, M.M. and Iyer, L.M. (2005) The many faces of the helix-turn-helix domain: transcription regulation and beyond. *FEMS Microbiol Rev* **29**, 231–262.

Arnone, M.I. and Davidson, E.H. (1997) The hardwiring of development: organization and function of genomic regulatory systems. *Development* **124**, 1851–1864.

Babu, M.M., Luscombe, N.M., Aravind, L., Gerstein, M. and Teichmann, S.A. (2004) Structure and evolution of transcriptional regulatory networks. *Curr Opin Struct Biol* **14**, 283–291.

Bailey, T.L. and Elkan, C. (1995) The value of prior knowledge in discovering motifs with MEME. *Proc Int Conf Intell Syst Mol Biol* **3**, 21–29.

Baldwin, D.A. and Gurley, W.B. (1996) Isolation and characterization of cDNAs encoding transcription factor IIB from Arabidopsis and soybean. *Plant J* **10**, 561–568.

Bao, X., Franks, R.G., Levin, J.Z. and Liu, Z. (2004) Repression of AGAMOUS by BELLRINGER in floral and inflorescence meristems. *Plant Cell* **16**, 1478–1489.

Barabasi, A.L. and Oltvai, Z.N. (2004) Network biology: understanding the cell's functional organization. *Nat Rev Genet* **5**, 101–113.

Basehoar, A.D., Zanton, S.J. and Pugh, B.F. (2004) Identification and distinct regulation of yeast TATA box-containing genes. *Cell* **116**, 699–709.

Beltran, A.C., Dawson, P.E. and Gottesfeld, J.M. (2005) Role of DNA sequence in the binding specificity of synthetic basic-helix-loop-helix domains. *Chembiochem* **6**, 104–113.

Benfey, P.N. and Chua, N.-H. (1990) The cauliflower mosaic virus 35S Promoter: combinatorial regulation of transcription in plants. *Science* **250**, 959–966.

Berk, A.J. (2000) TBP-like factors come into focus. *Cell* **103**, 5–8.

Berman, B.P., Nibu, Y., Pfeiffer, B.D., Tomancak, P., Celniker, S.E., Levine, M., *et al.* (2002) Exploiting transcription factor binding site clustering to identify cis-regulatory modules involved in pattern formation in the Drosophila genome. *Proc Natl Acad Sci USA* **99**, 757–762.

Birnbaum, K., Shasha, D.E., Wang, J.Y., Jung, J.W., Lambert, G.M., Galbraith, D.W., *et al.* (2003) A gene expression map of the Arabidopsis root. *Science* **302**, 1956–1960.

Blais, A. and Dynlacht, B.D. (2005) Constructing transcriptional regulatory networks. *Genes Dev* **19**, 1499–1511.

Blanc, G. and Wolfe, K.H. (2004) Functional divergence of duplicated genes formed by polyploidy during Arabidopsis evolution. *Plant Cell* **16**, 1679–1691.

Bodeau, J.P. and Walbot, V. (1992) Regulated transcription of the maize Bronze-2 promoter in electroporated protoplasts requires the C1 and R gene products. *Mol Gen Genet* **233**, 379–387.

Boyer, L.A., Lee, T.I., Cole, M.F., Johnstone, S.E., Levine, S.S., Zucker, J.P., *et al.* (2005) Core transcriptional regulatory circuitry in human embryonic stem cells. *Cell* **122**, 947–956.

Braun, E.L. and Grotewold, E. (1999a) Diversification of the *R2R3 Myb* gene family and the segmental allotetraploid origin of the maize genome. *Maize Genet Coop Newsl* **73**, 26–27.

Braun, E.L. and Grotewold, E. (1999b) Newly discovered plant *c-myb*-like genes rewrite the evolution of the plant *myb* gene family. *Plant Physiol* **121**, 21–24.

Bray, D. (2003) Molecular networks: the top-down view. *Science* **301**, 1864–1865.

Britten, R.J. and Davidson, E.H. (1969) Gene regulation for higher cells: a theory. *Science* **165**, 349–357.

Buchanan, B., Gruissem, W. and Jones, R. (2000) *Biochemistry and Molecular Biology of Plants* (Rockville: ASPP).

Buratowski, S., Hahn, S., Guarente, L. and Sharp, P.A. (1989) Five intermediate complexes in transcription initiation by RNA polymerase II. *Cell* **56**, 549–561.

Butler, J.E. and Kadonaga, J.T. (2002) The RNA polymerase II core promoter: a key component in the regulation of gene expression. *Genes Dev* **16**, 2583–2592.

Cao, X., Costa, L.M., Biderre-Petit, C., Kbhaya, B., Dey, N., Perez, P., *et al.* (2007) Abscisic Acid and stress signals induce viviparous1 expression in seed and vegetative tissues of maize. *Plant Physiol* **143**, 720–731.

Cardon, L.R. and Stormo, G.D. (1992) Expectation maximization algorithm for identifying protein-binding sites with variable lengths from unaligned DNA fragments. *J Mol Biol* **223**, 159–170.

Carninci, P., Sandelin, A., Lenhard, B., Katayama, S., Shimokawa, K., Ponjavic, J., *et al.* (2006) Genome-wide analysis of mammalian promoter architecture and evolution. *Nat Genet* **38**, 626–635.

Cheng, W.H., Taliercio, E.W. and Chourey, P.S. (1999) Sugars modulate an unusual mode of control of the cell-wall invertase gene (Incw1) through its 3' untranslated region in a cell suspension culture of maize. *Proc Natl Acad Sci USA* **96**, 10512–10517.

Clancy, M. and Hannah, L.C. (2002) Splicing of the maize Sh1 first intron is essential for enhancement of gene expression, and a T-rich motif increases expression without affecting splicing. *Plant Physiol* **130**, 918–929.

Clark, R.M., Wagler, T.N., Quijada, P. and Doebley, J. (2006) A distant upstream enhancer at the maize domestication gene tb1 has pleiotropic effects on plant and inflorescent architecture. *Nat Genet* **38**, 594–597.

Colinas, J., Birnbaum, K. and Benfey, P.N. (2002) Using cauliflower to find conserved non-coding regions in Arabidopsis. *Plant Physiol* **129**, 451–454.

Davidson, E.H. (2001) *Genomic Regulatory Systems: Development and Evolution* (San Diego: Academic Press).

Davuluri, R.V., Sun, H., Palaniswamy, S.K., Matthews, N., Molina, C., Kurtz, M., *et al.* (2003) AGRIS: Arabidopsis gene regulatory information server, an information resource of Arabidopsis cis-regulatory elements and transcription factors. *BMC Bioinformatics* **4**, 25.

De Bodt, S., Maere, S. and Van de Peer, Y. (2005) Genome duplication and the origin of angiosperms. *Trends Ecol Evol* **20**, 591–597.

Decary, S., Decesse, J.T., Ogryzko, V., Reed, J.C., Naguibneva, I., Harel-Bellan, A., *et al.* (2002) The retinoblastoma protein binds the promoter of the survival gene bcl-2 and regulates its transcription in epithelial cells through transcription factor AP-2. *Mol Cell Biol* **22**, 7877–7888.

Dekker, J. (2006) The three 'C's of chromosome conformation capture: controls, controls, controls. *Nat Methods* **3**, 17–21.

Dekker, J., Rippe, K., Dekker, M. and Kleckner, N. (2002) Capturing chromosome conformation. *Science* **295**, 1306–1311.

Dias, A.P., Braun, E.L., McMullen, M.D. and Grotewold, E. (2003) Recently duplicated maize *R2R3 Myb* genes provide evidence for distinct mechanisms of evolutionary divergence after duplication. *Plant Physiol* **131**, 610–620.

Djordjevic, M. and Sengupta, A.M. (2006) Quantitative modeling and data analysis of SELEX experiments. *Phys Biol* **3**, 13–28.

Djordjevic, M., Sengupta, A.M. and Shraiman, B.I. (2003) A biophysical approach to transcription factor binding site discovery. *Genome Res* **13**, 2381–2390.

Dostie, J., Richmond, T.A., Arnaout, R.A., Selzer, R.R., Lee, W.L., Honan, T.A., *et al.* (2006) Chromosome conformation capture carbon copy (5C): a massively parallel solution for mapping interactions between genomic elements. *Genome Res* **16**, 1299–1309.

Dynan, W.S. (1989) Modularity in promoters and enhancers. *Cell* **58**, 1–4.

Ehlert, A., Weltmeier, F., Wang, X., Mayer, C.S., Smeekens, S., Vicente-Carbajosa, J., *et al.* (2006) Two-hybrid protein-protein interaction analysis in Arabidopsis protoplasts: establishment of a heterodimerization map of group C and group S bZIP transcription factors. *Plant J* **46**, 890–900.

Eulgem, T., Rushton, P.J., Robatzek, S. and Somssich, I.E. (2000) The WRKY superfamily of plant transcription factors. *Trends Plant Sci* **5**, 199–206.

Fernandez, P.C., Frank, S.R., Wang, L., Schroeder, M., Liu, S., Greene, J., *et al.* (2003) Genomic targets of the human c-Myc protein. *Genes Dev* **17**, 1115–1129.

Florquin, K., Saeys, Y., Degroeve, S., Rouze, P. and Van de Peer, Y. (2005) Large-scale structural analysis of the core promoter in mammalian and plant genomes. *Nucleic Acids Res* **33**, 4255–4264.

Fujimori, S., Washio, T., Higo, K., Ohtomo, Y., Murakami, K., Matsubara, K., *et al.* (2003) A novel feature of microsatellites in plants: a distribution gradient along the direction of transcription. *FEBS Lett* **554**, 17–22.

Gao, Y., Li, J., Strickland, E., Hua, S., Zhao, H., Chen, Z., *et al.* (2004) An *Arabidopsis* promoter microarray and its initial usage in the identification of HY5 binding targets *in vitro*. *Plant Mol Biol* **54**, 683–699.

Garvie, C.W., Hagman, J. and Wolberger, C. (2001) Structural studies of Ets-1/Pax5 complex formation on DNA. *Mol Cell* **8**, 1267–1276.

Gidekel, M., Jimenez, B. and Herrera-Estrella, L. (1996) The first intron of the Arabidopsis thaliana gene coding for elongation factor 1 beta contains an enhancer-like element. *Gene* **170**, 201–206.

Goff, S.A., Cone, K.C. and Chandler, V.L. (1992) Functional analysis of the transcriptional activator encoded by the maize *B* gene: evidence for a direct functional interaction between two classes of regulatory proteins. *Genes Dev* **6**, 864–875.

Goff, S.A., Cone, K.C. and Fromm, M.E. (1991) Identification of functional domains in the maize transcriptional activator C1: comparison of wild-type and dominant inhibitor proteins. *Genes Dev* **5**, 298–309.

Goff, S.A., Klein, T.M., Roth, B.A., Fromm, M.E., Cone, K.C., Radicella, J.P., *et al.* (1990) Transactivation of anthocyanin biosynthetic genes following transfer of *B* regulatory genes into maize tissues. *EMBO J* **9**, 2517–2522.

Gold, L., Brown, D., He, Y., Shtatland, T., Singer, B.S. and Wu, Y. (1997) From oligonucleotide shapes to genomic SELEX: novel biological regulatory loops. *Proc Natl Acad Sci USA* **94**, 59–64.

Grace, M.L., Chandrasekharan, M.B., Hall, T.C. and Crowe, A.J. (2004) Sequence and spacing of TATA box elements are critical for accurate initiation from the beta-phaseolin promoter. *J Biol Chem* **279**, 8102–8110.

Greene, B., Walko, R. and Hake, S. (1994) Mutator insertions in an intron of the maize knotted 1 gene result in dominant suppressible mutations. *Genetics* **138**, 1275–1285.

Grotewold, E. (2005) Plant metabolic diversity: a regulatory perspective. *Trends Plant Sci* **10**, 57–62.

Grotewold, E. (2006) The genetics and biochemistry of floral pigments. *Annu Rev Plant Biol* **57**, 761–780.

Grotewold, E., Drummond, B.J., Bowen, B. and Peterson, T. (1994) The myb-homologous P gene controls phlobaphene pigmentation in maize floral organs by directly activating a flavonoid biosynthetic gene subset. *Cell* **76**, 543–553.

Guo, A., He, K., Liu, D., Bai, S., Gu, X., Wei, L., *et al.* (2005) DATF: a database of Arabidopsis transcription factors. *Bioinformatics* **21**, 2568–2569.

Guo, H. and Moose, S.P. (2003) Conserved noncoding sequences among cultivated cereal genomes identify candidate regulatory sequence elements and patterns of promoter evolution. *Plant Cell* **15**, 1143–1158.

Haberer, G., Mader, M.T., Kosarev, P., Spannagl, M., Yang, L. and Mayer, K.F. (2006) Large-scale cis-element detection by analysis of correlated expression and sequence conservation between Arabidopsis and Brassica oleracea. *Plant Physiol* **142**, 1589–1602.

Haerizadeh, F., Singh, M.B. and Bhalla, P.L. (2006) Transcriptional repression distinguishes somatic from germ cell lineages in a plant. *Science* **313**, 496–499.

Haga, N., Kato, K., Murase, M., Araki, S., Kubo, M., Demura, T., *et al.* (2007) R1R2R3-Myb proteins positively regulate cytokinesis through activation of KNOLLE transcription in Arabidopsis thaliana. *Development* **134**, 1101–1110.

Hake, S., Vollbrecht, E. and Freeling, M. (1989) Cloning knotted, the dominant morphological mutant in maize using Ds2 as a transposon tag. *EMBO J* **8**, 15–22.

Haralampidis, K., Milioni, D., Rigas, S. and Hatzopoulos, P. (2002) Combinatorial interaction of cis elements specifies the expression of the Arabidopsis AtHsp90-1 gene. *Plant Physiol* **129**, 1138–1149.

Hartmann, U., Sagasser, M., Mehrtens, F., Stracke, R. and Weisshaar, B. (2005) Differential combinatorial interactions of cis-acting elements recognized by R2R3-MYB, BZIP, and BHLH factors control light-responsive and tissue-specific activation of phenylpropanoid biosynthesis genes. *Plant Mol Biol* **57**, 155–171.

Haverty, P.M., Hansen, U. and Weng, Z. (2004) Computational inference of transcriptional regulatory networks from expression profiling and transcription factor binding site identification. *Nucleic Acids Res* **32**, 179–188.

Heine, G.F., Malik, V., Dias, A.P. and Grotewold, E. (2007) Expression and molecular characterization of ZmMYB-IF35 and related R2R3-MYB transcription factors. *Mol Biotechnol* **37**, 155–164.

Hernandez, J., Heine, G., Irani, N.G., Feller, A., Kim, M.-G., Matulnik, T., *et al.* (2004)

Different mechanisms participate in the R-dependent activity of the R2R3 MYB transcription factor C1. *J Biol Chem* **279**, 48205–48213.

Herr, A.J., Jensen, M.B., Dalmay, T. and Baulcombe, D.C. (2005) RNA polymerase IV directs silencing of endogenous DNA. *Science* **308**, 118–120.

Hertz, G.Z. and Stormo, G.D. (1999) Identifying DNA and protein patterns with statistically significant alignments of multiple sequences. *Bioinformatics* **15**, 563–577.

Higo, K., Ugawa, Y., Iwamoto, M. and Higo, H. (1998) PLACE: a database of plant cis-acting regulatory DNA elements. *Nucleic Acids Res* **26**, 358–359.

Higo, K., Ugawa, Y., Iwamoto, M. and Korenaga, T. (1999) Plant cis-acting regulatory DNA elements (PLACE) database: 1999. *Nucleic Acids Res* **27**, 297–300.

Hill, T.A., Day, C.D., Zondlo, S.C., Thackeray, A.G. and Irish, V.F. (1998) Discrete spatial and temporal cis-acting elements regulate transcription of the Arabidopsis floral homeotic gene APETALA3. *Development* **125**, 1711–1721.

Hiratsu, K., Matsui, K., Koyama, T. and Ohme-Takagi, M. (2003) Dominant repression of target genes by chimeric repressors that include the EAR motif, a repression domain, in *Arabidopsis*. *Plant J* **34**, 733–739.

Holt, K.E., Millar, A.H. and Whelan, J. (2006) ModuleFinder and CoReg: alternative tools for linking gene expression modules with promoter sequences motifs to uncover gene regulation mechanisms in plants. *Plant Methods* **2**, 8.

Huang, H., Mizukami, Y., Hu, Y. and Ma, H. (1993) Isolation and characterization of the binding sequence for the product of the Arabidopsis floral homeotic gene AGAMOUS. *Nucleic Acids Res* **21**, 4769–4776.

Huang, H., Tudor, M., Su, T., Zhang, Y., Hu, Y. and Ma, H. (1996) DNA binding properties of two arabidopsis MADS domain proteins: binding consensus and dimer formation. *Plant Cell* **8**, 81–94.

Inada, D.C., Bashir, A., Lee, C., Thomas, B.C., Ko, C., Goff, S.A., et al. (2003) Conserved noncoding sequences in the grasses. *Genome Res* **13**, 2030–2041.

Ito, M. (2005) Conservation and diversification of three-repeat Myb transcription factors in plants. *J Plant Res* **118**, 61–69.

Ito, M., Araki, S., Matsunaga, S., Itoh, T., Nishihama, R., Machida, Y., et al. (2001) G2/M-phase-specific transcription during the plant cell cycle is mediated by c-Myb-like transcription factors. *Plant Cell* **13**, 1891–1905.

Itzkovitz, S., Tlusty, T. and Alon, U. (2006) Coding limits on the number of transcription factors. *BMC Genomics* **7**, 239.

Jeong, Y.M., Mun, J.H., Lee, I., Woo, J.C. Hong, C.B. and Kim, S.G. (2006) Distinct roles of the first introns on the expression of Arabidopsis profilin gene family members. *Plant Physiol* **140**, 196–209.

Jia, J., Fu, J., Zheng, J., Zhou, X., Huai, J., Wang, J., et al. (2006) Annotation and expression profile analysis of 2073 full-length cDNAs from stress-induced maize (Zea mays L.) seedlings. *Plant J* **48**, 710–727.

Jiang, C., Gu, J., Chopra, S., Gu, X. and Peterson, T. (2004) Ordered origin of the typical two- and three-repeat Myb genes. *Gene* **326**, 13–22.

Jin, H. and Martin, C. (1999) Multifunctionality and diversity within the plant *MYB*-gene family. *Plant Mol Biol* **41**, 577–585.

Kanei-Ishii, C., Sarai, A., Sawazaki, T., Nakagoshi, H., He, D.-N., Ogata, K., et al. (1990) The tryptophan cluster: a hypothetical structure of the DNA-binding domain of the myb protooncogene product. *J Biol Chem* **265**, 19990–19995.

Kazan, K. (2006) Negative regulation of defence and stress genes by EAR-motif-containing repressors. *Trends Plant Sci* **11**, 109–112.

Keegan, L., Gill, G. and Ptashne, M. (1986) Separation of DNA binding from the transcription-activating function of a eukaryotic regulatory protein. *Science* **231**, 699–704.

Kiran, K., Ansari, S.A., Srivastava, R., Lodhi, N., Chaturvedi, C.P., Sawant, S.V., *et al.* (2006) The TATA-box sequence in the basal promoter contributes to determining light-dependent gene expression in plants. *Plant Physiol* **142**, 364–376.

Klempnauer, K.-H., Gonda, T.J. and Bishop, J.M. (1982) Nucleotide sequence of the retroviral leukemia gene v-myb and its cellular progenitor c-myb: the architecture of a transduced oncogene. *Cell* **31**, 453–463.

Klempnauer, K.-H., Ramsay, G., Bishop, J.M., Moscovici, M.G., Moscovici, C., Mc-Grath, J.P., *et al.* (1983) The product of the retroviral transforming gene v-myb is a truncated version of the protein encoded by the cellular oncogene c-myb. *Cell* **33**, 345–355.

Koch, M.A., Weisshaar, B., Kroymann, J., Haubold, B. and Mitchell-Olds, T. (2001) Comparative genomics and regulatory evolution: conservation and function of the Chs and Apetala3 promoters. *Mol Biol Evol* **18**, 1882–1891.

Kranz, H., Scholz, K. and Weisshaar, B. (2000) c-MYB oncogene-like genes encoding three MYB repeats occur in all major plant lineages. *Plant J* **21**, 231–235.

Kuhlemeier, C., Cuozzo, M., Green, P.J., Goyvaerts, E., Ward, K. and Chua, N.H. (1988) Localization and conditional redundancy of regulatory elements in rbcS-3A, a pea gene encoding the small subunit of ribulose-bisphosphate carboxylase. *Proc Natl Acad Sci USA* **85**, 4662–4666.

Lago, C., Clerici, E., Dreni, L., Horlow, C., Caporali, E., Colombo, L., *et al.* (2005) The Arabidopsis TFIID factor AtTAF6 controls pollen tube growth. *Dev Biol* **285**, 91–100.

Lagrange, T., Hakimi, M.A., Pontier, D., Courtois, F., Alcaraz, J.P., Grunwald, D., *et al.* (2003) Transcription factor IIB (TFIIB)-related protein (pBrp), a plant-specific member of the TFIIB-related protein family. *Mol Cell Biol* **23**, 3274–3286.

Lagrange, T., Kapanidis, A.N., Tang, H., Reinberg, D. and Ebright, R.H. (1998) New core promoter element in RNA polymerase II-dependent transcription: sequence-specific DNA binding by transcription factor IIB. *Genes Dev* **12**, 34–44.

Lam, E., Benfey, P.N., Gilmartin, P.M., Fang, R.X. and Chua, N.H. (1989) Site-specific mutations alter *in vitro* factor binding and change promoter expression pattern in transgenic plants. *Proc Natl Acad Sci USA* **86**, 7890–7894.

Lamb, R.S., Hill, T.A., Tan, Q.K. and Irish, V.F. (2002) Regulation of APETALA3 floral homeotic gene expression by meristem identity genes. *Development* **129**, 2079–2086.

Langham, R.J., Walsh, J., Dunn, M., Ko, C., Goff, S.A. and Freeling, M. (2004) Genomic duplication, fractionation and the origin of regulatory novelty. *Genetics* **166**, 935–945.

Lawit, S.J., O'Grady, K., Gurley, W.B. and Czarnecka-Verner, E. (2007) Yeast two-hybrid map of Arabidopsis TFIID. *Plant Mol Biol* **64**, 73–87.

Lee, J., He, K., Stolc, V., Lee, H., Figueroa, P., Gao, Y., *et al.* (2007) Analysis of transcription factor HY5 genomic binding sites revealed its hierarchical role in light regulation of development. *Plant Cell* **19**, 731–749.

Lee, T.I., Rinaldi, N.J., Robert, F., Odom, D.T., Bar-Joseph, Z., Gerber, G.K., *et al.* (2002) Transcriptional regulatory networks in Saccharomyces cerevisiae. *Science* **298**, 799–804.

Lescot, M., Dehais, P., Thijs, G., Marchal, K., Moreau, Y., Van de Peer, Y., *et al.* (2002) PlantCARE, a database of plant cis-acting regulatory elements and a portal to tools for in silico analysis of promoter sequences. *Nucleic Acids Res* **30**, 325–327.

Lesnick, M.L. and Chandler, V.L. (1998) Activation of the maize anthocyanin gene a2 is mediated by an element conserved in many anthocyanin promoters. *Plant Physiol* **117**, 437–445.

Lesnick, M.L. (1997) *Analysis of the cis-Acting Sequences Required for C1/B Activation of the Maize Anthocyanin Biosynthetic Pathway.* PhD Thesis, University of Oregon, Eugene, OR, pp. 32–53.

Levesque, M.P., Vernoux, T., Busch, W., Cui, H., Wang, J.Y., Blilou, I., *et al.* (2006) Whole-genome analysis of the SHORT-ROOT developmental pathway in Arabidopsis. *PLoS Biol* **4**, e143.

Levine, H.A. and Nilsen-Hamilton, M. (2007) A mathematical analysis of SELEX. *Comput Biol Chem* **31**, 11–35.

Lipsick, J.S. (1996) One billion years of Myb. *Oncogene* **13**, 223–235.

Locke, J.C., Kozma-Bognar, L., Gould, P.D., Feher, B., Kevei, E., Nagy, F., *et al.* (2006) Experimental validation of a predicted feedback loop in the multi-oscillator clock of Arabidopsis thaliana. *Mol Syst Biol* **2**, 59.

Locke, J.C., Southern, M.M., Kozma-Bognar, L., Hibberd, V., Brown, P.E., Turner, M.S., *et al.* (2005) Extension of a genetic network model by iterative experimentation and mathematical analysis. *Mol Syst Biol* **1**, 0013.

Luscombe, N.M., Austin, S.E., Berman, H.M. and Thornton, J.M. (2000) An overview of the structures of protein-DNA complexes. *Genome Biol* **1**, REVIEWS001.

Maere, S., De Bodt, S., Raes, J., Casneuf, T., Van Montagu, M., Kuiper, M., *et al.* (2005) Modeling gene and genome duplications in eukaryotes. *Proc Natl Acad Sci USA* **102**, 5454–5459.

Maniatis, T., Falvo, J.V., Kim, T.H., Kim, T.K., Lin, C.H., Parekh, B.S., *et al.* (1998) Structure and function of the interferon-beta enhanceosome. *Cold Spring Harb Symp Quant Biol* **63**, 609–620.

Maniatis, T. and Reed, R. (2002) An extensive network of coupling among gene expression machines. *Nature* **416**, 499–506.

Martin, C. and Paz-Ares, J. (1997) MYB transcription factors in plants. *Trends Genet* **13**, 67–73.

Massari, M.E. and Murre, C. (2000) Helix-loop-helix proteins: regulators of transcription in eucaryotic organisms. *Mol Cell Biol* **20**, 429–440.

Matys, V., Kel-Margoulis, O.V., Fricke, E., Liebich, I., Land, S., Barre-Dirrie, A., *et al.* (2006) TRANSFAC and its module TRANSCompel: transcriptional gene regulation in eukaryotes. *Nucleic Acids Res* **34**, D108–D110.

Menossi, M., Rabaneda, F., Puigdomenech, P. and Martinez-Izquierdo, J.A. (2003) Analysis of regulatory elements of the promoter and the 3′ untranslated region of the maize Hrgp gene coding for a cell wall protein. *Plant Cell Rep* **21**, 916–923.

Merika, M. and Thanos, D. (2001) Enhanceosomes. *Curr Opin Genet Dev* **11**, 205–208.

Milo, R., Shen-Orr, S., Itzkovitz, S., Kashtan, N., Chklovskii, D. and Alon, U. (2002) Network motifs: simple building blocks of complex networks. *Science* **298**, 824–827.

Molina, C. and Grotewold, E. (2005) Genome wide analysis of Arabidopsis core promoters. *BMC Genomics* **6**, 25.

Morishige, D.T., Childs, K.L., Moore, L.D. and Mullet, J.E. (2002) Targeted analysis of orthologous phytochrome A regions of the sorghum, maize, and rice genomes using comparative gene-island sequencing. *Plant Physiol* **130**, 1614–1625.

Morohashi, K., Zhao, M., Yang, M., Read, B., Lloyd, A., Lamb, R., *et al.* (2007) Participation of the *Arabidopsis* bHLH factor GL3 in trichome initiation regulatory events. *Plant Physiol* **145**, 736–746.

Murre, C., McCaw, P.S. and Baltimore, D. (1989) A new DNA binding and dimerization motif in immunoglobulin enhancer binding, daughterless, MyoD, and myc proteins. *Cell* **56**, 777–783.

Nakamura, M., Tsunoda, T. and Obokata, J. (2002) Photosynthesis nuclear genes generally lack TATA-boxes: a tobacco photosystem I gene responds to light through an initiator. *Plant J* **29**, 1–10.

Nakazono, M., Qiu, F., Borsuk, L.A. and Schnable, P.S. (2003) Laser-capture microdissection, a tool for the global analysis of gene expression in specific plant cell types: identification of genes expressed differentially in epidermal cells or vascular tissues of maize. *Plant Cell* **15**, 583–596.

Narusaka, Y., Nakashima, K., Shinwari, Z.K., Sakuma, Y., Furihata, T., Abe, H., *et al.* (2003) Interaction between two cis-acting elements, ABRE and DRE, in ABA-dependent expression of Arabidopsis rd29A gene in response to dehydration and high-salinity stresses. *Plant J* **34**, 137–148.

Ogata, K., Hojo, H., Aimoto, S., Nakai, T., Nakamura, H., Sarai, A., *et al.* (1992) Solution structure of a DNA-binding unit of Myb: a helix-turn-helix-related motif with conserved tryptophans forming a hydrophobic core. *Proc Natl Acad Sci USA* **89**, 6428–6432.

Ogata, K., Morikawa, S., Nakamura, H., Sekikawa, A., Inoue, T., Kanai., H., *et al.* (1994) Solution structure of a specific DNA complex of the Myb DNA-binding domain with cooperative recognition helices. *Cell* **79**, 639–648.

Ohler, U., Liao, G., Niemann, H. and Rubin, G.M. (2002) Computational analysis of core promoters in the *Drosophila* genome. *Genome Biol* **3**, 0087.1–0087.12.

Ohler, U. and Niemann, H. (2001) Identification and analysis of eukaryotic promoters: recent computational approaches. *Trends Genet* **17**, 56–60.

Ohtsu, K., Takahashi, H., Schnable, P.S. and Nakazono, M. (2006) Cell type-specific gene expression profiling in plants by using a combination of laser microdissection and high-throughput technologies. *Plant Cell Physiol* **48**, 3–7.

Onodera, Y., Haag, J.R., Ream, T., Nunes, P.C., Pontes, O. and Pikaard, C.S. (2005) Plant nuclear RNA polymerase IV mediates siRNA and DNA methylation-dependent heterochromatin formation. *Cell* **120**, 613–622.

Pabo, C.O. and Sauer, R.T. (1992) Transcription factors: structural families and principles of DNA recognition. *Annu Rev Biochem* **61**, 1053–1095.

Palaniswamy, K., James, S., Sun, H., Lamb, R., Davuluri, R.V. and Grotewold, E. (2006) AGRIS and AtRegNet: a platform to link cis-regulatory elements and transcription factors into regulatory networks. *Plant Phyisiol* **140**, 818–829.

Pan, S., Czarnecka-Verner, E. and Gurley, W.B. (2000) Role of the TATA binding protein-transcription factor IIB interaction in supporting basal and activated transcription in plant cells. *Plant Cell* **12**, 125–135.

Pennacchio, L.A., Ahituv, N., Moses, A.M., Prabhakar, S., Nobrega, M.A., Shoukry, M., *et al.* (2006) *in vivo* enhancer analysis of human conserved non-coding sequences. *Nature* **444**, 499–502.

Pooma, W., Gersos, C. and Grotewold, E. (2002) Transposon insertions in the promoter of the *Zea mays a1* gene differentially affect transcription by the Myb factors P and C1. *Genetics* **161**, 793–801.

Quattrocchio, F., Baudry, A., Lepiniec, L. and Grotewold, E. (2006) The regulation of flavonoid biosynthesis. In *The Science of Flavonoids*, E. Grotewold, ed. (New York, NY: Springer), pp. 97–122.

Quiros, C.F., Grellet, F., Sadowski, J., Suzuki, T., Li, G. and Wroblewski, T. (2001)

Arabidopsis and Brassica comparative genomics: sequence, structure and gene content in the ABI-Rps2-Ck1 chromosomal segment and related regions. *Genetics* **157**, 1321–1330.

Rabinowicz, P.D., Braun, E.L., Wolfe, A.D., Bowen, B. and Grotewold, E. (1999) Maize *R2R3 Myb* genes: sequence analysis reveals amplification in higher plants. *Genetics* **153**, 427–444.

Rand, D.A., Shulgin, B.V., Salazar, D. and Millar, A.J. (2004) Design principles underlying circadian clocks. *J R Soc Interface* **1**, 119–130.

Riano-Pachon, D.M., Ruzicic, S., Dreyer, I. and Mueller-Roeber, B. (2007) PlnTFDB: an integrative plant transcription factor database. *BMC Bioinformatics* **8**, 42.

Riechmann, J.L., Heard, J., Martin, G., Reuber, L., Jiang, C., Keddie, J., *et al.* (2000) Arabidopsis transcription factors: genome-wide comparative analysis among eukaryotes. *Science* **290**, 2105–2110.

Riechmann, J.L. and Ratcliffe, O.J. (2000) A genomic perspective on plant transcription factors. *Curr Opin Plant Biol* **3**, 423–434.

Roberts, S.G. (2000) Mechanisms of action of transcription activation and repression domains. *Cell Mol Life Sci* **57**, 1149–1160.

Rombauts, S., Florquin, K., Lescot, M., Marchal, K., Rouze, P. and Van de Peer, Y. (2003) Computational approaches to identify promoters and cis-regulatory elements in plant genomes. *Plant Physiol* **132**, 1162–1176.

Romero, I., Fuertes, A., Benito, M.J., Malpica, J.M., Leyva, A. and Paz-Ares, J. (1998) More than 80 *R2R3-MYB* regulatory genes in the genome of *Arabidopsis thaliana*. *Plant J* **14**, 273–284.

Rose, A.B. (2004) The effect of intron location on intron-mediated enhancement of gene expression in *Arabidopsis*. *Plant J* **40**, 744–751.

Rose, A.B. and Beliakoff, J.A. (2000) Intron-mediated enhancement of gene expression independent of unique intron sequences and splicing. *Plant Physiol* **122**, 535–542.

Roth, F.P., Hughes, J.D., Estep, P.W. and Church, G.M. (1998) Finding DNA regulatory motifs within unaligned noncoding sequences clustered by whole-genome mRNA quantitation. *Nat Biotechnol* **16**, 939–945.

Rubio, V., Shen, Y., Saijo, Y., Liu, Y., Gusmaroli, G., Dinesh-Kumar, S.P., *et al.* (2005) An alternative tandem affinity purification strategy applied to Arabidopsis protein complex isolation. *Plant J* **41**, 767–778.

Sablowski, R.W.M. and Meyerowitz, E.M. (1998) A Homolog of *NO APICAL MERISTEM* is an immediate target of the floral homeotic genes APETALA3/PISTILLATA. *Cell* **92**, 93–103.

Sainz, M.B., Goff, S.A. and Chandler, V.L. (1997a) Extensive mutagenesis of a transcriptional activation domain identifies single hydrophobic and acidic amino acids important for activation *in vivo*. *Mol Cell Biol* **17**, 115–122.

Sainz, M.B., Grotewold, E. and Chandler, V.L. (1997b) Evidence for direct activation of an anthocyanin promoter by the maize C1 protein and comparison of DNA binding by related Myb domain proteins. *Plant Cell* **9**, 611–625.

Seki, M., Narusaka, M., Kamiya, A., Ishida, J., Satou, M., Sakurai, T., *et al.* (2002) Functional annotation of a full-length *Arabidopsis* cDNA collection. *Science* **296**, 141–145.

Seki, M., Narusaka, M., Yamaguchi-Shinozaki, K., Carninci, P., Kawai, J., Hayashizaki, Y., *et al.* (2001) *Arabidopsis* encyclopedia using full-length cDNAs and its application. *Plant Physiol Biochem* **39**, 211–220.

Seoighe, C. and Gehring, C. (2004) Genome duplication led to highly selective expansion of the Arabidopsis thaliana proteome. *Trends Genet* **20**, 461–464.

Serfling, E., Jasin, M. and Schaffner, W. (1985) Enhancers and eukaryotic gene-transcription. *Trends Genet* **1**, 224–230.

Shi, W. and Zhou, W. (2006) Frequency distribution of TATA box and extension sequences on human promoters. *BMC Bioinformatics* **7**, S2.

Shin, B., Choi, G., Yi, H., Yang, S., Cho, I., Kim, J., et al. (2002) AtMYB21, a gene encoding a flower-specific transcription factor, is regulated by COP1. *Plant J* **30**, 23–32.

Shiu, S.H., Shih, M.C. and Li, W.H. (2005) Transcription factor families have much higher expansion rates in plants than in animals. *Plant Physiol* **139**, 18–26.

Sidorenko, L.V., Li, X., Cocciolone, S.M., Chopra, S., Tagliani, L., Bowen, B., et al. (2000) Complex structure of a maize Myb gene promoter: functional analysis in transgenic plants. *Plant J* **22**, 471–482.

Smale, S.T. and Kadonaga, J.T. (2003) The RNA polymerase II core promoter. *Annu Rev Biochem* **72**, 449–479.

Spelt, C., Quattrocchio, F., Mol, J.N. and Koes, R. (2000) Anthocyanin1 of petunia encodes a basic helix-loop-helix protein that directly activates transcription of structural anthocyanin genes. *Plant Cell* **12**, 1619–1632.

Springer, N.M. and Kaeppler, S.M. (2005) Evolutionary divergence of monocot and dicot methyl-CpG-binding domain proteins. *Plant Physiol* **138**, 92–104.

Springer, N.M., Napoli, C.A., Selinger, D.A., Pandey, R., Cone, K.C., Chandler, V.L., et al. (2003) Comparative analysis of SET domain proteins in maize and Arabidopsis reveals multiple duplications preceding the divergence of monocots and dicots. *Plant Physiol* **132**, 907–925.

Stam, M., Belele, C., Dorweiler, J.E. and Chandler, V.L. (2002a) Differential chromatin structure within a tandem array 100 kb upstream of the maize b1 locus is associated with paramutation. *Genes Dev* **16**, 1906–1918.

Stam, M., Belele, C., Ramakrishna, W., Dorweiler, J.E., Bennetzen, J.L. and Chandler, V.L. (2002b) The regulatory regions required for B' paramutation and expression are located far upstream of the maize b1 transcribed sequences. *Genetics* **162**, 917–930.

Stormo, G.D. (1990) Consensus patterns in DNA. *Methods Enzymol* **183**, 211–221.

Stormo, G.D. (2000) DNA binding sites: representation and discovery. *Bioinformatics* **16**, 16–23.

Stormo, G.D. and Hartzell, G.W., III. (1989) Identifying protein-binding sites from unaligned DNA fragments. *Proc Natl Acad Sci USA* **86**, 1183–1187.

Stracke, R., Werber, M. and Weisshaar, B. (2001) The R2R3 MYB gene family in *Arabidopsis thaliana*. *Curr Opin Plant Biol* **4**, 447–456.

Thibaud-Nissen, F., Wu, H., Richmond, T., Redman, J.C., Johnson, C., Green, R., et al. (2006) Development of Arabidopsis whole-genome microarrays and their application to the discovery of binding sites for the TGA2 transcription factor in salicylic acid-treated plants. *Plant J* **47**, 152–162.

Tirosh, I., Weinberger, A., Carmi, M. and Barkai, N. (2006) A genetic signature of interspecies variations in gene expression. *Nat Genet* **38**, 830–834.

Tompa, M., Li, N., Bailey, T.L., Church, G.M., De Moor, B., Eskin, E., et al. (2005) Assessing computational tools for the discovery of transcription factor binding sites. *Nat Biotechnol* **23**, 137–144.

Tuerk, C. and Gold, L. (1990) Systematic evolution of ligands by exponential enrichment: RNA ligands to bacteriophage T4 DNA polymerase. *Science* **249**, 505–510.

Ulyanov, A.V. and Stormo, G.D. (1995) Multi-alphabet consensus algorithm for identification of low specificity protein-DNA interactions. *Nucleic Acids Res* **23**, 1434–1440.

Vandepoele, K., Casneuf, T. and Van de Peer, Y. (2006) Identification of novel regulatory modules in dicot plants using expression data and comparative genomics. *Genome Biol* **7**, R103.

Vogel, J.M., Roth, B., Cigan, M. and Freeling, M. (1993) Expression of the two maize TATA binding protein genes and function of the encoded TBP proteins by complementation in yeast. *Plant Cell* **5**, 1627–1638.

Wang, D., Amornsiripanitch, N. and Dong, X. (2006) A genomic approach to identify regulatory nodes in the transcriptional network of systemic acquired resistance in plants. *PLoS Pathog* **2**, e123.

Wang, T. and Stormo, G.D. (2003) Combining phylogenetic data with co-regulated genes to identify regulatory motifs. *Bioinformatics* **19**, 2369–2380.

Wasserman, W.W. and Sandelin, A. (2004) Applied bioinformatics for the identification of regulatory elements. *Nat Rev Genet* **5**, 276–287.

Wellmer, F., Alves-Ferreira, M., Dubois, A., Riechmann, J.L. and Meyerowitz, E.M. (2006) Genome-wide analysis of gene expression during early Arabidopsis flower development. *PLoS Genet* **2**, e117.

West, A.G. and Fraser, P. (2005) Remote control of gene transcription. *Hum Mol Genet* **14** (Spec No 1), R101–R111.

Weston, K. (1998) Myb proteins in life, death and differentiation. *Curr Opin Gen Dev* **8**, 76–81.

William, D.A., Su, Y., Smith, M.R., Lu, M., Baldwin, D.A. and Wagner, D. (2004) Genomic identification of direct target genes of LEAFY. *Proc Natl Acad Sci USA* **101**, 1775–1780.

Williams, C.E. and Grotewold, E. (1997) Differences between plant and animal Myb domains are fundamental for DNA-binding, and chimeric Myb domains have novel DNA-binding specificities. *J Biol Chem* **272**, 563–571.

Wray, G.A., Hahn, M.W., Abouheif, E., Balhoff, J.P., Pizer, M., Rockman, M.V., *et al.* (2003) The evolution of transcriptional regulation in eukaryotes. *Mol Biol Evol* **20**, 1377–1419.

Yamaguchi-Shinozaki, K. and Shinozaki, K. (2005) Organization of cis-acting regulatory elements in osmotic- and cold-stress-responsive promoters. *Trends Plant Sci* **10**, 88–94.

Yuh, C.-H., Bolouri, H. and Davidson, E.H. (1998) Genomic cis-regulatory logic: experimental and computational analysis of a sea urchin gene. *Science* **279**, 1896–1902.

Zhang, L., Zuo, K., Zhang, F., Cao, Y., Wang, J., Zhang, Y., *et al.* (2006) Conservation of noncoding microsatellites in plants: implication for gene regulation. *BMC Genomics* **7**, 323.

Zhao, Z., Tavoosidana, G., Sjolinder, M., Gondor, A., Mariano, P., Wang, S., *et al.* (2006) Circular chromosome conformation capture (4C) uncovers extensive networks of epigenetically regulated intra- and interchromosomal interactions. *Nat Genet* **38**, 1341–1347.

Zhu, Q., Dabi, T. and Lamb, C. (1995) TATA box and initiator functions in the accurate transcription of a plant minimal promoter *in vitro*. *Plant Cell* **7**, 1681–1689.

Zhu, Q., Ordiz, M.I., Dabi, T., Beachy, R.N. and Lamb, C. (2002) Rice TATA binding protein interacts functionally with transcription factor IIB and the RF2a bZIP transcriptional activator in an enhanced plant *in vitro* transcription system. *Plant Cell* **14**, 795–803.

Annual Plant Reviews (2009) **35**, 229–242
doi: 10.1111/b.9781405175326.2009.00007.x

www.interscience.wiley.com

Chapter 7

THE RNA WORLD: IDENTIFYING miRNA-TARGET RNA PAIRS AS POSSIBLE MISSING LINKS IN MULTI-NETWORK MODELS

Pamela J. Green and Blake C. Meyers

Department of Plant and Soil Sciences and Delaware Biotechnology Institute, University of Delaware, Newark, DE, USA

Abstract: Plant cells contain highly complex and abundant populations of small RNAs that regulate gene expression at many levels, in diverse tissues, and show evidence of regulation under different growth conditions. These molecules have been recognized as important components of signalling networks, but investigations of their contributions to plant signalling pathways have only taken place within the past few years. As sequencing methods have improved, many new microRNAs (miRNAs) have been identified in the most well-studied species; the discovery of miRNAs in less-studied species is just beginning but has been greatly enabled by these new methods. Studies to compile and experimentally validate lists of miRNAs combined with high-throughput expression profiling of small RNAs will better define the extent of small RNA regulation in plant cellular processes. Validation of miRNAs often includes the prediction and verification of their mRNA targets and the identification of these targets can provide insight into the biological role of individual miRNAs and miRNA families. However, the power of this information could be farther reaching if it can be interpreted with other interaction data. The incorporation of pairs of miRNAs and their targets into multi-network models would accomplish this and would likely to provide the missing links in many regulatory pathways. Since approaches to address the function of individual miRNAs in mutants and transgenic plants are labour-intensive, the hypotheses resulting from the models could focus attention on the miRNAs that play pivotal roles in a number of key regulatory pathways.

Keywords: microRNA; small RNA; target mRNA; gene expression; gene regulation; RNA

7.1 Introduction

Nearly all eukaryotes produce small RNAs that function in gene regulation and/or chromatin structure. The first microRNA (miRNA) described was *lin-4*, a 22 nucleotide (nt) regulator of larval development in *Caenorhabditis elegans* (Lee *et al.*, 1993; Wightman *et al.*, 1993). The lack of other examples of ~21–25 nt small RNAs until 1999 (Hamilton and Baulcombe, 1999) or other miRNAs until 2000 (Pasquinelli *et al.*, 2000) prevented recognition of the broader implications of this finding for several years, but many subsequent studies have defined miRNA molecules as biologically important regulators of gene expression in plants, animals and algae. In plants, we now know that miRNAs play critical roles in various developmental, stress and signalling responses reviewed by Chen (2005) and Zhang *et al.* (2006). However, in plants, the small RNA population is particularly complex because plants also produce a complex set of small interfering RNAs (siRNAs) reviewed by Vaucheret (2006). As an example, in Arabidopsis, only 184 miRNA loci have been annotated in the Sanger miRNA registry (Griffiths-Jones, 2004), which is the primary curatorial resource for miRNAs. In contrast, more than 100 000 different endogenous 21–24 nt small interfering RNAs (siRNAs) have been sequenced from Arabidopsis (Lu *et al.*, 2005, 2006; Rajagopalan *et al.*, 2006; Fahlgren *et al.*, 2007), and this is likely to be just a fraction of the total diversity of siRNAs produced from the Arabidopsis genome. These siRNAs are primarily produced by repetitive sequences such as transposons and retrotransposons. Plant species with larger genomes, including nearly all crop plants and particularly many of the grasses are known to have higher contents of repetitive DNA. The complexity of the siRNA population is expected to increase roughly in proportion to genome size and repeat complexity, while miRNA complexity may increase only in proportion to the number of genes (although the latter has yet to be examined).

The notion that small RNAs may have a broader role in biology originated in part from plant research; Hamilton and Baulcombe (1999) identified small antisense RNAs associated with both transgene-induced and viral-induced post-transcriptional gene silencing. Sequencing-based approaches were soon launched to identify endogenous small RNAs from a number of organisms, including Arabidopsis, using standard methods to sequence size-fractionated RNA population (Llave *et al.*, 2002a; Lu *et al.*, 2005). The Arabidopsis miR-NAs that were cloned initially have turned out to be mainly well-conserved with other plant species, are highly expressed sequences and have biological roles primarily associated with growth and development (Llave *et al.*, 2002a; Aukerman and Sakai, 2003; Palatnik *et al.*, 2003; Chen, 2004, 2005; Han *et al.*, 2004; Juarez *et al.*, 2004; Mallory *et al.*, 2005). These were easily identified, primarily because their high abundance made them readily accessible to low-depth sequencing. As sequencing continued to dig deeper into the pools of small RNAs, several miRNAs with predicted roles in stress responses emerged (Jones-Rhoades and Bartel, 2004; Sunkar and Zhu, 2004), and a large number of miRNAs predicted to interact with genes important for other processes are now identified (Lu *et al.*, 2006; Rajagopalan *et al.*, 2006; Fahlgren *et al.*, 2007).

miRNAs and siRNAs are products of distinct biochemical pathways, and information about their origin and biogenesis can be used to categorize small RNAs into one of these two major categories, although several additional classes of small RNAs have been described, each with relatively few known members. miRNAs are generally produced from non-coding RNA precursors that are transcribed by RNA polymerase II (Xie *et al.*, 2005); these capped and polyadenylated 'pri-miRNA' precursors are predicted to form stem-loop structures, usually with an imperfectly complementary stem. Nearly all plant miRNAs require a specific RNAse III-like enzyme known as Dicer-like1 (DCL1) for biogenesis. In the case of the maturation of miR163, DCL1 is required at several steps during miRNA biogenesis in both Arabidopsis and rice (Kurihara and Watanabe, 2004; Liu *et al.*, 2005). According to current models, the stem-loop-like structure of the pri-miRNA is processed into a pre-miRNA and then into a mature miRNA duplex by DICER1 with the assistance of the HLY1 and SERRATE proteins which are associated with nuclear bodies (Fang and Spector, 2007; Fujioka *et al.*, 2007; Song *et al.*, 2007). This duplex is subsequently methylated by HEN1 presumably to stabilize the miRNA (Yu *et al.*, 2005).

In plants, miRNAs function to downregulate the product of the target gene, via a process of either mRNA cleavage and degradation, as is usually the case in plants, or translational attenuation, a phenomenon more frequently reported in animals (Llave *et al.*, 2002b; Aukerman and Sakai, 2003; Kasschau *et al.*, 2003; Chen, 2004; Souret *et al.*, 2004). However, both cases must result from binding of the mature single-stranded miRNAs to complementary sites in their mRNA 'targets'. The only known example of a positive regulatory function of a plant miRNA is the miRNAs that can stimulate production of trans-acting siRNAs. The ta-siRNAs are a special class of small RNA composed of relatively few 21 nt small RNAs, which are generated from at least five loci in Arabidopsis (Peragine *et al.*, 2004; Vazquez *et al.*, 2004; Allen *et al.*, 2005). ta-siRNAs have different genetic requirements, including Dicer-like4 (DCL4), RNA-dependent RNA-polymerase 6 (RDR6), as well as DCL1 and SGS3 (Peragine *et al.*, 2004; Vazquez *et al.*, 2004; Gasciolli *et al.*, 2005). Some of the 21 nt ta-siRNAs mediate the cleavage of specific mRNAs in *trans*, similar to miRNAs, hence the name ta-siRNAs. The ta-siRNAs are themselves the products of a regulatory cascade and its production requires an initial miRNA cleavage reaction (Peragine *et al.*, 2004; Vazquez *et al.*, 2004; Yoshikawa *et al.*, 2005). Because miRNA is required for ta-siRNA production, DCL1 is a requirement for both species of small RNAs. Neither miRNAs nor ta-siRNAs require RDR2, an enzyme essential for the production of most siRNAs, for their biosynthesis. It is likely that other small classes of siRNAs will be discovered, such as natural antisense siRNAs, a type thus far observed only under stress conditions (Borsani *et al.*, 2005).

siRNAs are generated in a manner that is quite different than miRNAs. According to current models, most are processed from perfectly complementary longer double-stranded RNA molecules that are the products of a plant specific RNA polymerase (Pol IV) (Herr *et al.*, 2005; Onodera *et al.*, 2005) and an RNA-dependent RNA polymerase (RDR2). It is believed that the dsRNA template that is produced by these enzymes is cleaved into ~24 nt small RNAs by a DICER enzyme, typically DCL3 (Xie *et al.*, 2004). However, it has not been shown that Pol IV is transcriptionally active, and its substrate has not been well-defined. Nonetheless, the Pol IV-dependent siRNAs, when mapped back to the genome, siRNA-producing loci are well represented on both strands of the DNA because of the dsRNA source. The function of siRNAs is probably heterochromatin formation via the guidance of nuclear complexes to modify histones and methylate DNA (Lippman and Martienssen, 2004; Verdel *et al.*, 2004). This typically results in the transcriptional silencing of transposons and retrotransposons and other types of potentially deleterious genomic sequences. The situation is further complicated by additional DICERs, such as DCL2 and DCL4, which are capable of generating 21 or 22 nt siRNAs or in the case of DCL4, can in some cases produce evolutionarily 'young' miRNAs (Rajagopalan *et al.*, 2006; Fahlgren *et al.*, 2007). This origin from a complex component of the plant genome is the reason for the tremendous diversity of siRNAs in plants.

7.2 Sequencing of small RNAs

The first comprehensive profiling of small RNAs was accomplished in our laboratories using deep sequencing methods, applied to Arabidopsis (Lu *et al.*, 2005). From this experiment, the extent of the complexity of plant small RNAs became clear. These data were obtained using a technology that was impressive in its time, an unconventional (non-Sanger) sequencing method that utilized a series of cycles involving adapter ligation, hybridization and enzymatic digestion. Known as massively parallel signature sequencing (MPSS) (Brenner *et al.*, 2000; Meyers *et al.*, 2004), this method had several disadvantages, including a limit of 17 nucleotides for each template molecule, and a complex methodology that limited the practice to the company that developed it. The partial length (17 nt) of the MPSS data made it hard to distinguish miRNAs from siRNAs without informed bioinformatic strategies, while a full-length sequence might have helped distinguish these classes as primarily 21 nt and 24 nt small RNAs, respectively. On the other hand, the depth of MPSS was tremendous compared to the alternative, conventional methods available at that time, with the application of MPSS to small RNA sequencing exceeding by more than an order of magnitude the number of small RNAs discovered by all prior conventional sequencing approaches (Llave *et al.*, 2002a; Reinhart *et al.*, 2002; Sunkar and Zhu, 2004; Gustafson *et al.*, 2005; Meyers *et al.*, 2006). However, even at depth of approximately 500 000 to a million sequences per library, the MPSS analysis was not saturating in Arabidopsis, and larger genomes would still require even greater depth.

Newer sequencing methods with advantages over MPSS for small RNA sequencing have since been described and are now more commonly employed for miRNA discovery. For example, 454 Life Sciences' pyrosequencing method can sequence in parallel hundreds of thousands of template molecules, while offering a longer read length than MPSS (Margulies et al., 2005). This permits the full-length sequencing of small RNAs. Illumina, Inc., the company that now includes the original developer of MPSS, has released a powerful new technology called 'sequencing-by-synthesis' (SBS) that sequences tens of millions of 35+ nucleotide tags, potentially offering the richest source of small RNA data. Because these methods all avoid the expense and complication of bacterial cloning by using direct amplification of templates from DNA, the price of sequencing a single cDNA comes out to a fraction of a cent. This offers advantages in both cost and throughput of several orders of magnitude over traditional sequencing methods. SBS offers dramatic cost savings and significant technical improvements over other methods, including much deeper sequencing, a simple sequencing process and a relatively low cost. These advantages have, for now, made this technology of choice for small RNA sequencing and miRNA discovery, although a number of other companies are on the verge of launching competing methods. The product of any of these methods, when applied to plant small RNAs, is typically a large number of repeat-associated siRNAs combined with a small proportion of miRNAs, ta-siRNAs and other minor classes of siRNAs. Indeed, with SBS and future short-read technologies, the sequencing of small RNAs is a solved problem, and there is no technical limitation to the discovery of novel miRNAs from any plant species. However, because these small RNAs do not come with names attached, and because they have no distinguishing characteristics at the sequence level, the challenge is then to map them to the genome or other available sequence resources and determine which sequences are the miRNAs of interest. In poorly described plant genomes, the development of a 'parts list' of miRNAs will remain a challenge, simply because the plethora of siRNAs will prove a challenge to miRNA identification.

7.3 Identification of miRNAs

The number of known, annotated plant miRNAs is now in the hundreds, with at least 186 loci identified from Arabidopsis that comprise at least 55 families as listed in the Sanger miRNA registry (Griffiths-Jones, 2004; Xie et al., 2005). To designate an miRNA as distinct from an siRNA or other small RNA, a set of criteria were proposed in 2003 that include distinguishing features of expression (identification by cloning) and biogenesis (a hairpin-forming precursor transcript, phylogenetic conservation and increased accumulation of the precursor when Dicer activity is reduced; Ambros et al., 2003). While these criteria are now somewhat out of date in terms of their sufficiency and general applicability, they are still a good first step toward identification of miRNAs. In addition, the availability of mutants in small RNA biogenesis pathways of Arabidopsis often makes categorizing small RNAs a more straightforward task reviewed by Vaucheret (2006). Work in our labs, and others has demonstrated that parallel sequencing of small RNA populations using key genetic mutants in small RNA biogenesis pathways will help to characterize large numbers of small RNAs (Lu *et al.*, 2005; 2006; Rajagopalan *et al.*, 2006; Fahlgren *et al.*, 2007).

Prior to the widespread use of deep sequencing, *a priori* miRNA prediction using purely bioinformatics-based methods was a popular way to identify miRNA candidates. These approaches typically combine both RNA secondary structure analyses and a cross-genome comparison to identify conserved sequences and secondary structures. At least three groups applied such an approach to the Arabidopsis genome with comparisons to rice (Bonnet *et al.*, 2004; Jones-Rhoades and Bartel, 2004; Wang *et al.*, 2004). A different miRNA prediction strategy was developed by the Sundaresan laboratory, which requires only a single-genomic sequence and relies on the complementarity between plant miRNAs and their intragenomic targets to identify candidate sequences; these candidates are then assessed for their potential to fold into a miRNA-like precursor (Adai *et al.*, 2005). However, the output of all these programmes includes many false positives, and the data are best combined with substantial experimental data (deep sequence data and target cleavage data) to validate these results and avoid both false positives and false negatives in the miRNA identification process (Jones-Rhoades and Bartel, 2004; Jones-Rhoades *et al.*, 2006).

Comparative genomics approaches are also quite promising, but these have yet to be fully exploited. The cloning of miRNAs from moss and Chlamydomonas indicates that miRNA functionality is conserved over more than 400 million years of evolution (Floyd and Bowman, 2004; Arazi *et al.*, 2005; Axtell and Bartel, 2005; Molnar *et al.*, 2007; Zhao *et al.*, 2007). Although the miRNAs in Chlamydomonas are not homologous to those in plants, conserved miRNA sequences representing several families have been found that span eudicots, monocots, gymnosperms and even mosses (Arazi *et al.*, 2005; Axtell and Bartel, 2005; Zhang *et al.*, 2006). However, these highly conserved miRNAs represent only a portion of the complete set in Arabidopsis, and numerous miRNAs have been identified in Arabidopsis for which putative orthologs in other plant genomes have yet to be described (Lu *et al.*, 2006; Rajagopalan *et al.*, 2006; Fahlgren *et al.*, 2007). Among these miRNAs, at least two have evolved recently, suggesting that other lineage-specific miRNAs are likely to be found (Allen *et al.*, 2004). With the increasing availability of both deep small RNA data and plant genome sequences from diverse species, comparative genomics approaches are likely to be powerful methods for the discovery of miRNAs with different conservation patterns, such as specific plant orders, families, genera, species, etc.

Both the validation and verification steps required to shift a miRNA from the 'predicted' to the 'known' category has proven to be the bottleneck in miRNA discovery. These steps currently require low-throughput and time-consuming experimental approaches to satisfy a set of validation criteria for miRNAs. These criteria began with the early consensus for plants and animals mentioned above (Ambros *et al.*, 2003) and have evolved to be more rigorous and flexible as our tools and understanding has grown. Currently, plant miRNA validation most often includes several of the following steps: detection on northern blots, confirming the expected effects of small RNA biosynthetic mutants (e.g. DCL1 dependence, RDR2 independence), sequencing of the miRNA * (the opposite strand of the miRNA duplex), eliminating sequences in dense small RNA clusters or identification of the cleavage product of the target (see below). The weight and number of these criteria satisfied in various studies varies greatly and none can be viewed as the gold standard for all miRNAs. For example, not all miRNAs are dependent on DCL1; miR168 is controlled by a feedback loop (Vaucheret *et al.*, 2006) and some, possibly young miRNAs, appear to be dependent on DCL4 rather than DCL1(Rajagopalan *et al.*, 2006). Most consider validation of cleavage of the predicted target as the ultimate proof that a small RNA is a miRNA, but this is not always possible to demonstrate; for example, in cases where the cleaved target is present at such low abundances or very few cells such that it is undetectable. Additionally, some cleavage events may be due to the action of minor classes of endogenous siRNAs (reviewed by Chapman and Carrington, 2007, and Mallory and Vaucheret, 2006). Finally, some miRNAs repress translation as their primary mode of action, and protein levels are rarely assayed. Nevertheless, it is safe to say that the field has evolved to use more sequences, evidence and knowledge for miRNA annotation, and this should lead to a more reliable parts list of miRNAs than was available previously. Yet, a parts list can only begin to provide insight about the biology of those sequences or fit them into a specific regulatory or signalling pathway. This latter activity is in fact the long-term goal of most miRNA discovery projects.

7.4 Identification of miRNA targets

In plants, the mRNA targets of miRNAs are commonly first predicted using computational approaches because of the near perfect complementary nature of most miRNA-mRNA target pairs (Llave *et al.*, 2002a; Reinhart *et al.*, 2002; Kasschau *et al.*, 2003; Jones-Rhoades and Bartel, 2004). This requires a scan of the miRNA sequence against the genome, with a number of rules to guide and score the types of allowed matches and mismatches (Jones-Rhoades and Bartel, 2004; Schwab *et al.*, 2006). MicroRNA-guided cleavage of the mRNA target typically occurs in the middle of the region of complementarity between the miRNAs and their target mRNAs. To experimentally validate potential targets, and the 5′ terminus of the 3′ RNA cleavage, products can be amplified by RNA ligase mediated 5' RACE (Llave *et al.*, 2002a, 2002b; Kasschau *et al.*, 2003). This method involves ligating an RNA adapter to the 5' end of RNAs in a total or poly(A)+ RNA preparation, reverse transcription with an oligo dT primer and amplification with oligos matching the adapter and an internal primer of the predicted target. Cloning and sequencing of these cleavage products indicates the exact cleavage location(s) and their representation in the RNA population. The ligation step requires a 5' phosphate, like that generated from miRNA-directed cleavage by Argonaute; intact mRNAs or those with other types of 5' ends will not be substrates for adapter ligation and are therefore unavailable as amplification templates. The cleavage products of some bona fide targets sometimes are hard to detect with this method if they are low abundance due to low expression levels or RNA instability, although sometimes sampling in a different tissue may address these issues. Another approach is to work in an *Atxrn4* background which lacks the activity of the 5'–3' exoribonuclease AtXRN4 (Kastenmayer and Green, 2000), a homolog of the yeast cytoplasmic enzyme XRN1. The cleavage products of many miRNA targets accumulate in *Atxrn4* mutants because these molecules are substrates for the enzyme (Souret *et al.*, 2004).

Another way to validate miRNA-target mRNA pairs is to carry out overexpression and underexpression experiments. One prime example is to engineer or identify within a mutant an insensitive target gene in which the sequence that is predicted to pair with the miRNA has been disrupted. The disruption should prevent the miRNA-directed cleavage of the target with the result that the target mRNA levels are elevated. In this mutant, phenotypic changes may be predicted depending on what is known about the role of the mRNA target (Chen, 2004; Zhong and Ye, 2004; Mallory *et al.*, 2005). Alternatively, the miRNA may be overexpressed or ectopically expressed (Palatnik *et al.*, 2003; Mallory *et al.*, 2004; Fujii *et al.*, 2005). If this results in cleavage and decay of the target and the expected physiological phenotype, then this would also provide evidence for function. Finally, validation of miRNA activities and the impact on putative targets can be performed for individual miRNAs using mutants for instances in which a T-DNA insertion is available which alters or eliminates the expression of a miRNA gene (Palatnik *et al.*, 2003; Guo *et al.*, 2005). These and other methods are important for the identification and validation of miRNA-mRNA target pairs, and are a necessary step to determine the biology of individual miRNAs in plant signalling.

7.5 MicroRNA-target mRNA pairs: missing links in multi-network models?

A step toward understanding the biology of specific miRNAs is to include miRNA-target mRNA pairs in multi-network models. These models typically comprise other interaction data, such as protein–protein and protein–DNA interactions, and the development of these models with the inclusion of miRNA–mRNA interactions will lead to logical hypotheses of how specific miRNAs fit into different cellular networks. Testing of these hypotheses can be based on or include experiments using inactivation or overexpression of the miRNA and/or target, understanding the expression dynamics of the miRNA and target mRNA and other experiments suggested by the model for a given miRNA. As a first step towards a 'wiring diagram' of the plant cell, Gutierrez *et al.* (2007) generated an integrated 'multi-network' for Arabidopsis; the network has nodes (genes) connected by 'edges' supported by a variety of data types. This model includes metabolism and regulatory interactions from KEGG, AraCyc, Transfac, inferred protein–protein interactions using the interolog and interactions obtained from mining the literature with the GeneWays system (Kanehisa, 2002; Matys *et al.*, 2003; Mueller *et al.*, 2003; Rzhetsky *et al.*, 2004; Yu *et al.*, 2004). As a proof-of-principle for the use of this multi-network model to hypothesize miRNA gene function, we initially added to the current model a total of 61 miRNA–target interactions. These comprised 14 known miRNAs, 59 different mRNA targets and two common mRNA targets. Figure 7.1 shows part of the subnetwork that was found when querying the multi-network for genes that are linked by up to seven nodes from miR156 and miR157 which are in the same family. Two of the targets of miR156 and miR157, the squamosa promoter binding protein (At2g33810) and a transcription factor-related protein (At1g53160) show protein–DNA interactions with *apetala1* (At1g69120), a key component in flower development. This suggests that miR156 and miR157 are involved in flower development, a prediction that is consistent with, but derived independently of, recent literature. Indeed, the regulation of miR156 during flowering and the role of SPL genes (targets of miR156) in floral induction have been reported by several groups (Cardon *et al.*, 1997; Schmid *et al.*, 2003; Kidner and Martienssen, 2005). This initial analysis of a small group of miRNAs is encouraging because it demonstrates the utility of the model to provide testable hypotheses about miRNA roles and modes of action. As new and more highly developed models are created (e.g. Gifford *et al.*, 2008; Gutierrez *et al.*, 2008), the incorporation of miRNA-target mRNA pairs may provide the missing links in many regulatory pathways.

Acknowledgements

We are grateful to members of the Green and Meyers laboratories for many stimulating discussions. Work in the Green and Meyers laboratories is supported by NSF, USDA and DOE awards. We thank Rodrigo Gutiérrez and Gloria Coruzzi for providing Fig. 7.1.

References

Adai, A., Johnson, C., Mlotshwa, S., Archer-Evans, S., Manocha, V., Vance, V., *et al.* (2005) Computational prediction of miRNAs in *Arabidopsis thaliana*. *Genome Res* **15**, 78–91.

Allen, E., Xie, Z., Gustafson, A.M. and Carrington, J.C. (2005) MicroRNA-directed phasing during trans-acting siRNA biogenesis in plants. *Cell* **121**, 207–221.

Allen, E., Xie, Z., Gustafson, A.M., Sung, G.H., Spatafora, J.W. and Carrington, J.C. (2004) Evolution of microRNA genes by inverted duplication of target gene sequences in *Arabidopsis thaliana*. *Nat Genet* **36**, 1282–1290.

Ambros, V., Bartel, B., Bartel, D.P., Burge, C.B., Carrington, J.C., Chen, X., *et al.* (2003) A uniform system for microRNA annotation. *RNA* **9**, 277–279.

Arazi, T., Talmor-Neiman, M., Stav, R., Riese, M., Huijser, P. and Baulcombe, D.C. (2005) Cloning and characterization of micro-RNAs from moss. *Plant J* **43**, 837–848.

Aukerman, M.J. and Sakai, H. (2003) Regulation of flowering time and floral organ identity by a microRNA and its APETALA2-like target genes. *Plant Cell* **15**, 2730–2741.

Axtell, M.J. and Bartel, D.P. (2005) Antiquity of microRNAs and their targets in land plants. *Plant Cell* **17**, 1658–1673.

Bonnet, E., Wuyts, J., Rouze, P. and Van de Peer, Y. (2004) Detection of 91 potential conserved plant microRNAs in *Arabidopsis thaliana* and *Oryza sativa* identifies important target genes. *Proc Natl Acad Sci USA* **101**, 11511–11516.

Borsani, O., Zhu, J., Verslues, P.E., Sunkar, R. and Zhu, J.K. (2005) Endogenous siRNAs derived from a pair of natural cis-antisense transcripts regulate salt tolerance in Arabidopsis. *Cell* **123**, 1279–1291.

Brenner, S., Johnson, M., Bridgham, J., Golda, G., Lloyd, D.H., Johnson, D., *et al.* (2000) Gene expression analysis by massively parallel signature sequencing (MPSS) on microbead arrays. *Nat Biotechnol* **18**, 630–634.

Cardon, G.H., Hohmann, S., Nettesheim, K., Saedler, H. and Huijser P. (1997) Functional analysis of the *Arabidopsis thaliana*, S.B.P-box gene SPL3: a novel gene involved in the floral transition. *Plant J* **12**, 367–377.

Chapman, E.J. and Carrington, J.C. (2007) Specialization and evolution of endogenous small RNA pathways. *Nat Rev Genet* **8**, 884–896.

Chen, X. (2004) A microRNA as a translational repressor of APETALA2 in Arabidopsis flower development. *Science* **303**, 2022–2025.

Chen, X. (2005) MicroRNA biogenesis and function in plants. *FEBS Lett* **579**, 5923–5931.

Fahlgren, N., Howell, M.D., Kasschau, K.D., Chapman, E.J., Sullivan, C.M., Cumbie, J.S., *et al.* (2007) High-throughput sequencing of Arabidopsis microRNAs: evidence for frequent birth and death of miRNA genes. *PLoS ONE* **2**, e219.

Fang, Y. and Spector, D.L. (2007) Identification of nuclear dicing bodies containing proteins for microRNA biogenesis in living Arabidopsis plants. *Curr Biol* **17**, 818–823.

Floyd, S.K. and Bowman, J.L. (2004) Gene regulation: ancient microRNA target sequences in plants. *Nature* **428**, 485–486.

Fujii, H., Chiou, T.J., Lin, S.I., Aung, K. and Zhu, J.K. (2005) A miRNA involved in phosphate-starvation response in Arabidopsis. *Curr Biol* **15**, 2038–2043.

Fujioka, Y., Utsumi, M., Ohba, Y. and Watanabe, Y. (2007) Location of a possible miRNA processing site in SmD3/SmB nuclear bodies in Arabidopsis. *Plant Cell Physiol* **48**, 1243–1253.

Gasciolli, V., Mallory, A.C., Bartel, D.P. and Vaucheret, H. (2005) Partially redundant functions of Arabidopsis DICER-like enzymes and a role for DCL4 in producing trans-acting siRNAs. *Curr Biol* **15**, 1494–1500.

Gifford, M.L., Dean, A., Gutierrez, R.A., Coruzzi, G.M. and Birnbaum, K.D. (2008)

Cell-specific nitrogen responses mediate developmental plasticity. *Proc Natl Acad Sci USA* **105**, 803–808.

Griffiths-Jones, S. (2004) The microRNA registry. *Nucleic Acids Res* **32**, D109–D111.

Guo, H.S., Xie, Q., Fei, J.F. and Chua, N.H. (2005) MicroRNA directs mRNA cleavage of the transcription factor NAC1 to downregulate auxin signals for Arabidopsis lateral root development. *Plant Cell* **17**, 1376–1386.

Gustafson, A.M., Allen, E., Givan, S., Smith, D., Carrington, J.C. and Kasschau, K.D. (2005) ASRP: the Arabidopsis small RNA project database. *Nucleic Acids Res* **33**, D637–D640.

Gutierrez, R.A., Lejay, L.V., Dean, A., Chiaromonte, F., Shasha, D.E. and Coruzzi, G.M. (2007) Qualitative network models and genome-wide expression data define carbon/nitrogen-responsive molecular machines in Arabidopsis. *Genome Biol* **8**, R7.

Gutierrez, R.A., Stokes, T.L., Thum, K., Xu, X., Obertello, M., Katari, M.S., *et al.* (2008) Systems approach identifies an organic nitrogen-responsive gene network that is regulated by the master clock control gene CCA1. *Proc Natl Acad Sci USA* **105**, 4939–4944.

Hamilton, A.J. and Baulcombe, D.C. (1999) A species of small antisense RNA in posttranscriptional gene silencing in plants. *Science* **286**, 950–952.

Han, M.H., Goud, S., Song, L. and Fedoroff, N. (2004) The Arabidopsis double-stranded RNA-binding protein HYL1 plays a role in microRNA-mediated gene regulation. *Proc Natl Acad Sci USA* **101**, 1093–1098.

Herr, A.J., Jensen, M.B., Dalmay, T. and Baulcombe, D.C. (2005) RNA polymerase IV directs silencing of endogenous DNA. *Science* **308**, 118–120.

Jones-Rhoades, M.W. and Bartel, D.P. (2004) Computational identification of plant microRNAs and their targets, including a stress-induced miRNA. *Mol Cell* **14**, 787–799.

Jones-Rhoades, M.W., Bartel, D.P. and Bartel, B. (2006) MicroRNAs and their regulatory roles in plants. *Annu Rev Plant Biol* **57**, 19–53.

Juarez, M.T., Kui, J.S., Thomas, J., Heller, B.A. and Timmermans, M.C. (2004) MicroRNA-mediated repression of rolled leaf1 specifies maize leaf polarity. *Nature* **428**, 84–88.

Kanehisa, M. (2002) The KEGG database. *Novartis Found Symp* **247**, 91–101.

Kasschau, K.D., Xie, Z., Allen, E., Llave, C., Chapman, E.J., Krizan, K.A., *et al.* (2003) P1/HC-Pro, a viral suppressor of RNA silencing, interferes with Arabidopsis development and miRNA function. *Dev Cell* **4**, 205–217.

Kastenmayer, J.P. and Green, P.J. (2000) Novel features of the XRN-family in Arabidopsis: evidence that AtXRN4, one of several orthologs of nuclear Xrn2p/Rat1p, functions in the cytoplasm. *Proc Natl Acad Sci USA* **97**, 13985–13990.

Kidner, C.A. and Martienssen, R.A. (2005) The developmental role of microRNA in plants. *Curr Opin Plant Biol* **8**, 38–44.

Kurihara, Y. and Watanabe, Y. (2004) Arabidopsis micro-RNA biogenesis through Dicer-like 1 protein functions. *Proc Natl Acad Sci USA* **101**, 12753–12758.

Lee, R.C., Feinbaum, R.L. and Ambros V. (1993) The C. elegans heterochronic gene lin-4 encodes small RNAs with antisense complementarity to lin-14. *Cell* **75**, 843–854.

Lippman, Z. and Martienssen, R. (2004) The role of RNA interference in heterochromatic silencing. *Nature* **431**, 364–370.

Liu, B., Li, P., Li, X., Liu, C., Cao, S., Chu, C., *et al.* (2005) Loss of function of OsDCL1 affects microrna accumulation and causes developmental defects in rice. *Plant Physiol* **139**, 296–305.

Llave, C., Kasschau, K.D., Rector, M.A. and Carrington, J.C. (2002a) Endogenous and silencing-associated small RNAs in plants. *Plant Cell* **14**, 1605–1619.

Llave, C., Xie, Z., Kasschau, K.D. and Carrington, J.C. (2002b) Cleavage of scarecrow-like mRNA targets directed by a class of Arabidopsis miRNA. *Science* **297**, 2053–2056.

Lu, C., Kulkarni, K., Souret, F.F., MuthuValliappan, R., Tej, S.S., Poethig, R.S., *et al.* (2006) MicroRNAs and other small RNAs enriched in the Arabidopsis RNA-dependent RNA polymerase-2 mutant. *Genome Res* **16**, 1276–1288.

Lu, C., Tej, S.S., Luo, S., Haudenschild, C.D., Meyers, B.C. and Green, P.J. (2005) Elucidation of the small RNA component of the transcriptome. *Science* **309**, 1567–1569.

Mallory, A.C. and Vaucheret, H. (2006) Functions of microRNAs and related small RNAs in plants. *Nat Genet* **38**(Suppl), S31–S36.

Mallory, A.C., Bartel, D.P. and Bartel B. (2005) MicroRNA-directed regulation of Arabidopsis AUXIN RESPONSE FACTOR17 is essential for proper development and modulates expression of early auxin response genes. *Plant Cell* **17**, 1360–1375.

Mallory, A.C., Reinhart, B.J., Jones-Rhoades, M.W., Tang, G., Zamore, P.D., Barton, M.K., *et al.* (2004) MicroRNA control of PHABULOSA in leaf development: importance of pairing to the microRNA 5' region. *EMBO J* **23**, 3356–3364.

Margulies, M., Egholm, M., Altman, W.E., Attiya, S., Bader, J.S., Bemben, L.A., *et al.* (2005) Genome sequencing in microfabricated high-density picolitre reactors. *Nature* **437**, 376–380.

Matys, V., Fricke, E., Geffers, R., Gossling, E., Haubrock, M., Hehl, R., *et al.* (2003) TRANSFAC: transcriptional regulation, from patterns to profiles. *Nucleic Acids Res* **31**, 374–378.

Meyers, B.C., Souret, F.F., Lu, C. and Green, P.J. (2006) Sweating the small stuff: microRNA discovery in plants. *Curr Opin Biotechnol* **17**, 139–146.

Meyers, B.C., Tej, S.S., Vu, T.H., Haudenschild, C.D., Agrawal, V., Edberg, S.B., *et al.* (2004) The use of MPSS for whole-genome transcriptional analysis in Arabidopsis. *Genome Res* **14**, 1641–1653.

Molnar, A., Schwach, F., Studholme, D.J., Thuenemann, E.C. and Baulcombe, D.C. (2007) MiRNAs control gene expression in the single-cell alga *Chlamydomonas reinhardtii*. *Nature* **447**, 1126–1129.

Mueller, L.A., Zhang, P. and Rhee, S.Y. (2003) AraCyc: a biochemical pathway database for Arabidopsis. *Plant Physiol* **132**, 453–460.

Onodera, Y., Haag, J.R., Ream, T., Nunes, P.C., Pontes, O. and Pikaard, C.S. (2005) Plant nuclear RNA polymerase IV mediates siRNA and DNA methylation-dependent heterochromatin formation. *Cell* **120**, 613–622.

Palatnik, J.F., Allen, E., Wu, X., Schommer, C., Schwab, R., Carrington, J.C., *et al.* (2003) Control of leaf morphogenesis by microRNAs. *Nature* **425**, 257–263.

Pasquinelli, A.E., Reinhart, B.J., Slack, F., Martindale, M.Q., Kuroda, M.I., Maller, B., *et al.* (2000) Conservation of the sequence and temporal expression of let-7 heterochronic regulatory RNA. *Nature* **408**, 86–89.

Peragine, A., Yoshikawa, M., Wu, G., Albrecht, H.L. and Poethig, R.S. (2004) SGS3 and SGS2/SDE1/RDR6 are required for juvenile development and the production of trans-acting siRNAs in Arabidopsis. *Genes Dev* **18**, 2368–2379.

Rajagopalan, R., Vaucheret, H., Trejo, J. and Bartel, D.P. (2006) A diverse and evolutionarily fluid set of microRNAs in *Arabidopsis thaliana*. *Genes Dev* **20**, 3407–3425.

Reinhart, B.J., Weinstein, E.G., Rhoades, M.W., Bartel, B. and Bartel, D.P. (2002) MicroRNAs in plants. *Genes Dev* **16**, 1616–1626.

Rzhetsky, A., Iossifov, I., Koike, T., Krauthammer, M., Kra, P., Morris, M., *et al.* (2004) GeneWays: a system for extracting, analyzing, visualizing, and integrating molecular pathway data. *J Biomed Inform* **37**, 43–53.

Schmid, M., Uhlenhaut, N.H., Godard, F., Demar, M., Bressan, R., Weigel, D., *et al.* (2003) Dissection of floral induction pathways using global expression analysis. *Development* **130**, 6001–6012.

Schwab, R., Ossowski, S., Riester, M., Warthmann, N. and Weigel, D. (2006) Highly specific gene silencing by artificial microRNAs in Arabidopsis. *Plant Cell* **18**, 1121–1133.

Song, L., Han, M.H., Lesicka, J. and Fedoroff, N. (2007) Arabidopsis primary microRNA processing proteins HYL1 and DCL1 define a nuclear body distinct from the Cajal body. *Proc Natl Acad Sci USA* **104**, 5437–5442.

Souret, F.F., Kastenmayer, J.P. and Green, P.J (2004). AtXRN4 degrades mRNA in Arabidopsis and its substrates include selected miRNA targets. *Mol Cell* **15**, 173–183.

Sunkar, R. and Zhu, J.K. (2004) Novel and stress-regulated microRNAs and other small RNAs from Arabidopsis. *Plant Cell* **16**, 2001–2019.

Vaucheret, H. (2006) Post-transcriptional small RNA pathways in plants: mechanisms and regulations. *Genes Dev* **20**, 759–771.

Vaucheret, H., Mallory, A.C. and Bartel, D.P. (2006) AGO1 homeostasis entails coexpression of MIR168 and AGO1 and preferential stabilization of miR168 by AGO1. *Mol Cell* **22**, 129–136.

Vazquez, F., Vaucheret, H., Rajagopalan, R., Lepers, C., Gasciolli, V., Mallory, A.C., *et al.* (2004) Endogenous trans-acting siRNAs regulate the accumulation of Arabidopsis mRNAs. *Mol Cell* **16**, 69–79.

Verdel, A., Jia, S., Gerber, S., Sugiyama, T., Gygi, S., Grewal, S.I., *et al.* (2004) RNAi-mediated targeting of heterochromatin by the RITS complex. *Science* **303**, 672–676.

Wang, X.J., Reyes, J.L., Chua, N.H. and Gaasterland, T. (2004) Prediction and identification of *Arabidopsis thaliana* microRNAs and their mRNA targets. *Genome Biol* **5**, R65.

Wightman, B., Ha, I. and Ruvkun, G. (1993) Posttranscriptional regulation of the heterochronic gene lin-14 by lin-4 mediates temporal pattern formation in C. elegans. *Cell* **75**, 855–862.

Xie, Z., Allen, E., Fahlgren, N., Calamar, A., Givan, S.A. and Carrington, J.C. (2005) Expression of Arabidopsis MIRNA genes. *Plant Physiol* **138**, 2145–2154.

Xie, Z., Johansen, L.K., Gustafson, A.M., Kasschau, K.D., Lellis, A.D., Zilberman, D., *et al.* (2004) Genetic and functional diversification of small RNA pathways in plants. *PLoS Biol* **2**, E104.

Yoshikawa, M., Peragine, A., Park, M.Y. and Poethig, R.S. (2005) A pathway for the biogenesis of trans-acting siRNAs in Arabidopsis. *Genes Dev* **19**, 2164–2175.

Yu, B., Yang, Z., Li, J., Minakhina, S., Yang, M., Padgett, R.W., *et al.* (2005) Methylation as a crucial step in plant microRNA biogenesis. *Science* **307**, 932–935.

Yu, H., Luscombe, N.M., Lu, H.X., Zhu, X., Xia, Y., Han, J.D., *et al.* (2004) Annotation transfer between genomes: protein-protein interologs and protein-DNA regulogs. *Genome Res* **14**, 1107–1118.

Zhang, B., Pan, X., Cannon, C.H., Cobb, G.P. and Anderson, T.A. (2006) Conservation and divergence of plant microRNA genes. *Plant J* **46**, 243–259.

Zhao, T., Li, G., Mi, S., Li, S., Hannon, G.J., Wang, X.J., *et al.* (2007) A complex system of small RNAs in the unicellular green alga *Chlamydomonas reinhardtii. Genes Dev* **21**, 1190–1203.

Zhong, R. and Ye, Z.H. (2004) Amphivasal vascular bundle 1, a gain-of-function mutation of the IFL1/REV gene, is associated with alterations in the polarity of leaves, stems and carpels. *Plant Cell Physiol* **45**, 369–385.

Annual Plant Reviews (2009) **35**, 243–257
doi: 10.1111/b.9781405175326.2009.00008.x

www.interscience.wiley.com

Chapter 8

PROTEOMICS: SETTING THE STAGE FOR SYSTEMS BIOLOGY

Scott C. Peck

Division of Biochemistry, University of Missouri-Columbia, Columbia, MO, USA

Abstract: Proteomics is a broadly defined discipline. Loosely, the early definition referred to the goal of studying 'all proteins produced by a genome'. The reality is that this goal has never been achieved because of the complexity of the proteome, and most studies actually deal with sub-proteomes – for example, proteins from a particular organelle or with a specific modification. This distinction, however, does not detract from the increasing importance of proteomics in systems biology. Rather, the study of sub-proteomes merely emphasizes the combinatorial complexity that can be achieved by the variety of protein forms produced by the genome. The underlying paradigm is that each form of a protein may be functionally distinct, meaning that a more detailed understanding of how a protein changes in form, location or activity during development or a response is required to understand its role in a biological process. This level of understanding is the current goal of proteomic studies.

Keywords: proteomics; 2D gel; LC-MS/MS; post-translational modification; quantitative proteomics; isotopic labelling

8.1 Introduction: the need for proteomics in systems biology

Proteomic studies provide unique information about changes in the cell. Although analysis of changes in mRNA abundance (e.g. using microarrays) provides useful information about potential rates of transcription and/or mRNA stability, the level of mRNA correlates very poorly with the level of protein (Gygi *et al.*, 1999; Tian *et al.*, 2004). Differences in the translational efficiency of the mRNA as well as the stability of the protein result in tremendous variation in the ratio of protein:mRNA between different gene products. Therefore, the presence of a transcript provides little information about the abundance of the translation product. Furthermore, transcriptome studies provide no information on the form of a protein or its location within the cell. A single gene may give rise to a variety of protein forms. Alternative transcriptional initiation, splicing or translation can give rise to multiple protein products. Moreover, these individual products can undergo a wide array of post-translational modifications (PTMs) that can alter a protein's activity, location or binding partners. Although empirical proteomic analyses in plants have improved bioinformatic programmes, predicting PTMs or subcellular localization based on primary sequence information remains problematic. Moreover, alternative transcription or translation may remove an accurately predicted chloroplast targeting sequence or myristoylation motif, relocating the protein to the cytoplasm contrary to expectations. Similarly, a phosphorylation site may be predicted on a protein, but the reversible nature of protein phosphorylation precludes predicting whether the potentially phosphorylated residue exists during the response being studied. Therefore, even if prediction programmes are refined sufficiently, the dynamic nature of the proteome imposes the requirement of obtaining empirical evidence of protein form and location during a response to a stimulus or at a particular stage of development.

8.2 Determination of protein location in the cell

From the efforts of many groups, the proteomic analyses of subcellular compartments in plants have greatly increased our knowledge of the dynamic composition of organelles (Millar, 2004). The breadth of these studies requires nearly an individual chapter for each organelle to do justice to the field and is beyond the scope of this chapter. Instead, only two examples of organelle proteomic studies will be discussed to illustrate two major concerns in subcellular proteomic studies: problems in detecting the full dynamic range of proteins due to the presence of highly abundant proteins and problems with contamination obscuring definitive conclusions from these studies. As these problems begin to be resolved, the next step will be quantitative comparisons to examine how organelles change during development or in response to various biotic or abiotic stimuli (see Section 'Quantitation' in this chapter).

8.2.1 Chloroplasts: problems of abundant proteins

Because of the importance of its role in the plant cell, the chloroplast arguably has been the most thoroughly studied of the plant organelle proteomes. These studies have resulted in both large amounts of empirical confirmation of protein localization as well as significant refinement of targeting prediction programmes (Baginski and Gruissem, 2004). Moreover, these studies corrected database annotations of false N- and C-termini as well as exon/intron boundary predictions, demonstrating the value of thorough proteomics experiments when they provide sufficient sequence coverage. In-depth studies of the chloroplast proteome, however, must deal with the presence of a small number of extremely abundant proteins, such as RuBISCO (ribulose-1,5-bisphosphate carboxylase/oxygenase). On 2D (two-dimensional) gels, extremely abundant proteins not only obscure large areas of the gel which they occupy, but the presence of these abundant proteins also distorts the migration of proteins in surrounding regions of the gel. In LC-MS/MS (liquid chromatography-tandem mass spectrometry) analyses, extremely abundant proteins are problematic because high concentrations of individual peptides from these proteins do not fractionate in tight peaks during LC separation, meaning that the same peptide is repeatedly resequenced. These problems interfere with the ability to study/identify proteins at the lower end of the dynamic range. In mammalian serum profiling, the presence of serum albumin and a few other proteins account for more than half of the total protein (Anderson and Anderson, 2002). In this case, commercial antibody columns have been produced to affinity deplete these highly abundant protein to allow a complete analysis of the dynamic complement of proteins. In plants, a method describing FPLC (fast protein liquid chromatography) anion-exchange chromatography to deplete RuBISCO from crude leaf extracts provides a working solution (Wienkoop *et al.*, 2004), and a few companies are beginning to produce RuBISCO antibody columns for affinity depletion. Similar problems of abundant proteins extend to the study of specialized tissues (e.g. highly abundant seed storage proteins), and in each case, it may be necessary to develop effective depletion strategies in order to study the full dynamic complement of an organelle or tissue.

8.2.2 Vacuoles: questions of contamination

In all studies of organelles, a major concern is the cross-contamination of other organelle compartments. Even in highly enriched isolations of a particular organelle, some degree of contamination, particularly by extremely abundant proteins, is not unexpected. However, the presence of contaminants raises questions about accurately assigning the location of unknown proteins to the compartment being studied. This potential problem is particularly complicated with proteomic studies of the vacuole (Carter *et al.*, 2004; Shimaoka *et al.*, 2004; Szponarski *et al.*, 2004). Being a lytic compartment involved in turnover of proteins from other organelles (or perhaps even whole organelles), potential 'contaminating' proteins may actually be present in the vacuole. Moreover, via alternative transcription/splicing/translation, the same gene may target proteins to two different compartments. Therefore, proteomic localization must be interpreted as evidence, not proof, of a protein's location. As such, the need for independent validation such as by GFP (green fluorescent protein) localization of proteins (Cutler *et al.*, 2000; Koroleva *et al.*, 2005) is becoming increasingly important to provide accurate predictions of subcellular localization. Synthesis of many independent lines of evidence have been used to generate the Arabidopsis subcellular database

(SUBA) which has compiled information from a variety of sources to provide subcellular locations for more than half of the proteins encoded by expressed sequence tags in Arabidopsis (Heazlewood *et al.*, 2007).

8.3 Identification of different protein forms

Modification of proteins by phosphorylation, glycosylation and ubiquitylation add to the complexity of the proteome in the cell. The reality, however, is that more than 200 PTMs have been reported (Aebersold and Goodlett, 2001). Because a protein may contain multiple different PTMs at any given time and these PTMs may change throughout the life of the plant, the proteome is estimated to be at least 1–2 orders of magnitude larger than the number of open reading frames predicted by the genome. It is this level of dynamic complexity that makes studying the proteome so challenging.

Currently, bioinformatic predictions are not reliably accurate for most PTMs. Moreover, because many PTMs are reversible, even an accurate prediction of a potential modification will require empirical determination of when a PTM is present. These analyses are not straightforward as the modified forms of the protein are often present at sub-stoichiometric levels, so these modifications will often escape detection in a general proteome survey. Therefore, the development of selective enrichment techniques for various PTMs has been a major focus of proteomic studies.

8.3.1 Phosphorylation

Dynamic protein phosphorylation is a nearly ubiquitous regulatory process during biological responses and, therefore, is one of the most-studied of the PTMs. Phosphorylation can modulate protein activity, target proteins for degradation or lead to the formation of new protein complexes. Until recently, the analysis of protein phosphorylation remained a challenging task, primarily because the phosphorylated form of the protein often represents a relatively small proportion of the total protein but also because analysis of phosphorylated peptides by LC-MS/MS is not straightforward (as discussed below).

Traditionally, the simplest method for detecting a phosphorylated protein was to look for the appearance of new protein isoforms on 2D gels. With 2D gels, proteins are separated in the first dimension based on their isoelectric point (pI) and then in the second dimension based on their apparent molecular weight using SDS–PAGE. When a protein becomes phosphorylated, its pI becomes more acidic, and the phosphoprotein shifts to a new position on the 2D gel. This method only works, of course, if the protein is of sufficient abundance to be detected by general protein stains. A more sensitive alternative is using ^{32}P- or ^{33}P-orthophosphate to radioactively label phosphoproteins in the cell. After separation by 2D gels, the radioactive image of the gel indicates which proteins are phosphorylated. This approach has proven to be effective in studying rapid phosphorylation changes in Arabidopsis suspension-cultured cells in response to microbial elicitors, such as the flagellin peptide (Peck *et al.*, 2001; Nühse *et al.*, 2003a). However, this approach is only amenable to biological systems in which the radioactive pulse labelling of cells can occur fairly rapidly (e.g. in less than 10 min). Otherwise, the radioactivity accumulates in constitutively phosphorylated proteins, making detection of differentially phosphorylated proteins extremely difficult. An alternative method to avoid this problem is staining the gels with Pro-Q Diamond (Molecular Probes). This fluorescent stain has a strong preference for the phosphorylated form of proteins and provides linear and sensitive detection of phosphoproteins. This approach successfully identified substrates of the SnRK2.8 kinase in a proteomic analysis of plants overexpressing this kinase (Shin *et al.*, 2007).

Both strategies above will identify candidate phosphoproteins. For identification, these proteins will be excised from the gel, digested with trypsin and identified by mass spectrometry. However, ionization of phosphopeptides is often suppressed in the presence of non-phosphopeptides, meaning that the phosphopeptide is not sequenced. A standard method for circumventing this problem is using immobilized metal affinity chromatography (IMAC) to enrich for phosphopeptide(s). The strong positive charge of the transition metal, usually Fe^{3+} or Ga^{3+}, binds the negatively charged phosphate group and enriches the phosphopeptide. This pre-enrichment step often improves the success of sequencing the phosphopeptide by mass spectrometry.

A large advance in analyzing the phosphoproteome, the complement of phosphorylated proteins in a cell, came from the development of methods for using IMAC to enrich hundreds to thousands of phosphopeptides from complex mixtures of trypsin digests of total cellular protein. A concern about using IMAC was that it may also bind peptides containing many acidic residues. As a solution to this potential problem, chemical methylation of acidic residues prior to IMAC was found to improve the specific binding of phosphopeptides from yeast (Ficarro *et al.*, 2002), and this approach was employed to identify eight phosphopeptides from Arabidopsis thylakoid membranes, including three new phosphopeptides (Hansson and Vener, 2003). Another approach was based on the fact that the pKa's of acidic residues or PO_4^- are significantly different. Therefore, binding at sufficiently acidic conditions should result in the specific protonation of the acidic residues, leaving only the phosphopeptides able to bind to the IMAC column. Indeed, this strategy was successful in identifying more than 300 phosphorylation sites from about 200 Arabidopsis plasma membrane proteins (Nühse *et al.*, 2003b, 2004). A further recent option is the use of cation exchange columns under very specific pH conditions that greatly enrich phosphopeptides, resulting in the identification of 2000 phosphorylation sites from more than 900 proteins from HeLa cell nuclei (Beausoleil *et al.*, 2004). At this time, the technology platforms for phosphoproteomic analyses have matured sufficiently that large-scale studies are being performed in plants (De la Fuente van Bentem *et al.*, 2006; Benschop *et al.*, 2007; Niittyla *et al.*, 2007; Nühse *et al.*, 2007).

8.3.2 Glycosylphosphatidylinositol (GPI)

GPI is a PTM on the C-terminus of some proteins, tethering them to the extracellular membrane. These GPI-anchored proteins (GPI-APs) have greater flexibility than true integral membrane proteins and the possibility of dynamic protein release to the extracellular space by specific phospholipase cleavage of the GPI moiety. Although the C-terminal attachment motif allows prediction of GPI-APs (Borner *et al.*, 2002), prediction programmes are not always in agreement. Thus, empirical evidence was obtained using the phospholipase, Pi-PLC, to 'shave' GPI-APs from plasma membrane preparations of Arabidopsis (Borner *et al.*, 2003; Elorteza *et al.*, 2003). Interestingly, only 19 proteins overlapped between the two sets of data, perhaps reflecting the large differences in age of the cells used in the two studies. Because of the potential role of GPI-APs in growth and development, quantitative comparisons of GPI-AP populations from different stages of growth or hormone treatment could be very informative.

8.3.3 Ubiquitylation

Components of the ubiquitylation machinery have been genetically defined in a variety of biological responses in plants, implicating targeted degradation of ubiquitylated proteins as an important regulatory mechanism in plant biology (Smalle and Vierstra, 2004). However, identification of the actual ubiquitylated protein is challenging. Recently, a method has been described for using ubiquitin-binding domains to affinity capture ubiquitylated proteins from complex mixtures of plant proteins (Maor *et al.*, 2007). From the 294 proteins enriched by this method, 85 ubiquitylated lysine residues from 56 proteins were sequenced directly, confirming the specific enrichment of the target population. This method provides the technical framework to perform comparative studies between different genotypes or biological responses.

8.4 Quantitation

The topics above generally described static proteomic analyses such as determining the protein complement of an organelle, but the full exploitation of proteomics in systems biology requires the ability to perform comparative studies. Therefore, quantitative methods are required to investigate dynamic changes in the proteome. A number of quantitative methods are currently available for proteomic studies. Importantly, all the methods described below have been employed successfully to gain new insights into plant biology. Rather than considering these methods as competing for superiority, it is perhaps better to consider them as complementary alternatives. In a cross-comparison of methods in plants, 2D gels, isotope-coded affinity tags (ICAT) and label-free quantitative methods were used to investigate differential accumulation of maize chloroplast proteins in bundle sheath versus mesophyll cells (Majeran et al., 2005). Only 20 of the 125 proteins quantified in these experiments were identified by all three methods. This example demonstrates that many methods provide useful information and that pursuing multiple strategies is likely to provide a deeper understanding of the biology in question. In many cases, the final decision on selecting a method is likely to be determined by the researcher's access to local technology to support the method.

8.4.1 Gel-based quantitation: two-dimensional gel electrophoresis (2DE)

The most established method for proteomic comparisons relies on protein isoelectric focusing coupled to SDS–PAGE, usually referred to as 2DE, followed by comparisons of stained protein patterns in the gels.

Although this traditional approach has been used successfully for many years, the reproducibility of 2DE gels and limits in sensitivity for gel staining are problematic. Streaks and poorly focused spots can arise from sample overloading as one tries to increase the percentage of the proteome that can be visualized. Ultra-sensitive, quantitative detection methods such as fluorescent dyes (e.g. Sypro Ruby (Invitrogen) and Deep Purple (GE Healthcare)) address some of these sample loading issues (Miller et al., 2006), but matching of 2DE spots between multiple gels containing >1000 spots each remains highly challenging even with advanced 2DE analysis software. Therefore, even though the method works, traditional 2DE comparisons are still laborious and time consuming because of the number of technical and biological replicates necessary to obtain conclusive results.

An approach that improves on both sensitivity and gel variability problems of traditional 2DE is difference gel electrophoresis (DIGE). In DIGE, protein samples are labelled with distinct fluorescent dyes, typically on lysine residues (Tonge et al., 2001). Three different charge-matched, Lys-reactive dyes are commercially available at present (Cy 2, 3, 5; GE Healthcare). Generally, the typical design is to label one sample with Cy3 and the other with Cy5 (Cy2 is used as a pooled internal control if multiple gels are to be used), mix the samples and separate them in a single 2D gel (Alban et al., 2003). Once resolved in the gel, the labelled proteins are visualized using a laser fluorescent scanner capable of detecting each dye in subsequent scans. Analyzing multiple samples in a single 2DE gel eliminates gel-to-gel variation and greatly simplifies spot matching. In addition, the sensitivity of the dyes enables the routine detection of >1000 protein spots with only 50 μg of protein sample. Of course, the sensitivity of detection may exceed the ability to sequence the protein by mass spectrometry. Therefore, it may be necessary to resolve a separate, preparative 2DE gel containing Cy-labelled protein mixed with larger amounts of unlabelled protein to match fluorescent and Coomassie-stained images for protein identification.

In plants, DIGE has been used for a variety of applications. For example, near isogenic sunflower lines varying in achene oil content were compared by DIGE to identify markers associated with the variations in oil content (Hajduch *et al.*, 2007). As expected for near isogenic lines, a small percentage of the total 2DE spots were differential. Of the differential proteins, many were involved in glycolysis and protein synthesis or storage, supporting a relationship between oil and protein content. These experiments also demonstrate that DIGE is amenable to comparisons within plant species not supported by extensive genome information. DIGE has also been used in studies to characterize changes in the proteome in response to biotic and abiotic stresses. In Arabidopsis cell cultures treated with fungal elicitors, approximately 10% of the proteins (154 of 1500) changed in abundance by at least 20% after 24 h of elicitation (Chivasa *et al.*, 2006). Of these, 45 were of sufficient abundance to be identified by mass spectrometry. Similarly, DIGE successfully identified proteins differentially accumulated in maize leaves in response to UV-B treatment (Casati *et al.*, 2005) and in Arabidopsis plants in response to cold stress (Amme *et al.*, 2006).

For many researchers, the requirement of only a laser imager and analysis software to perform DIGE comparisons may make this a more affordable approach for quantitative proteomics compared to mass spectrometry-based quantitative approaches. Despite these advantages, however, DIGE suffers from the same problems as traditional 2DE including under-representation of proteins with high/low molecular weights or with basic/acidic pI as well as hydrophobic membrane proteins. Still, the simplicity of analysis and the ability to employ this method for studies in plants with less than complete genome sequences will outweigh the disadvantages for many researchers.

8.4.2 Quantitation by mass spectrometry

As discussed above, 2D gel-based quantification is limited by the properties of intact proteins. Quantification by mass spectrometry, MS-based approaches, works from a different premise because it relies on the abundance of peptides as surrogates for intact proteins. Although a variety of mass spectrometers exist with differences in how they detect and fragment peptides (Domon and Aebersold, 2005), the workflow for MS analyses is fairly similar. After digestion of proteins with a protease (typically trypsin), the complex peptide mixture is separated by chromatography either prior to MS analysis or directly coupled to the mass spectrometer. The peptides are ionized to acquire the initial MS scan, a spectrum of the mass to charge ratio of peptide ions in that sample. Selected peptides from the MS scan are then individually fragmented for the MS/MS scan to determine the amino acid sequence of the peptide. In many types of machines, some form of dynamic exclusion is used to maximize the number of peptides sequenced. Individual peptides elute from a reversed-phase column coupled to the mass spectrometer in the time scale of minutes while a mass spectrometer collects data in seconds or milliseconds. Therefore, rather than repeatedly resequencing an abundant peptide, dynamic exclusion will cause the machine to ignore ions for which MS/MS spectra have already been acquired within a certain time frame.

To avoid variations between sample runs on the mass spectrometer, a number of different labelling strategies have been developed that allow direct comparisons of peaks, corresponding to the peptide abundance in different samples, within the same MS or MS/MS scan. At a basic level, these strategies are variations on a similar theme: peptides are modified with inert, stable isotopic tags such that the ionization and chromatographic properties of the tagged peptides are similar. After the analysis, the sample origin can be deciphered based upon a signature mass shift either in the MS or MS/MS spectrum, and the peak intensity reflects the relative abundance of that peptide in each sample. As discussed below, the main differences between these labelling methods are when the tag is introduced into the protein/peptide and how the quantitative data are extracted.

8.4.3 In vivo isotopic labelling

In vivo metabolic labelling of proteins with isotopes involves growing one set of samples on a natural nitrogen source while the comparative sample is grown in the presence of the heavy isotope. The isotopic label is introduced either using ^{15}N as the sole nitrogen source or using isotopically labelled amino acids, termed stable isotopic labelling by amino acids in cell culture (SILAC). The end result in both cases is that the mass of a peptide from a control or labelled population will be different allowing for a direct comparison of MS peak intensities between the two samples. In theory, the differential samples can be mixed very early in the experiment, virtually eliminating potential variation that might arise from technical variation during sample handling. However, only two or three samples can be compared at one time, limited by both the ability to introduce distinguishing isotopic tags into the cells as well as the resulting increase in complexity in the MS scan. Also, because all peptides in the MS scan are not always sequenced, machines with extremely high mass accuracy in the MS mode are necessary to ensure that the peaks being compared are the same peptide sequence and not one with a very similar mass. Therefore, some researchers may not be able to employ these methods depending on the type of mass spectrometer available.

8.4.3.1 Stable isotopic labelling by amino acids in cell culture (SILAC)

Usually, a single isotopically labelled amino acid is used for SILAC analyses. If the supplied amino acid is lysine or arginine, analysis of peptides from a trypsin digest (which cleaves immediately after these two amino acids) will result in peptides containing only a single, reproducible mass difference. Gruhler *et al.* (2005) used this method to identify subsets of glutathione-S-transferases and 14–3-3 proteins that were differentially regulated in response to treatment of Arabidopsis cell cultures with salicylic acid. Because plants can synthesize all amino acids from inorganic nitrogen, however, exogenous amino acid feeding of Arabidopsis cell cultures achieved on average only 70–80% efficiency (Gruhler *et al.*, 2005). Therefore, a portion of the peptide from the heavy isotopic-labelled sample is not labelled and will always contaminate the unlabelled peptide. Although this contamination can be back-calculated, it complicates the analysis. The other disadvantage of SILAC is the expense of labelled amino acids. The cost restrictions and efficiency of labelling are likely to limit the use of this method to plant cell cultures.

8.4.3.2 ^{15}N metabolic labelling

Metabolic labelling with ^{15}N, on the other hand, is much more efficient, achieving >98% incorporation in both intact plants (Ippel *et al.*, 2004) and cell cultures (Engelsberger *et al.*, 2006). In addition, it is more cost-effective than SILAC. The trade-off is that all amino acids will incorporate the isotopic label, meaning that the mass difference between labelled and unlabelled peptides will be different for each peptide depending on its amino acid sequence. Therefore, for each peak comparison, the peptide must be sequenced before the location of the paired MS peak can be calculated and located. Although software performs these tasks, comparisons could be complicated with highly complex samples. In addition, it is not yet clear whether all plant tissues can be efficiently labelled with this method.

8.4.4 In vitro isotopic labelling of peptides

An alternative approach to in vivo labelling is to introduce the label into the peptides after isolation and digestion of the proteins. Therefore, in vitro labelling of peptides can be used with any sample. Of course, because the isotopic label is introduced after the proteins are isolated and digested, greater care is needed to control technical variation introduced during the isolation of protein (e.g. during subcellular fractionation).

8.4.4.1 Isotope-coded affinity tags (ICAT)

One method for labelling peptides relies on covalent modification of cysteine residues with chemically identical tags that differ only in mass because of inclusion of heavy and light isotopes. The tags include biotin which facilitates the rapid enrichment of tagged peptides using streptavidin. Using a strategy based on cysteine chemistry is both an advantage and disadvantage. Most proteins only contain a few cysteine residues. Therefore, each protein will be represented by a small number of peptides, greatly reducing the sample complexity and increasing the probability that a greater dynamic range of proteins will be analyzed. However, about one in seven proteins do not contain cysteine residues guaranteeing limitations in the completeness of the analysis. Furthermore, comparisons of peptides with PTMs will be limited to only modified peptides containing cysteine residues, greatly reducing the value of ICAT for these types of experiments.

8.4.4.2   Isobaric-tags for relative and absolute quantitation (iTRAQ)

An alternative method is similar to the concept for ICAT but involves chemical modification of all primary amines with isobaric-tags for relative and absolute quantitation (iTRAQ) (Ross *et al.*, 2004). Therefore, each sample population receives its own unique tag, but iTRAQ reagents label (nearly) all tryptic peptides instead of only those containing cysteine residues. In addition, iTRAQ currently allows comparison of up to four samples in the same experiment (new 8-plex reagents are set to be released by Applied Biosystems in 2008). Another aspect that makes iTRAQ unique from other methods is that quantitation occurs in the MS/MS scan after fragmentation of the peptide. Peptide fragmentation releases the iTRAQ mass reporter of 114, 115, 116 or 117 Daltons, and the intensity of this reporter peak reflects the relative quantity of the peptide in each sample. Because the comparison occurs in the MS/MS scan (i.e. when the peptide is sequenced), relative quantitative comparisons are obtained unambiguously for each peptide sequenced.

The possibility of labelling at least four samples allows analysis of time-course experiments. This approach was employed in conjunction with chromatographic enrichment of phosphopeptides to identify proteins undergoing differential phosphorylation in response to microbial elicitation of Arabidopsis suspension cultured cells (Nühse *et al.*, 2007). These studies also demonstrated that phosphorylation of the differentially regulated residues were mechanistically required for the activation of rbohD, the NADPH oxidase required for the early oxidative burst during defence responses. Another application of iTRAQ in plants was to quantitate information from 2D gel regions where spots had not resolved sufficiently to provide unequivocal results (Rudella *et al.*, 2006). While investigating the proteomic changes in chloroplasts of *clpr2–1*, a mutant of a plastidial protease complex, the ClpP/R/S subunit proteins of the Clp core complex ran as overlapping spots on 2D gels, obscuring accurate quantitation. Isolation of this complex gel region followed by iTRAQ labelling of peptides from gels containing control and mutant protein samples demonstrated that most of the subunits of Clp complex were decreased in the mutant but to different degrees.

8.5 Conclusion

Proteomics has matured as a field and is now producing exciting new insights into a range of questions in plant biology. As discussed at the beginning of this chapter, the true value of a proteomic study is in its ability to provide unique information about cellular responses. Valuable information does not necessarily require a massive investment in mass spectrometry sequencing time to generate lists of thousands of proteins. As with many other functional genomic approaches, good experimental design to address focused questions will have the greatest immediate impact on plant biology, and these types of studies will include integrating quantitative proteomic methods with the use of genetic mutants and/or tightly controlled treatment conditions. One of the greatest challenges we face in the near future is learning how to effectively integrate the information from proteomic studies with that produced by other functional genomic studies. With the successes in functional genomic platforms we have witnessed in the past decade, however, one can be confident that this challenge too will be handled in time.

References

Aebersold, R. and Goodlett, D.R. (2001) Mass spectrometry in proteomics. *Chem Rev* **101**, 269–295.

Alban, A., David, S.O., Bjorkesten, L., Andersson, C., Sloge, E., Lewis, S., *et al.* (2003) A novel experimental design for comparative two-dimensional gel analysis: two-dimensional difference gel electrophoresis incorporating a pooled internal standard. *Proteomics* **3**, 36–44.

Amme, S., Matros, A., Schlesier, B. and Mock, H.P. (2006) Proteome analysis of cold stress response in *Arabidopsis thaliana* using DIGE-technology. *J Exp Bot* **57**, 1537–1546.

Anderson, N.L. and Anderson, N.G. (2002) The human plasma proteome: history, character, and diagnostic prospects. *Mol Cell Proteomics* **1**, 845–867.

Baginski, S. and Gruissem, W. (2004) Chloroplast proteomics: potentials and challenges. *J Exp Bot* **55**, 1213–1220.

Beausoleil, S.A., Jedrychowski, M., Schwartz, D., Elias, J.E., Villén, J., Li, J., *et al.* (2004) Large-scale characterization of HeLa cell nuclear phosphoproteins. *Proc Natl Acad Sci USA* **101**, 12130–12135.

Benschop, J.J., Mohammed, S., O'Flaherty, M., Heck, A.J.R., Slijper, M. and Menke, F.L.H. (2007) Quantitative phospho-proteomics of early elicitor signalling in Arabidopsis. *Mol Cell Proteomics* **6**, 1198–1214.

Borner, G.H.H., Lilley, K.S., Stevens, T.J. and Dupree, P. (2003) Identification of glycosylphosphatidylinositol-anchored proteins in Arabidopsis. A proteomic and genomic analysis. *Plant Physiol* **132**, 568–577.

Borner, G.H.H., Sherrier, D.J., Stevens, T.J., Arkin, I.T. and Dupree, P. (2002) Prediction of glycosylphosphatidylinositol-anchored proteins in Arabidopsis: a genomic analysis. *Plant Physiol* **129**, 486–499.

Carter, C., Pan, S., Zouhar, J., Avila, E.L., Girke, T. and Raikhel, N.V. (2004) The vegetative vacuole proteome of *Arabidopsis thaliana* reveals predicted and unexpected proteins. *Plant Cell* **16**, 3285–3303.

Casati, P., Zhang, X., Burlingame, A.L. and Walbot, V. (2005) Analysis of leaf proteome after UV-B irradiation in maize lines differing in sensitivity. *Mol Cell Proteomics* **4**, 1673–1685.

Chivasa, S., Hamilton, J.M., Pringle, R.S., Ndimba, B.K., Simon, W.J., Lindsey, K., *et al.* (2006) Proteomic analysis of differentially expressed proteins in fungal elicitor-treated cultures. *J Exp Bot* **57**, 1553–1562.

Cutler, S.R., Ehrhardt, D.W., Griffitts, J.S. and Somerville, C.R. (2000) Random GFP::cDNA fusions enable visualization of subcellular structures in cells of Arabidopsis at a high frequency. *Proc Natl Acad Sci USA* **97**, 3718–3723.

De la Fuente van Bentem, S., Anrather, D., Roitinger, E., Djamei, A., Hufnagl, T., Barta, A., *et al.* (2006) Phosphoproteomics reveals extensive in vivo phosphorylation of Arabidopsis proteins involved in RNA metabolism. *Nucleic Acids Res* **34**, 3267–3278.

Domon, B. and Aebersold, R. (2005) Mass spectrometry and protein analysis. *Science* **312**, 212–217.

Elorteza, F., Nühse, T.S., Stansballe, A., Peck, S.C. and Jensen, O.N. (2003) Proteomic analysis of glycosylphosphatidylinositol-anchored membrane proteins. *Mol Cell Proteomics* **2**, 1261–1270.

Engelsberger, W.R., Erban, A., Kopka, J. and Schulze, W.X. (2006) Metabolic labeling of plant cell cultures with $K^{15}NO_3$ as a tool for quantitative analysis of proteins and metabolites. *BMC Plant Methods* **2**, 14.

Ficarro, S.B., McCleland, M.L., Stukenber, P.T., Burke, D.J., Ross, M.M., Shabanowitz, J., *et al.* (2002) Phosphoproteome analysis by mass spectrometry and its application to Saccharomyces cerevisiae. *Nat Biotechnol* **20**, 301–305.

Gruhler, A., Schulze, W.X., Matthiesen, R., Mann, M. and Jensen, O.N. (2005) Stable isotope labeling of *Arabidopsis thaliana* cells and quantitative proteomics by mass spectrometry. *Mol Cell Proteomics* **4**, 1697–1709.

Gygi, S.P., Rochon, Y., Franza, B.R. and Aebersold, R. (1999) Correlation between protein and mRNA abundance in yeast. *Plant Physiol* **19**, 1720–1730.

Hajduch, M., Casteel, J.E., Tang, S., Hearne, L.B., Knapp, S. and Thelen, J.J. (2007) Proteomic analysis of near isogenic sunflower varieties differing in seed oil traits. *J Proteome Res* **6**, 3232–3241.

Hansson, M. and Vener, A.V. (2003) Identification of three previously unknown in vivo protein phosphorylation sites in thylakoid membranes of *Arabidopsis thaliana*. *Mol Cell Proteomics* **2**, 550–559.

Heazlewood, J.L., Verboom, R.E., Tonti-Filippini, J., Small, I. and Millar, A.H. (2007) SUBA: the arabidopsis subcellular database. *Nucleic Acids Res* **35**, D213–D218.

Ippel, J.H., Pouvreau, L., Kroef, T., Gruppen, H., Versteeg, G., Van Den Putten, P., *et al.* (2004) *In vivo* uniform ^{15}N-isotope labeling of plants: using the greenhouse for structural proteomics. *Proteomics* **4**, 226–234.

Koroleva, O.A., Tomlinson, M.L., Leader, D., Shaw, P. and Doonan, J.H. (2005) High-throughput protein localization in Arabidopsis using agrobacterium-mediated transient expression of GFP-ORF fusions. *Plant J* **41**, 162–174.

Majeran, W., Cai, Y., Sun, Q. and van Wijk, K.J. (2005) Functional differentiation of bundle sheath and mesophyll maize chloroplasts determined by comparative proteomics. *Plant Cell* **17**, 3111–3140.

Maor, R., Jones, A., Nühse, T.S., Studholme, D.J., Peck, S.C. and Shirasu, K. (2007)

Multidimensional protein identification technology (MudPIT) analysis of ubiquitylated proteins in plants. *Mol Cell Proteomics* **6**, 601–610.

Millar, A.H. (2004) Location, location, location: surveying the intracellular real estate through proteomics in plants. *Funct Plant Biol* **31**, 563–571.

Miller, I., Crawford, J. and Gianazza, E. (2006) Protein stains for proteomic applications: which, when, why? *Proteomics* **6**, 5385–5408.

Niittyla, T., Fuglsang, A.T., Palmgren, M.G., Frommer, W.F. and Schulze, W.X. (2007) Temporal analysis of sucrose-induced phosphorylation changes in plasma membrane proteins of Arabidopsis. *Mol Cell Proteomics* **6**, 1711–1726.

Nühse, T.S., Boller, T. and Peck, S.C. (2003a) A plasma membrane syntaxin is phosphorylated in response to the bacterial elicitor flagellin. *J Biol Chem* **278**, 45248–45254.

Nühse, T.S., Bottrill, A.R., Jones, A.M.E. and Peck, S.C. (2007) Quantitative phosphoproteomic analysis of plasma membrane proteins reveals regulatory mechanisms of plant innate immune responses. *Plant J* **51**, 931–940.

Nühse, T.S., Stansballe, A., Jensen, O.N. and Peck, S.C. (2003b) Large-scale analysis of in vivo phosphorylated membrane proteins by immobilized metal ion affinity chromatography and mass spectrometry. *Mol Cell Proteomics* **2**, 1234–1243.

Nühse, T.S., Stansballe, A., Jensen, O.N. and Peck, S.C. (2004) Phosphoproteomics of the Arabidopsis plasma membrane and a new phosphorylation site database. *Plant Cell* **16**, 2394–2405.

Peck, S.C., Nühse, T.S., Iglesias, A., Hess, D., Meins, F. and Boller, T. (2001) Directed proteomics identifies a plant-specific protein rapidly phosphorylated in response to bacterial and fungal elicitors. *Plant Cell* **13**, 1467–1475.

Ross, P.L., Huang, Y.N., Marchese, J.N., Williamson, B., Parker, K., Hattan, S., *et al.* (2004) Multiplexed protein quantitation in *Saccharomyces ceravisiae* using amine-reactive isobaric tagging reagents. *Mol Cell Proteomics* **3**, 1154–1169.

Rudella, A., Friso, G., Alonso, J.M., Ecker, J.R. and van Wijk, K.J. (2006) Downregulation of ClpR2 leads to reduced accumulation of the ClpPRS protease complex and defects in chloroplast biogenesis in Arabidopsis. *Plant Cell* **18**, 1704–1721.

Shimaoka, T., Ohnishi, M., Sazuka, T., Mitsuhashi, N., Hara-Nishimura, I., Shimazaki, K.I., *et al.* (2004) Isolation of intact vacuoles and proteomic analysis of tonoplast from suspension-cultured cell of *Arabidopsis thaliana*. *Plant Cell Physiol* **45**, 672–683.

Shin, R., Alvarez, S., Burch, A.Y., Jez, J.M. and Schachtman, D.P. (2007) Phosphoproteomic identification of targets of the Arabidopsis sucrose nonfermenting-like kinase SnRK2.8 reveals a connection to metabolic processes. *Proc Natl Acad Sci USA* **104**, 6460–6465.

Smalle, J. and Vierstra, R.D. (2004) The ubiquitin 26S proteasome proteolytic pathway. *Annu Rev Plant Biol* **55**, 555–590.

Szponarski, W., Sommere, N., Boyer, J.C., Rossignol, M. and Gibrat, R. (2004) Large-scale characterization of integral proteins from Arabidopsis vacuolar membrane by two-dimensional liquid chromatography. *Proteomics* **4**, 397–406.

Tian, Q., Stepaniants, S.B., Mao, M., Weng, L., Feetham, M.C., Doyle, M.J., *et al.* (2004) Integrated genomic and proteomic analyses of gene expression in mammalian cells. *Mol Cell Proteomics* **3**, 960–969.

Tonge, R., Shaw, J., Middleton, B., Rowlinson, R., Rayner, S., Young, J., *et al.* (2001) Validation and development of fluorescent two-dimensional gel electrophoresis proteomics technology. *Proteomics* **1**, 277–296.

Wienkoop, S., Glinski, M., Tanaka, N., Tolstikov, V., Fiehn, O. and Weckwerth, W. (2004) Linking protein fractionation with multidimensional monolithic reversed-phase peptide chromatography/mass spectrometry enhances protein identification from complex mixtures even in the presence of abundant proteins. *Rapid Commun Mass Spectrom* **18**, 643–650.

Annual Plant Reviews (2009) **35**, 258–289
doi: 10.1111/b.9781405175326.2009.00009.x

www.interscience.wiley.com

Chapter 9

METABOLOMICS: INTEGRATING THE METABOLOME AND THE PROTEOME FOR SYSTEMS BIOLOGY

Wolfram Weckwerth

Department of Molecular Plant Physiology and Systems Biology, University of Vienna, Vienna, Austria
GoFORSYS, Potsdam, Germany
Max Planck Institute of Molecular Plant Physiology, Potsdam, Germany

Abstract: The term 'metabolome' was coined in analogy to 'genome' and 'proteome' and relates to the metabolism. The metabolome contains all the characteristic metabolic activities of a cell or tissue and can be described on the basis of conversion rates, enzyme activities and intermediates, the set of metabolites. In order to study a cell, tissue or an organism in its entirety it is not sufficient to merely consider its metabolome alone. Only the interactions and synergisms between the transcriptome, the proteome and finally the metabolome will furnish the key to a functioning living system. This approach does not only promise to provide an improved picture of the metabolic processes and the interaction. In this way, we also hope to understand the function of hitherto unknown genes representing a large proportion of all so far published genomes. Finally, computer models for making predictions of the metabolic network and their iterative linkage to experimental data via the interaction of proteins and metabolic processes may contribute to studying fundamental mechanisms of metabolism.

Keywords: plant systems biology; multivariate statistics; sample pattern recognition; mass spectrometry; gas chromatography coupled to time-of-flight mass spectrometry (GC-TOF-MS); liquid chromatography coupled to mass spectrometry (LC/MS); shotgun proteomics; label-free

9.1 The molecular hierarchy in biochemical networks, the concept of systems biology and functional genomics in the post-genome era

The central dogma of biology defines a molecular hierarchy in living systems from a gene to a protein (Crick, 1970). Consequently, important progress towards a complete understanding of biological systems was achieved by sequencing whole genomes of a variety of prokaryotes and eukaryotes (http://www.ncbi.nlm.nih.gov/Genomes/). These large-scale sequencing projects have also yielded the complete genomic sequence for a number of plant species, including *Chlamydomonas reinhardtii, Oryza sativa, Popolus trichocarpa* and *Arabidopsis thaliana*. Interestingly, gene annotation and homology searches have revealed that only a small fraction are functionally characterized experimentally either directly or through gene homology with other species (Arabidopsis Genome Initiative, 2000). Thus, analysis of gene function or 'functional genomics' is the prior 'post-genome' discipline (Trethewey *et al.*, 1999; Somerville and Dangl, 2000). The first step towards functional analysis is to quantify differential gene expression. Several technology developments have addressed whole genome analysis using techniques such as oligonucleotide and cDNA microarray analysis (Schena *et al.*, 1995; Chee *et al.*, 1996; Aharoni and Vorst, 2002) and RtPCR (real-time-PCR; Czechowski *et al.*, 2004) to systematically measure gene expression at the mRNA level. These methods have provided a vast amount of genomic and transcriptomic information about plants as well as other organisms in molecular, developmental and stress physiology (http://www.arabidopsis.org/info/expression/ATGenExpress.jsp). However, measuring mRNA dynamics is only a very early step in the assessment of biological systems. In Fig. 9.1, the different molecular levels and their interactions including gene expression by transcription and translation are shown. In several studies dealing with holistic views of biological systems, it became increasingly obvious that transcriptional and translational organizations are not necessarily correlated with all other subsequent responses of a biological system (Gygi *et al.*, 1999; ter Kuile and Westerhoff, 2001; MacCoss *et al.*, 2003; Gibon *et al.*, 2004; Baginsky *et al.*, 2005). This is exemplified by metabolic control which is executed at a post-transcriptional and translational level and results in tremendous phenotypic plasticity, which is especially important in plant systems. Plants are sessile systems who cannot escape environmental pressures. Thus, they have developed methods to survive under highly variable conditions and evolved a molecular flexibility to cope with these pressures. These mechanisms include metabolite sensing, signal cascades via protein kinase pathways and, the more recently recognized mechanisms of post-transcriptional regulation (Koch, 1996; Jang and Sheen, 1997; MacKintosh, 1998; Romeis, 2001; Hamilton *et al.*, 2002; Jonak *et al.*, 2002; Rolland *et al.*, 2002; Apel and Hirt, 2004).

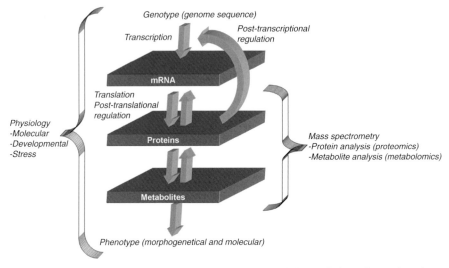

Figure 9.1 The 'genotype–phenotype relationship' and the underlying hierarchy of biochemistry and physiology. The analysis of molecular levels of proteins and metabolites is bound to bioanalytical and mass spectrometry-based techniques.

From such complexity it becomes clear that gene function can only be resolved with measurements of the actively expressed molecular phenotype comprising mRNA expression, protein concentrations, protein activities, protein modifications and metabolite fluctuations (see Fig. 9.1). The assessment of the whole interacting molecular system depicted in Fig. 9.1 makes it in principle possible, for the first time, to reveal a complete picture of metabolism combined with statistical and mathematical modelling of the data thereby unravelling the driving mechanisms of biological systems. Thus, the term 'systems biology' best encapsulates our efforts to undertake this challenge (Kitano, 2000; Ideker *et al.*, 2001a; Weckwerth, 2003; Ideker, 2004).

Somerville and Dangl (2000) defined a catalogue of things 'to do' at the completion of the *Arabidopsis* genome sequencing project. The following items are relevant to this review:

'Representative goals of the 2010 project are as follows:

1. *To 3-year goals.* Describe global protein profiles at organ, cellular and subcellular levels under various environmental conditions.
2. *To 6-year goals.* Develop global understanding of post-translational modification.
3. *To 6-year goals.* Undertake global metabolic profiling at organ, cellular and subcellular levels under various environmental conditions (Somerville and Dangl, 2000)'.

At the time 'global profiling' methods were at very early stages of development and these comments thus represent very visionary ideas. However, the following years have seen many of these technologies develop and emerge and have brought these visions to reality.

9.2 Metabolomics and proteomics: post-genome disciplines intimately bound to mass spectrometric techniques

The analysis of the quantitative molecular phenotype demands the step towards post-genomic techniques such as 'metabolomics' and 'proteomics' for the processes of unbiased identification and state-specific quantification of all metabolites and proteins in a cell, tissue or whole organism. These large-scale analyses of proteins and metabolites are intimately bound to advancements in mass spectrometric technologies (see Fig. 9.1 and the following sections) and emerged in parallel with the development of novel mass analyzers and hyphenated techniques, for instance gas chromatography coupled to time-of-flight mass spectrometry (GC-TOF-MS) for metabolomics (see Section 9.3.1) and multidimensional LC-MS for proteomics (see Section 9.5.1). The following sections describe the development and applications of these technologies in molecular plant physiology and biochemistry and their significance in the post-genome era.

9.3 Metabolomics: global analysis of rapid metabolic responses combined with computer-aided iterative metabolic modelling

The concept of metabolomics was introduced as the global analysis of all metabolites in a sample (Oliver *et al.*, 1998; Tweeddale *et al.*, 1998; Nicholson *et al.*, 1999; Trethewey *et al.*, 1999). It is an extension of existing methods looking for target compounds and their biochemical transformation. Novel developments for mass detectors and hyphenated techniques such as LC-MS, GC-MS and CE-MS (capillary electrophoresis coupled to mass spectrometry) techniques promote the challenge to measure the metabolome. The most important difference to conventional protocols of metabolite analysis is the concept to reveal metabolite dynamics in an unbiased manner, according to the general definition of 'omics-technologies' (Weckwerth, 2003). The aim is to study gene function and fundamental metabolic principles in biochemical networks. Metabolomic technology enables the rapid, accurate and precise analysis of metabolites and subsequent pattern recognition of biological samples, identification of biomarkers and the direct analysis of the biochemical consequences of gene mutations and environmental perturbations. In contrast to other post-genomic technologies, such as proteomics and transcriptomics, metabolomics has developed into a routine process with respect to sample throughput and robustness within a few years (Weckwerth and Morgenthal, 2005). The analytical techniques are capable of detecting hundreds of individual chemical structures, ranging from metabolite fingerprinting to an extensive metabolite profiling (see next Section 9.3.1).

For the extraction of biologically meaningful information from such huge data sets, the application of biostatistics and novel mathematical frameworks is essential. A whole new field of biostatistics is emerging according to the classical concept of chemometrics in analytical chemistry. For multivariate data analysis, problems arise from small sample numbers in contrast to the high number of measured metabolites and the resulting high dimensionality of the data matrix. Novel algorithms and statistical analysis have to be established for test cases and learning sets.

After a general introduction to metabolomic techniques with respect to diagnostics, pattern recognition and biomarker discovery, this chapter focuses on developments in biostatistics and biological interpretation of metabolite profiles as an 'in vivo' fingerprint of metabolic networks. A stochastic model of metabolic networks is introduced leading to a novel understanding of co-regulation in biochemical networks and direct consequences for statistical interpretation and computer simulation of metabolomic data.

9.3.1 Metabolomic techniques

In view of the physicochemical diversity of small biological molecules the challenge remains to develop one or more protocols to gather the whole 'metabolome', the complete set of small molecules present in a sample. The general estimation of size and dynamic range of a species-specific metabolome is at a preliminary stage. In the plant kingdom the structural diversity is enormous revealing new compounds on a daily basis. Estimates exceed 500 000 putative structures (Hadacek, 2002). No single technique is suitable for the analysis of all these different chemical structures, which is why a mixture of techniques has to be used and the data have to be combined (Weckwerth, 2003).

Goodacre et al. (2004) give a valid overview of metabolomic techniques. Metabolic fingerprinting with high sample throughput but decreased dynamic range and deconvolution of individual components achieve a global view on in vivo dynamics of metabolic networks (Nicholson et al., 1999, 2002; Castrillo and Oliver, 2004; Goodacre et al., 2004; Kell, 2004; Dunn and Ellis, 2005). A lower sample throughput but unassailable identification and quantitation of individual compounds in a complex sample is achieved by GC-MS and LC-MS technology.

Owing to major steps forward in these technologies, it is possible to match specific demands with specific instruments and novel developments in the performance of mass analyzers (see Table 9.1). However, it is important to

Table 9.1 Mass analyzers, hyphenated techniques and their performances

Mass analyzer	Ionization technique	Chromatography	Scan modes	Speediness/sensitivity/mass accuracy
Quadrupole	ESI, EI, CI, FI; APCI, APPI	GC, CE, LC	Full scan; SIM	Scan speed slow, faster with SIM mode, but mass range restricted
Triple quadrupole	ESI; APCI, APPI; MALDI	GC, CE, LC	SIM; Full scan; MS^2; SRM/MRM	Full scan slow; MRM very fast and sensitive; Exact masses with internal calibration
Triple quadrupole linear trap	ESI; APCI, APPI; MALDI	CE, LC	Full scan; MS^2, MS^n; SRM/MRM	Full scan medium; MS^n possible
Ion trap	ESI, EI, CI; APCI, APPI; MALDI	GC, CE, LC	Full scan; SIM; MS^2, MS^n	As for above
Linear ion trap	ESI; APCI, APPI; MALDI	CE, LC	Full scan; SIM; MS^2, MS^n	Very fast and sensitive full scan; Rest as for above
ToF	ESI, EI, FI; APCI, APPI; MALDI	GC, CE, LC	Full scan; Source fragmentation	Very sensitive full scan; Exact masses with internal calibration
Quadrupole ToF	ESI; APCI, APPI; MALDI	CE, LC	Full scan; MS^2	Most sensitive full scan; Exact masses with internal calibration; Resolution 20 000
OrbitrapR	ESI, EI,; APCI, APPI; MALDI	LC, CE	Full scan; MS^2, MS^n	Exact masses ($<$ 2 ppm) without internal calibration; Resolution 100 000
FTICR	ESI, EI, FI; APCI, APPI; MALDI	LC	Full scan; MS^2, MS^n	Exact masses ($<$1 ppm) without internal calibration; Resolution 1 000 000

Note: For further details see text.
ToF, time of flight; FTICR, Fourier transform ion cyclotron resonance; ESI, electrospray ionization; EI, electron impact; FI, field ionization; APCI, atmospheric pressure chemical ionization; APPI, atmospheric pressure photoionization; MALDI, matrix assisted laser desorption ionization; LC, liquid chromatography; GC, gas chromatography; CE, capillary electrophoresis; SIM, single ion monitoring; SRM, single reaction monitoring; MRM, multiple reaction monitoring.

consider that each technology will have a bias towards certain compound classes, mostly due to ionization techniques, chromatography and detector capabilities. GC-MS has evolved as an imperative technology for metabolomics owing to its comprehensiveness and sensitivity (Webb *et al.*, 1986; Sauter *et al.*, 1988; Fiehn *et al.*, 2000a, b; Roessner *et al.*, 2000, 2001; Weckwerth *et al.*, 2001, 2004a, b; Wagner *et al.*, 2003; Jonsson *et al.*, 2004; Boldt *et al.*, 2005; Broeckling *et al.*, 2005; Morgenthal *et al.*, 2005, 2006; Weckwerth, 2006). In 2001, the first application of plant metabolome analysis using gas chromatography coupled to a time-of-flight mass analyzer GC-TOF-MS was presented (Weckwerth *et al.*, 2001). The idea was to improve the conventional approach using quadrupole mass analyzer coupled to gas chromatography (Fiehn *et al.*, 2000a; Roessner *et al.*, 2000). In this study, two basic questions had to be resolved: (1) Can one increase the sensitivity which results in better signal-to-noise, more low-abundance compounds identified and less starting material for extraction? (2) Can one decrease measurement time resulting in higher sample throughput? Usually, one has to cope with a high dynamic range of abundance and co-elution in complex metabolite fractions. Thus, accurate deconvolution of chromatogram peaks – identification and quantification of analytes out of the raw data – demands high-quality spectra and peak shapes. The quality of peak shape and deconvolution is proportional to the scan speed and signal-to-noise of the mass analyzer. With respect to these demands the GC-TOF-MS instrument was a clear improvement over GC-quadrupole-MS. The time-of-flight mass analyzer can provide two basic features, one is mass resolution and accuracy and the other is high sensitivity in full scan mode. Mass resolution and accuracy is inversely related to sensitivity. High sensitivity but loss in resolution in the full scan mode is achieved by time-array detection using integrated transient recorder technology (ICR^{TM}) (Watson *et al.*, 1990; Veriotti and Sacks, 2001). The signal-to-noise ratio is increased by summing up several 100 'microscans' per scan making the search for low-abundance analytes in complex samples possible. Additionally, the TOF detector divides the mass ranges into small mass windows, which accelerate data transfer to the computer. Very high scan speeds of up to 500 full spectra/s are achieved improving the spectral deconvolution process.

In comparison to conventional GC-quadrupole-MS, these high scan speeds also enable very fast chromatography (Veriotti and Sacks, 2001). These features together provide an improvement over conventional GC-MS analysis with respect to the analysis of complex samples like in the metabolomic approach (Weckwerth *et al.*, 2001, 2004a, b; Boldt *et al.*, 2005; Morgenthal *et al.*, 2005, 2006; Weckwerth, 2006). For the convenient analysis of metabolite concentrations in many replicate plant samples, a novel protocol has been developed which is capable of analyzing the polar and the lipophilic fraction without phase separation in one chromatographic step on GC-TOF-MS (Weckwerth *et al.*, 2004b). A typical GC-TOF-MS analysis of *A. thaliana* leaf extract with representative metabolites can be found in this study (Weckwerth *et al.*, 2004b). For different plant tissues and species, technical and biological variability was compared in another study (Morgenthal *et al.*, 2006).

GC-TOF-MS analysis of complex plant tissue samples was also applied to distinguish a 'silent' plant phenotype from its wild type (WT) (Weckwerth *et al.*, 2004a). In this example, it was possible to extract ~1000 compounds from the data using the full potential of spectral deconvolution of the TOF-mass analyzer. However, this process is very time consuming as it is only semi-automated and due to the necessary manual interpretation. In the meantime, GC-TOF-MS has developed into a routine application in plant metabolomics, as well as in biomedical applications (Goodacre *et al.*, 2004; Dunn and Ellis, 2005; Shellie *et al.*, 2005; Underwood *et al.*, 2006; Weckwerth, 2006).

9.3.2 Metabolomics as an integral part of systems biology

Statistical data mining and integration of complex molecular data is one of the critical goals of systems biology. Ideker *et al.* (2001a) writes in '*A new approach to decoding life: Systems Biology*': 'Systems biology studies biological systems by systematically perturbing them (biologically, genetically, or chemically); monitoring the gene, protein, and informational pathway responses; integrating these data; and ultimately, formulating mathematical models that describe the structure of the system and its response to individual perturbations.' This is a very valid definition of systems biology as an iterative process by generating data, building mathematical models to explain these, generate hypotheses and testing these hypotheses to come at the end to a holistic picture of life. However, metabolomics has developed very rapidly into a well-accepted systems-biology discipline (Weckwerth, 2003; Fernie *et al.*, 2004). One could even argue that due to high sample throughput, comprehensiveness and accuracy, metabolomics is the superior technology for systems biology (Weckwerth, 2003). Consequently, one has to monitor metabolite responses alongside with protein and transcript analysis and consider these integrative dynamics for the assessment of biological systems and gene function analysis.

9.3.3 Interpretation of metabolomic data: a stochastic computer-aided model of metabolism, correlation networks and the underlying pathway structure in biochemical networks

Most of the current metabolite profiling approaches rely on the measurement of 'steady state' levels comparing for instance state A with state B, a mutant and a WT plant or stress perturbation (see Section 9.4). The tests for significant changes in averages or mean levels of specific metabolites will then reveal alterations in the regulation of plant metabolism. This comparative analysis relies on the statistically significant detection of differences between sample groups. Often a high biological variation of individual compounds is observed within a set of samples from the same background. This high biological variability of independent biological replicates can be exploited to go beyond the classical 'two state-differences-question' and can reveal systemic behaviour and biochemical regulations (see the following sections and Section 9.6). Metabolite profiling using GC-TOF-MS was applied to various plant systems such as *Solanum*, *Arabidopsis* and *Nicotiana* (Weckwerth *et al.*, 2001, 2004a, b; Morgenthal *et al.*, 2005, 2006), as well as *Chlamydomonas and Synechocystis*. In all these systems, significant pairwise correlations were observed between specific metabolites, termed co-regulation. These correlations showed conserved or altered structures between different species (Morgenthal *et al.*, 2006) and provided the basis for constructing connectivity networks of metabolites based on the Pearson's correlation coefficient. This coefficient was then facilitated to quantify the distance of the connectivity of all the measured metabolites and enabled the construction of metabolite distance maps visualized as differential metabolite correlation networks (Weckwerth *et al.*, 2001, 2004a; Weckwerth, 2003). Using this approach, significant alterations of these network structures were found in all plant systems investigated so far depending on the genotype and environmental perturbations (Kose *et al.*, 2001; Weckwerth *et al.*, 2001, 2004a, b; Steuer *et al.*, 2003a, b; Morgenthal *et al.*, 2005, 2006; Weckwerth and Steuer, 2005). A trend in these networks is a high connectivity of only a few nodes (metabolites), whereas many nodes have only a low connectivity (Weckwerth *et al.*, 2004a). The degree distribution of these networks is not clear-cut power law, coinciding with the observations of Ihmels *et al.* (2004) on gene co-expression/correlation networks in yeast. Thus, a refined analysis of real-world network structures is most important to define biochemical modules in these co-regulations.

Based on these empirical observations, a stochastic model of metabolism was developed that was able to explain these phenomena and to provide a reasonable framework for multivariate data mining and biological interpretation of huge metabolomic experiments (Weckwerth, 2003). Early work of Arkin and Ross (Arkin *et al.*, 1997, 1998; Samoilov *et al.*, 2001; Rao *et al.*, 2002; Vance *et al.*, 2002) and Rascher and Lüttge (Rascher *et al.*, 2001) demonstrated the need to introduce stochastic models for the interpretation of metabolic networks. By analogy, by introducing metabolite fluctuation using stochastic differential equations for a glycolytic pathway system, the putative origin of correlations in metabolomic data was proposed in order to connect these correlations to the underlying enzymatic pathway structure (Steuer *et al.*, 2003b). Using these correlation networks one is capable of revealing alterations in enzymatic activity and alterations in the differential analysis of various metabolic states (Weckwerth *et al.*, 2004a; Camacho *et al.*, 2005; Morgenthal *et al.*, 2005). Changes in the network topology point to regulatory hubs in the biochemical network because the correlation matrix of all metabolite pairs is a fingerprint of the enzymatic and regulatory reaction network as discussed above. It is further possible to compare the measured correlation network with the proposed underlying reaction network and the corresponding numerically resolved correlation network (Steuer *et al.*, 2003a, b; Morgenthal *et al.*, 2006). Here, it becomes evident that correlations cannot be predicted only on the basis of pathway connectivity. Most pairs of metabolites neighboured in the reaction network show a low correlation, whereas other metabolite pairs that are far apart from each other in the reaction network exhibit a strong correlation (Steuer *et al.*, 2003b; Weckwerth, 2003; Morgenthal *et al.*, 2005; Weckwerth and Steuer, 2005). Regulatory properties, for instance the modulation of enzyme activity serves as a source of changes in the topology of the correlation network (Weckwerth, 2003; Weckwerth *et al.*, 2004a; Camacho *et al.*, 2005; Weckwerth and Steuer, 2005; Morgenthal *et al.*, 2006). In other studies, metabolite correlation network analyses were adapted to yeast metabolism and enzyme concentration fluctuations (Camacho *et al.*, 2005), *Medicago truncatula* cell cultures and their response to methyljasmonate as an elicitor (Broeckling *et al.*, 2005) or lipid metabolism in a transgenic mouse model (Clish *et al.*, 2004). From all these studies, it became clear that the analysis of dynamic metabolic networks gives the opportunity to observe in vivo regulation of dynamic biochemical networks that are otherwise not accessible. Single components of the biochemical network function as harmonic oscillators or effectors, and it will be a challenge for future applications to compare experimental fluctuations of a perturbed system with computer simulations of fluctuating complex reaction pathway networks.

However, the interrelation of an enzymatic reaction network and the resulting correlation matrix is still difficult to interpret (Steuer *et al.*, 2003b; Weckwerth, 2003; Camacho *et al.*, 2005; Levine and Hwa, 2007; Mueller-Linow *et al.*, 2007). Any alteration in the reaction network, inhibition of enzyme activity, genetic suppression or enhancement of a reaction, or addition of new pathways, will result in specific metabolite patterns (Weckwerth, 2003; Weckwerth *et al.*, 2004a; Morgenthal *et al.*, 2005, 2006; Weckwerth and Steuer, 2005). The following section will demonstrate that these patterns are systematically explained by their covariance. Because most of the classical algorithms for unsupervised or supervised data mining (principal components or independent component analysis) look for optimal variance discrimination of sample groups in data sets, this stochastic model of metabolism provides a fundamental relationship between 'biometric' analyses, metabolite profiling and biochemical regulation (Weckwerth and Morgenthal, 2005) (see the following section).

9.3.4 Implications of the stochastic model of metabolism for statistical analysis of metabolomic data

The stochastic model of metabolism discussed above gives an analytical description linking the observed correlation matrix and the underlying biochemical reaction pathway network. Given the elasticities of the rate equations of an arbitrary enzymatic reaction network (called the Jacobian **J** of the system) and the additive fluctuation matrix $\mathbf{D}$, the resulting covariance matrix Γ (hence the correlation matrix $\mathbf{C}$ and the Pearson correlation coefficient $\mathbf{C}_{ij}$)

$$C_{ij} = \frac{\Gamma_{ij}}{\sqrt{\Gamma_{ij}\Gamma_{jj}}}$$

with Γ_{ij} as the covariance of two metabolite concentrations S_i and S_j

$$\Gamma_{ij} = \langle S_i S_j \rangle - \langle S_i \rangle \langle S_j \rangle \; i, j = 1, K, ...M$$

is given as the solution of a linear equation (vanKampen, 1992; Steuer *et al.*, 2003b):

$$J\Gamma + \Gamma J^T = -2D \tag{9.1}$$

Eq. (9.1) establishes a fundamental relationship between the observed covariance Γ, and thus the experimentally observed metabolite correlations, and the underlying reaction network in the form of the linear approximation of the Jacobian $\mathbf{J}$. According to this hypothesis, the emergent pattern of correlations within a metabolic system can thus be interpreted as a specific 'fingerprint' of that system. In this way, measuring an ensemble of identical genotypes under identical experimental conditions (snapshots of the system as seen in Fig. 9.2) exploits the intrinsic flexibility and variability in the concentrations to gain information about the current state of the system. However, we cannot calculate the complete covariance matrix of a plant metabolic network because we do not know all the enzymatic rate equations and in vivo constants of the system, nor can we recalculate the complete Jacobian (Steuer *et al.*, 2003b). Therefore, comparative studies are necessary to look for differentially regulated metabolite correlations to identify the most important hubs for biological interpretation (Weckwerth, 2003; Weckwerth *et al.*, 2004a; Morgenthal *et al.*, 2005, 2006; Mueller-Linow *et al.*, 2007). This differentiation process of variable states of independently genetically or environmentally perturbed systems can best be performed by one of the most widely used chemometric methods principal components analysis (PCA). PCA looks for the highest variance in the data thereby optimizing sample pattern recognition according to the covariance of the variables, in our case the metabolites. The samples will obtain new 'eigenvectors', and can be visualized in this new coordinate system because in most cases the first three components are able to explain more than 95% of total variance in the data. The sample group orientation depends on the weights of the corresponding variable, the metabolite. In other words, if some metabolites show a strong covariance within a sample group they will have high weights on the corresponding 'eigenvector', whereas the same metabolites possibly have no covariance in the other sample groups, thus leading to a clear separation of both sample groups. This process coincides with the search for differential metabolite correlation network connectivities and regulatory hubs within a set of sample replicate measurements (see Fig. 9.2 and Section 9.3.3). The more the sample groups of different experimental perturbations are separated in PCA, the more discriminatory biochemical regulation is to be expected, indicated by a set of highly 'co-regulated' metabolites. However, the interpretation strongly relies on the optimal separation and orientation of sample groups along the new components. A complicating fact is that different data transformations and different statistical procedures will change the output of such an experiment.

Systematic studies, using different data transformations such as log, sample max, sample range, dividing by median, or Z-transformation in combination with PCA, independent components analysis (ICA), hierarchical clustering analysis (HCA), partial least square analysis (PLS) or PCA/DFA (discriminatory function analysis) are necessary to unravel the optimal conditions and information extraction for multivariate data mining 'biometrics' in biological systems. For instance, if data are standardized to 0 mean and 1 variance, PCA is actually performed on a correlation matrix (Vantongeren *et al.*, 1992; Cao *et al.*, 1999); thus, following our proposed model to explain metabolite correlations based on the Jacobian of the underlying biochemical network, differentiates the sample groups according to their correlation network topology as an in vivo fingerprint. This is a novel and very intriguing concept for the interpretation of the 'eigenvectors' and in the future, this approach will be extended for biological interpretation (Weckwerth and Morgenthal, 2005). However, as a result, an important property of PCA will no longer apply, that is the 'eigenvalue' of an axis will not indicate the percentage of total variance.

According to this outcome, one has to decide during or before the process of statistical evaluation what kind of properties will give an optimal interpretation of the experiment. Our experience with most of the data analysis so far, is that log transformation in combination with PCA is a useful standard procedure with which to begin. An extended strategy is discussed in Section 9.6.

Another very important extension to consider is the fact that the outcome of such huge biometric studies on multivariate data sets is strongly dependent on the experimental design. In other words: the experimental design of the study defines in advance which sort of results will be obtained. Thus, there is not only an iterative and cyclic process for model generation and testing hypothesis but also for experimental design and all subsequent analyses (King *et al.*, 2004).

9.4 Application of metabolomics in molecular plant physiology and biochemistry

Determining metabolite levels as a measure of metabolic fine and coarse control of pathways has a long tradition in plant biochemistry and physiology (Preiss, 1980; Stitt *et al.*, 1988). These measurements enable the detection of processes such as diurnal rhythms in enzyme regulation (Stitt *et al.*, 1988), and serve as clues to understanding pathway organization. Changes at the metabolite level are closely related to the microenvironment. Metabolic reaction chains are able to sense environmental stimuli within seconds and milliseconds. The results are high metabolic fluctuations. It is possible to exploit this biological variance to investigate pathway structures or the regulation of different genotypes as discussed above. The application of metabolomics in plant physiology gives the unique opportunity to investigate whole metabolic networks instead of single pathways as response to various environmental or developmental stimuli or gene function analysis. Here, one can use metabolic marker as controls and correlate these to other processes in plant metabolism. For that purpose, we have performed several studies in which metabolomics is applied to problems in plant physiology and plant biochemistry. Examples include sucrose and starch metabolism (Weckwerth *et al.*, 2004a; Morgenthal *et al.*, 2005), diurnal rhythm (Morgenthal *et al.*, 2005), photosynthesis and photorespiration (Boldt *et al.*, 2005), abiotic stress (Wienkoop *et al.*, 2006b) (see Section 9.6) and interspecies comparison of different plant tissues like potato and tobacco leaf versus potato tuber (Morgenthal *et al.*, 2006). In the following section, some aspects of metabolite profiling and molecular plant physiology are briefly reviewed.

9.4.1 Metabolite profiles cover many physiological processes in the plant

In Fig. 9.3, typical metabolites are shown which can be easily identified and quantified in plant tissue samples by GC-TOF-MS. In principle, these metabolites overlap with many physiological processes in the plant and thus could function as specific metabolic markers. However, this is not always the case. It appears that pathway regulation is much more complex and distinct from classical textbook knowledge. The great promise is that gene function analysis in *A. thaliana* and other model plants reveals new pathways and regulatory loops. Pathways are emerging which were unknown before adding new links and explaining observations on the metabolite level. Recently, for instance, starch degradation during the night was discovered to be controlled by a novel maltose transporter (Niittyla *et al.*, 2004). Homologous genes in plants including rice and potato indicated that maltose export is of widespread significance. Consequently, maltose is a metabolic marker for alterations in starch metabolism in plants (see the following section).

9.4.2 Metabolite profiling for gene function discovery: discrimination of mutant and wild type

The development of metabolomics technology is intimately bound to whole-genome sequencing and the post-genome era of functional genomics. In a very visionary review, metabolic profiling was foreseen as an important tool for functional studies on genes (Trethewey *et al.*, 1999). However, studies which distinctly point to a specific gene function using metabolomics technology are quite rare, if not absent, in current literature. This is partly due to the fact that the complete spectrum of small molecules which can be measured with different techniques need to be assigned to specific metabolic processes or biomarkers. Once a metabolite is identified as a marker for a specific pathway, it can be used in future studies to investigate unknown gene functions using forward and reverse genetics. In that respect, systematic programmes profiling all the knockout/conditional knockout mutants for ~22 000 genes in *A. thaliana* will be crucial to provide the first hints on gene functions related to specific metabolite markers. However, pleiotropic effects and very complex metabolic regulation multiplies the effects of gene disruption on the metabolite level. In an example, GC-TOF-MS metabolite profiling was applied to investigate a starchless mutant (*pgm*) in a diurnal rhythm (see Fig. 9.4) (Morgenthal *et al.*, 2005). Besides strong sugar accumulation, severe alterations in asparagine metabolism were observed in the *pgm* mutant. Interestingly, levels of photorespiratory intermediates such as glycerate and glycine best characterized the phases of diurnal rhythm in both the WT and the mutant. Obviously, they were not influenced by the primary effect of high sugar accumulation in *pgm* mutant plants (Morgenthal *et al.*, 2005). In contrast to WT plants, *pgm* mutant plants showed an inversely regulated cluster of N-rich amino acid metabolites, asparagine as the strong discriminator and carbohydrates indicating a shift in C/N partitioning. This observation corresponds to the observed utilization of urea cycle intermediates and polyamines in *pgm* mutant plants (see Fig. 9.4c) suggesting enhanced protein degradation and carbon utilization due to growth inhibition and a dwarf phenotype (Morgenthal *et al.*, 2005).

Maltose is the export sugar of starch degradation during the night as discussed above (Section 9.4.1 Niittyla *et al.*, 2004). This is nicely seen in Fig. 9.4d in the WT plants where maltose accumulation is observed in the 8 h dark samples. The *pgm* mutant is severely disturbed in maltose metabolism compared to WT. Maltose accumulation which is tightly linked to sucrose/glucose/fructose concentrations is altered in the *pgm* mutant. Taken together, the lack of starch synthesis during the day results in severe reorganization of metabolism in the *pgm* mutant compared to the WT plant. This indicates that the diurnal rhythm of starch synthesis and degradation in the night is essential for normal development in plants. The most prominent effects in the *pgm* mutant are observed at the carbohydrate and N-rich metabolite level.

Another example of systematic metabolite profiling for gene function discovery is the analysis of photorespiratory mutants. This pathway is well described in the literature; however the linkage to housekeeping processes is still a matter of debate. We analyzed several photorespiratory mutants at the metabolome level and identified specific metabolic markers for these processes and differences depending on where the pathway is disrupted.

D-glycerate 3-kinase (GLYK) is a key enzyme in photorespiration catalyzing the final step in glycolate recycling (Boldt *et al.*, 2005). Although enzymatic activity and kinetic properties of the enzyme were described in different plants, neither the structure of the protein nor the encoding gene(s) were known in plants thus far. Boldt *et al.* (2005) purified the glycerate kinase (GLYK) to homogeneity from *A. thaliana* leaves. The protein was digested with trypsin, analyzed using nanoflow LC/MS, and identified as the *A. thaliana* gene At1g80380, formerly falsely annotated as a putative phosphoribulo-kinase protein. The Arabidopsis GLYK does not show any significant similarity to other proteins encoded by the Arabidopsis genome and is apparently encoded by a single gene. A GC-TOF-MS-based analysis of metabolites in leaves of WT and GLYK mutant plants grown under elevated CO_2 reveals an up to 200-fold accumulation of glycerate in the mutants and a fivefold increase in the downstream metabolites of the C2 cycle, hydroxypyruvate and serine, thus providing conclusive evidence on the nature of the final enzymatic step of the C2 cycle (Boldt *et al.*, 2005).

Hydroxypyruvate reductase (HPR) is another essential photorespiratory enzyme in the C2 cycle. Kolukisaoglu *et al.* isolated homozygote HPR knockout mutant plants (Timm *et al.*, 2008). Unlike other *Arabidopsis* photorespiratory mutants, both HPR1-1 and HPR1-2 null mutants are fully viable in normal air which is also visible in the PCA plot of GC-TOF-MS metabolite profiles in Fig. 9.5.

As expected, after growth in normal air, the leaf content of hydroxypyruvate is about tenfold elevated in the HPR mutants relative to WT plants. Interestingly, this enhancement is very moderate in comparison with the metabolic effects in the GLYK photorespiratory mutant described above, the up to 200-fold higher leaf glycerate content of the very air-sensitive *A. thaliana* glycerate kinase mutants (Boldt *et al.*, 2005). This relatively low increase in hydroxypyruvate levels corresponds with minor effects of the HPR1-knockout on growth and development. The leaf glycerate content is not reduced in the mutant but, instead, even twofold increased relative to the level observed in WT plants, an unexpected effect in this mutant. Serine content is elevated to approximately the same degree as that of hydroxypyruvate, which indicates that serine:glyoxylate aminotransferase operates at perfect equilibrium in the photorespiratory cycle. Ethanolamine is increased 15-fold in the GLYK mutant plants. It has been shown that plants produce most ethanolamine by decarboxylation of free serine mediated by a recently identified serine decarboxylase (Rontein *et al.*, 2001, 2003). Accordingly, the increased levels of serine are probably responsible for the increase in ethanolamine. This compound is essential for the synthesis of membrane components such as phosphatidylethanolamine and phosphatidylcholine as well as of free choline and, in some species, the osmoprotectant glycine betaine. The data provide evidence for the existence of a relatively tight linkage between leaf serine concentrations and the biosynthesis rate of ethanolamine.

However, the conclusion from these studies is that it is almost impossible to dissect primary and secondary effects and causality of gene function using high-throughput profiling without any time resolution. Therefore, the future of metabolomics, especially for causation studies, lies in the combination of high sample throughput metabolite profiling with very high kinetic resolution.

9.5 Measuring the key players: proteomics

Metabolomic technologies enable the very rapid non-targeted analysis of metabolites and provide an advanced diagnostic tool for pattern recognition and gene function discovery in biological systems (Trethewey, 2001; Goodacre *et al.*, 2004). Pattern recognition based on metabolite profiles, however, is only the first part to understand biological systems and their physiology. For the systems-level approach, it is straightforward to combine the superior diagnostic properties of metabolite profiling with the other post-genome disciplines transcriptomics and proteomics. Several studies demonstrate the parallel analysis of metabolites and large-scale transcript expression (Urbanczyk-Wochniak *et al.*, 2003; Hirai *et al.*, 2005). However, protein analysis has been ignored in these studies. While some information about relative protein expression levels may be obtained by analyzing the transcriptome, translational mechanisms that control the rate of synthesis, the half-life of proteins and post-translational modifications such as phosphorylation, glycosylation and myristoylation indicate that mRNA levels do not necessarily correlate with corresponding protein activity or abundance (Gygi *et al.*, 1999; MacCoss *et al.*, 2003; Baginsky *et al.*, 2005) demanding that actual protein levels have to be determined for protein function analysis (Gygi *et al.*, 1999; Ideker *et al.*, 2001b).

Following this line of reasoning, the analysis of the plant proteome and the complementation with metabolite profiling data is described in the next sections. First, we will review the most recent technologies coping with the large-scale identification and quantification of proteins. Secondly, examples are described integrating metabolite, protein data and multivariate statistics (see Section 9.6).

9.5.1 Shotgun proteomics

In the mid of the 1990s, two-dimensional polyacrylamid gel electrophoresis (2D-PAGE) in combination with MS was the approach of choice for protein profiling. As MS techniques advanced and 'soft' ionization methods such as matrix assisted laser desorption ionisation (MALDI) (Karas and Hillenkamp, 1988; Tanaka *et al.*, 1988; Hillenkamp and Karas, 1990) and electrospray

Complex protein sample (heterogeneous sample)

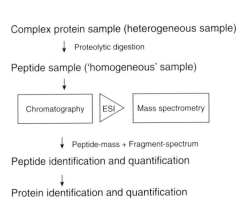

Peptide sample ('homogeneous' sample)

Peptide identification and quantification

Protein identification and quantification

Figure 9.6   Scheme of shotgun proteomics.

ionization (ESI) (Whitehouse *et al.*, 1985; Fenn *et al.*, 1989) were further developed, an alternative complementary approach for the analysis of protein mixtures was introduced to address a number of the technical limitations inherent to 2D-PAGE/MS/MS in proteome analysis. This approach called 'shotgun proteomics' is based on tryptic digestion of a complex protein sample and subsequent classical peptide LC/MS analysis with coupling of orthogonal chromatographic separation methods to cope with the sample complexity (Eng *et al.*, 1994; Link *et al.*, 1999; Washburn *et al.*, 2001; Wolters *et al.*, 2001) (see Fig. 9.6 and Wienkoop and Weckwerth, 2006).

In the so-called multidimensional protein identification strategy (MudPIT), the sample containing the tryptic peptides is loaded onto a strong cation exchange chromatography and stepped onto a reversed-phase (RP) chromatography in a series of salt steps that increase in concentration (Washburn *et al.*, 2001). A subsequent RP gradient separates the eluting peptides in relation to their hydrophobicities and delivers them into the mass spectrometer after each salt step. The theoretical peak capacity of the total MudPIT system was calculated to be ca. 23 000 (Wolters *et al.*, 2001) making the MudPIT strategy a powerful tool for proteomics. This process has the potential to fully automate the separation and identification of proteins from very complex samples (see also Section 9.5.2) in contrast to 2D-PAGE/MS/MS. Shotgun proteomics makes use of data-dependent MS/MS to determine the amino acid sequence of individual peptides. With these sequences an automated protein identification is possible by database searching using several algorithms, for example the SEQUEST (Eng *et al.*, 1994) or the Mascot algorithm (Perkins *et al.*, 1999). Each protein sequence from genomic database is virtually digested according to the specificity of the used protease. The resulting peptides that match the measured mass of the peptide ion are identified. In the next step, the experimentally derived tandem mass spectrum of the peptide ion is compared to the theoretical MS/MS spectra obtained by virtual fragmentation of candidate peptide sequences. Finally, a score is calculated for each peptide sequence by matching the predicted fragment ions to the ions observed in the experimental spectrum. When a preponderance of the fragment ions matches, it is considered a good fit. For a detailed description of shotgun proteomics see Glinski and Weckwerth, 2006.

With the MudPIT technique, very complex protein mixtures such as whole cell lysates can be analyzed directly. 2D-PAGE and MudPIT give overlapping but also complementary data on complex samples (Koller *et al.*, 2002; Schmidt *et al.*, 2004; Breci *et al.*, 2005; Kubota *et al.*, 2005). It should be noted that there are very critical issues regarding the use of these techniques. Database searches are prone to generate hundreds of false positives and false negatives depending on the parameters used. Clear rules are missing, and protein lists in the literature still provide empirical evaluation of the data. Comparisons among data sets are often limited by the parameters used. False positive identifications and protein/peptide modifications (resulting in unreliable identification of high-quality spectra) are liable to be the biggest hurdle.

In contrast to the traditional 2D-PAGE/MS/MS approach, the shotgun method for protein identification is largely unbiased providing a strategy for the efficient detection of low-abundance and hydrophobic proteins. The major improvement over 2D-PAGE systems is that the resolution of peptides and the generation of tandem mass spectra can be achieved simultaneously using the same sample. For analysis of protein dynamics (identification and quantification) and complementation of metabolic networks in plant biochemistry and physiology, LC/MS-based 'shotgun proteomics' have been implemented for qualitative and quantitative protein analysis (Weckwerth *et al.*, 2004b; Wienkoop *et al.*, 2004a, b, 2006b; Morgenthal *et al.*, 2005; Wienkoop and Weckwerth, 2006).

9.5.2 Protein fractionation, signature peptides and databases to cope with the proteome complexity

The dynamic of a biological system can only be addressed if large-scale and high-throughput methods are available. However, the enormous dynamic range and the number of protein species in a proteome sample are far beyond the technical practicability for large-scale protein analysis today. To cope with the complexity of the dynamic range and the number of protein species in a proteome, novel techniques for quantitative coverage are arising. These methods are based on a targeted quantification approach using proteotypic peptides (Wienkoop and Weckwerth, 2006; Brunner *et al.*, 2007; Hummel *et al.*, 2007). A fast and reproducible strategy has been developed to enhance protein identification in *A. thaliana* leaf and *M. truncatula* root nodule extracts by protein prefractionation via FPLC (fast performance liquid chromatography) (Wienkoop *et al.*, 2004a; Larrainzar *et al.*, 2007). This strategy is described in detail in Wienkoop and Weckwerth (2006). The application of anion exchange chromatography resulted in the separation of the high-abundance protein RUBISCO and photosystem components, as well as in the efficient fractionation of the other proteins. The use of online strong cation exchange chromatography and RP-monolithic columns, instead of classical microparticulate sorbents, enhanced the separation efficiency (Wienkoop et al., 2004a).

With the use of a 10 port valve coupling two peptide trap columns in front of a RP-monolithic microcapillary column, we were able to automate the loading and analysis process. The high number of co-eluting tryptic peptides was decreased in order to reduce disturbing ion suppression. The process allowed the identification of 1032 unique proteins from a single leaf protein sample (see Fig. 9.7 for protein classification and protein match in 'Mapman'; Thimm et al., 2004). In addition, protein prefractionation enables a further confidence level for the identification process in shotgun protein sequencing by assigning proteins to different fractions (Wienkoop et al., 2004a). This is an important feature in the face of the major problems of false positive and false negative identification rates. The proteins are identified by two or more tryptic peptides. These peptides are specific signature peptides representative for the corresponding protein. Currently, we establish an A. thaliana, a Chlamydomonas reinhardtii, and other general plant protein master lists with peptide signatures. The A. thaliana, the C. reinhardtii, as well as other plant proteins, are stored in a mass spectral reference library called ProMEX (Hummel et al., 2007). The ProMEX protein/peptide/phosphopeptide database represents a mass spectral reference library with the capability of matching unknown samples for protein and phosphorylation site identification. The database allows further text searches based on metadata such as experimental information of the samples, mass spectrometric instrument parameters or unique protein identifier like AGI codes. ProMEX integrates proteomics data with other levels of molecular organization including metabolite, pathway and transcript information and may thus become a useful resource for plant systems biology studies and genome annotation for newly sequenced species (May et al, 2008). The ProMEX mass spectral library is available at http://promex.mpimp-golm.mpg.de/. The whole approach is embedded in an international effort to measure the A. thaliana proteome and to provide proteomic databases about protein interaction, subcellular localization, tissue-specific protein localization, post-translational modification such as protein phosphorylation and other topics (http://www.masc-proteomics.org/) (Weckwerth et al. 2008). The A. thaliana proteomics consortium in turn is embedded in the MASC consortium (Multinational Arabidopsis Steering Committee) which has established several working groups for the development of plant systems biology on the basis of A. thaliana as model system (for further information see http://www.arabidopsis.org/portals/masc/MASC_members.jsp).

The selection of signature peptides from these databases allows for targeted analysis of interesting and low-abundance proteins in complex samples (Wienkoop and Weckwerth, 2006), as well as whole gene families and their differential expression in different plant tissues. We call this methodology a 'Mass Western' because of the analogy to Western Blotting (see next section).

9.5.3 Quantitative shotgun proteomics for systems biology

Mass spectrometry offers the opportunity to generate large amounts of protein sequence-dependent data giving the potential for high-throughput analysis of the plant proteome (Zivy and de Vienne, 2000; Roberts, 2002; Whitelegge, 2004; Agrawal and Rakwal, 2005; Glinski and Weckwerth, 2006). However, beside this qualitative process the ultimate benefit to biology will be determined by relative and absolute protein concentration measurements. Currently, three techniques are mainly used for quantitation in proteomics: 2D-PAGE, stable-isotope labelling and stable-isotope-free shotgun proteomics, respectively (see Fig. 9.8). A common argument in favour of 2D-PAGE is that comparison can readily be made between two gels so that proteome differences can be detected. However, 2D-PAGE is still problematic because of reproducibility, varying staining efficiency of individual gels and bias against protein classes such as membrane proteins making a comparison of different states unreliable.

In shotgun proteomics, a complex protein sample is subject to tryptic digestion and analyzed using LC-MS techniques (Yates, 2004). The heterogeneous sample with proteins of different chemical and physical behaviour is broken down to a mixture of peptides easily adaptable to classical reversed-phase LC-MS techniques (see Fig. 9.6). The increased complexity of the sample is a challenge to chromatographic and mass spectrometric resolution. Most quantitative shotgun proteomics techniques rely on modifying one of the samples with a stable isotope, which changes the molecular mass but not chromatographic behaviour. Quantitative differences are then determined directly as the difference in peak area between the two peptides in the mixed sample. There are several approaches for labelling peptides with stable isotopes: metabolic labelling using isotope-enriched or -depleted media (Oda et al., 1999; Ong et al., 2002; Whitelegge et al., 2004), proteolytic labelling of peptides using $H_2^{16}O/H_2^{18}O$ (Yao et al., 2001), protein/peptide derivatization with ICAT (isotope-coded affinity (tag)-technology) (Smolka et al., 2001) or methanol/HCl (Goodlett et al., 2001), and for instance ITRAQ^{TM}, that uses a multiplex set of four amine-specific isobaric reagents (Ross et al., 2004) allowing four-way relative and absolute quantitation.

However, only limited access to quantitative data has been demonstrated for these techniques using metabolic or chemical stable-isotope labelling techniques due to problems with quantitative and homogenous labelling (Smolka et al., 2001). Another drawback is that experiments using differential stable-isotope labelling are not providing statistical confidence of the data. Thus, current trends are the development of direct LC/MS analysis of many replicates without using stable-isotope labelling (Old *et al.*, 2005) or stable-isotope dilution with internal standards (Barr *et al.*, 1996).

In this context, one of the very early studies using label-free quantitative shotgun proteomics based on LC/MS analysis was published (Weckwerth *et al.*, 2004b). In this study metabolite profiling was accompanied with label-free LC/MS protein analysis. Based on the integration of metabolite and protein profiling in combination with multivariate data mining, two *A. thaliana* accessions were characterized. General correlations between metabolites and proteins were identified (see next Section 9.6).

Recently, a novel technique for quantitative label-free shotgun proteomics by using spectral counting and multivariate data mining was introduced (Liu *et al.*, 2004; Wienkoop *et al.*, 2006a). The application of independent component analysis enables the visualization of the dynamic behaviour of the proteome in dependence of genotype, environmental stress or developmental stages (see Weckwerth, 2008).

Besides the relative quantification of proteins, there is a strong need for the analysis of low-abundance proteins and the determination of absolute quantities of proteins in ultra complex mixtures. Here, stable-isotope dilution techniques, distinct from stable-isotope labelling approaches like ICAT, in combination with shotgun proteomics can be used (Barr *et al.*, 1996). Proteins of interest are digested with trypsin in the presence of synthetic peptide standards of known concentration with an incorporated stable isotope (^{13}C, ^{15}N). These standards are identical to the analyte peptides of interest but are distinguished by mass difference. Stable-isotope-labelled and unlabelled peptides co-migrate during chromatography. Absolute quantification is achieved by comparison of the peak area abundances of the internal standard peptide with the corresponding native counterpart. Using the highly selective and sensitive scan features of a triple quadrupole mass spectrometer, a strategy was developed for absolute quantification of proteins out of a complex mixture (Wienkoop and Weckwerth, 2006). A triple quadrupole mass spectrometer was used in the multiple reaction monitoring mode (MRM). In this mode, the mass spectrometer is tuned to the target peptide and a corresponding fragment ion thereby increasing selectivity and sensitivity of the analysis. Several single reaction monitoring modes can be performed in one chromatographic separation (so called MRM). Thus, the simultaneous analysis of dozens of target peptides can be performed in a single run which we demonstrated in the analysis of phosphopeptides (Glinski and Weckwerth, 2005). Using internal standard peptides, it is also possible to generate calibration curves for the absolute quantitation of peptides. Based on the analysis of a signature peptide for low-abundance sucrose synthase (Susy) isoform encoded by At3g43190, this protein was detected and quantified out of an ultra complex protein mixture of entire *A. thaliana* tissue. The Susy protein is usually not detectable in a typical non-targeted 1D shotgun analysis of a complex plant protein extract (Wienkoop and Weckwerth, 2006). The targeted analysis allows both quantification of low-abundance proteins and enables the analysis

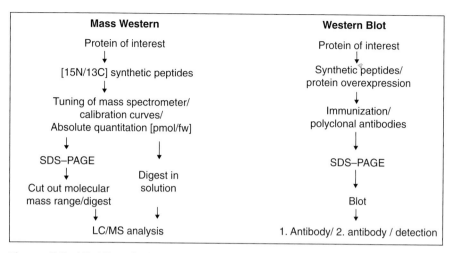

Figure 9.9 Workflows for 'Mass Western' and Western Blot.

of many samples. The proposed method, 'Mass Western', can be expanded for quantitative examinations of many proteins in a single run. For a complete strategy overview see Fig. 9.9.

9.6 Combining metabolomics, proteomics and multivariate data mining: a systems biology approach

For the integration of metabolite profiles with quantitative protein profiles, a method was required capable of a comparable sample throughput. For this purpose, we have implemented the shotgun proteomics approach using LC-iontrap-MS for identification and quantification of tryptic peptides from unique proteins in a complex protein digest (see Section 9.5.2 and Weckwerth *et al.*, 2004b). Multivariate statistics are applied to examine pattern recognition and biomarker identification. In particular, protein dynamics and metabolite dynamics alone were compared with a combined data matrix. The integration of the data revealed different patterns of samples and multiple biomarkers giving evidence for an increase of information in such holistic approaches (Weckwerth *et al.*, 2004b; Morgenthal *et al.*, 2005; Wienkoop *et al.*, 2006b). In combination with a novel kind of component analysis (ICA, see below), the identification of correlative sets of multiple biomarkers for distinct physiological processes such as day–night rhythm, sugar metabolism and temperature stress was possible. In a study of two *A. thaliana* accessions, differential metabolite–protein networks were revealed. An interesting discovery was the co-regulation of L-ascorbate peroxidase and inositol, pointing to a relationship between ascorbate metabolism and myo-inositol (Weckwerth *et al.*, 2004b). This pathway is only known in animals, but was recently uncovered in plants (Lorence *et al.*, 2004). This is one of the first examples of how integrative metabolite–protein data can reveal new hypotheses for protein function.

A further example is the investigation of an Arabidopsis *pgm* starchless mutant during a diurnal rhythm. Here we applied the strategy depicted in Fig. 9.10. For improved sample pattern recognition, we performed an ICA of the first most important principal components (for details see Scholz *et al.*, 2004, 2005). This method makes use of not only variance discrimination but also introduces a 'kurtosis' measure, thereby looking for the highest non-parametric distribution in the data. After combining the metabolite and protein profiles it was possible to separate day and night sample groups unambiguously (see Fig. 9.11, Section 9.4.2 and Morgenthal *et al.*, 2005). This led to the conclusion that photorespiratory intermediates were the best marker

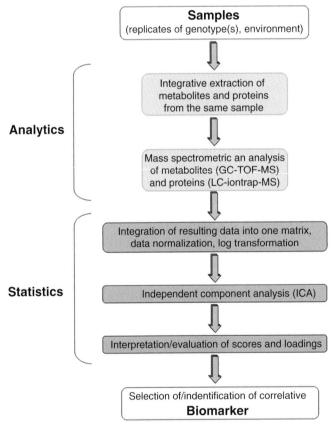

Figure 9.10 Strategy for the integration of metabolite and protein profiling in combination with multivariate data mining and subsequent biomarker identification. Due to the superior diagnostic metabolomics technology, it is possible to embed protein function analysis into this process.

for day–night samples of WT and the *pgm* mutant, although the *pgm* mutant had strong levels of accumulated sugar. Thus, the photorespiratory pathway was more or less not influenced by the constitutive synthesis of sugars in the *pgm* mutant. This observation was only possible due to an improved sample pattern recognition including both the metabolites as well as the protein dynamics (Morgenthal *et al.*, 2005). The conclusions from these initial studies became even clearer with a further experiment on temperature acclimation in *A. thaliana*. Again, we applied the same strategy as before except that protein quantification was improved enabling the detection of a much higher protein number. Proteins and metabolites demonstrated different sample pattern recognition in variance discrimination analysis and the combined data improved the biological interpretation leading to novel protein markers, RNA-binding proteins, for temperature acclimation in the plant (Wienkoop *et al.*, 2008). These proteins show high homology to RNA-binding proteins in animals known to control specific cell cycle processes as response to cold stress. Now, it becomes obvious that from these independent experiments information gain by data integration is indeed possible. The underlying reason is that data with different dynamic patterns are combined. The summation of these characteristics results in a new data matrix with novel and synergistic features. These features can be extracted with ICA. Most important is the fact that we can exploit the superior diagnostic metabolomics technique (see Section 9.3.1) for protein function analysis (see Figs. 9.10 and Figs. 9.11).

9.7 Conclusion

We are at the beginning of the integration of high-throughput profiling methods with systematic screening for biological processes and metabolic modelling. While the current methods are very limited, the first initial studies of integration are promising. Heterogeneity of the data structure – like metabolite and protein profiles – indicates that data integration will reveal synergistic effects. Refinement of statistical methods and biophysical theories such as stochastic processes in dynamic systems will allow iterative processing for matching experimental and computer-simulated data. This will lead to the refinement of metabolic models and test systems which can be ultimately validated. In parallel, high-throughput genome sequencing prepares the groundwork to ask similar and comparative questions in almost all biological systems. Especially, the interrelation of the metabolic components of the biological systems can be studied in such holistic approaches. Bottlenecks for the techniques used are simply the number of components which are detected in conventional analytical approaches, the throughput of the methods, the accuracy of quantification, data management and design of databases, the speed for compound identification in huge databases and algorithms for data integration, normalization and data transformation, statistics and metabolic modelling.

Despite these shortcomings, the first examples that touch only the tip of the iceberg of a complex biological system raise high expectations of the future of systems biology.

References

Agrawal, G.K. and Rakwal, R. (2005) Rice proteomics: a cornerstone for cereal food crop proteomes. *Mass Spectrom Rev.*

Aharoni, A. and Vorst, O. (2002) DNA microarrays for functional plant genomics. *Plant Mol Biol* **48** (1–2), 99.

Apel, K. and Hirt, H. (2004) Reactive oxygen species: metabolism, oxidative stress, and signal transduction. *Annu Rev Plant Biol* **55**, 373.

Preiss, T. (1980). In *The Biochemistry of Plants*, Vol. **3**, J. Preiss, ed. (New York: Academic Press), p. 1.

Arabidopsis Genome Initiative (2000) Analysis of the genome sequence of the flowering plant *Arabidopsis thaliana*. *Nature* **408** (6814), 796.

Arkin, A., Ross, J. and McAdams, H.H. (1998) Stochastic kinetic analysis of developmental pathway bifurcation in phage lambda-infected Escherichia coli cells. *Genetics* **149** (4), 1633.

Arkin, A., Shen, P.D. and Ross, J. (1997) A test case of correlation metric construction of a reaction pathway from measurements. *Science* **277** (5330), 1275.

Baginsky, S., Kleffmann, T., von Zychlinski, A. and Gruissem, W. (2005) Analysis of Shotgun Proteomics and RNA Profiling Data from *Arabidopsis thaliana* Chloroplasts. *J Proteome Res* **4** (2), 637.

Barr, J.R., Maggio, V.L., Patterson, D.G., Cooper, G.R., Henderson, L.O., Turner, W.E., *et al.* (1996) Isotope dilution mass spectrometric quantification of specific proteins: Model application with apolipoprotein A-I. *Clin Chem* **42** (10), 1676.

Boldt, R., Edner, C., Kolukisaoglu, U., Hagemann, M., Weckwerth, W., Wienkoop, S., *et al.* (2005) D-GLYCERATE 3-KINASE, the last unknown enzyme in the photorespiratory cycle in Arabidopsis, belongs to a novel kinase family. *Plant Cell* **17** (8), 2413.

Breci, L., Hattrup, E., Keeler, M., Letarte, J., Johnson, R. and Haynes, P.A. (2005) Comprehensive proteomics in yeast using chromatographic fractionation, gas phase fractionation, protein gel electrophoresis, and isoelectric focusing. *Proteomics* **5** (8), 2018.

Broeckling, C.D., Huhman, D.V., Farag, M.A., Smith, J.T., May, G.D., Mendes, P., *et al.* (2005) Metabolic profiling of Medicago truncatula cell cultures reveals the effects of biotic and abiotic elicitors on metabolism. *J Exp Bot* **56** (410), 323.

Brunner, E., Ahrens, C.H., Mohanty, S., Baetschmann, H., Loevenich, S., Potthast, F., *et al.* (2007) A high-quality catalog of the *Drosophila melanogaster* proteome. *Nat Biotechnol* **25** (5), 576.

Camacho, D., Fuente, A. and Mendes, P. (2005) The origin of correlations in metabolomics data. *Metabolomics* **1** (1), 53.

Cao, Y., Williams, D.D. and Williams, N.E. (1999) Data transformation and standardization in the multivariate analysis of river water quality. *Ecol Appl* **9** (2), 669.

Castrillo, J.O. and Oliver, S.G. (2004) Yeast as a touchstone in post-genomic research: strategies for integrative analysis in functional genomics. *J Biochem Mol Biol* **37** (1), 93.

Chee, M., Yang, R., Hubbell, E., Berno, A., Huang, X.C., Stern, D., *et al.* (1996) Accessing genetic information with high-density DNA arrays. *Science* **274** (5287), 610.

Clish, C.B., Davidov, E., Oresic, M., Plasterer, T.N., Lavine, G., Londo, T., *et al.* (2004) Integrative biological analysis of the APOE[*3]-Leiden transgenic mouse. *OMICS* **8** (1), 3.

Crick, F. (1970) Central dogma of molecular biology. *Nature* **227** (5258), 561.

Czechowski, T., Bari, R.P., Stitt, M., Scheible, W.R. and Udvardi, M.K. (2004) Real-time RT-PCR profiling of over 1400 Arabidopsis transcription factors: unprecedented sensitivity reveals novel root- and shoot-specific genes. *Plant J* **38** (2), 366.

Dunn, W.B. and Ellis, D.I. (2005) Metabolomics: current analytical platforms and methodologies. *TRAC Trends Anal Chem* **24** (4), 285.

Eng, J.K., Mccormack, A.L. and Yates, J.R. (1994) An approach to correlate tandem mass-spectral data of peptides with amino-acid-sequences in a protein database. *J Am Soc Mass Spectr* **5** (11), 976.

Fenn, J.B., Mann, M., Meng, C.K., Wong, S.F. and Whitehouse, C.M. (1989) Electrospray ionization for mass spectrometry of large biomolecules. *Science* **246** (4926), 64.

Fernie, A.R., Trethewey, R.N., Krotzky, A.J. and Willmitzer, L. (2004) Innovation – metabolite profiling: from diagnostics to systems biology. *Nat Rev Mol Cell Bio* **5** (9), 763.

Fiehn, O., Kopka, J., Dormann, P., Altmann, T., Trethewey, R.N. and Willmitzer, L. (2000a) Metabolite profiling for plant functional genomics. *Nat Biotechnol* **18** (11), 1157.

Fiehn, O., Kopka, J., Trethewey, R.N. and Willmitzer, L. (2000b) Identification of uncommon plant metabolites based on calculation of elemental compositions using gas chromatography and quadrupole mass spectrometry. *Anal Chem* **72** (15), 3573.

Gibon, Y., Blasing, O.E., Palacios-Rojas, N., Pankovic, D., Hendriks, J.H.M, Fisahn, J., *et al.* (2004) Adjustment of diurnal starch turnover to short days: depletion of sugar during the night leads to a temporary inhibition of carbohydrate utilization, accumulation of sugars and post-translational activation of ADP-glucose pyrophosphorylase in the following light period. *Plant J* **39** (6), 847.

Glinski, M. and Weckwerth, W. (2005) Differential multisite phosphorylation of the trehalose-6-phosphate synthase gene family in Arabidopsis thaliana – a mass spectrometry-based process for multiparallel peptide library phosphorylation analysis. *Mol Cell Proteomics* **4** (10), 1614.

Glinski, M. and Weckwerth, W. (2006) The role of mass spectrometry in plant systems biology. *Mass Spectrom Rev* **25** (2), 173.

Goodacre, R., Vaidyanathan, S., Dunn, W.B., Harrigan, G.G. and Kell, D.B. (2004) Metabolomics by numbers: acquiring and understanding global metabolite data. *Trends Biotechnol* **22** (5), 245.

Goodlett, D.R., Keller, A., Watts, J.D., Newitt, R., Yi, E.C., Purvine, S., *et al.* (2001) Differential stable isotope labeling of peptides for quantitation and de novo sequence derivation. *Rapid Commun Mass Spectrom* **15** (14), 1214.

Gygi, S.P., Rochon, Y., Franza, B.R. and Aebersold, R. (1999) Correlation between protein and mRNA abundance in yeast. *Mol Cell Biol* **19** (3), 1720.

Hadacek, F. (2002) Secondary metabolites as plant traits: current assessment and future perspectives. *Crit Rev Plant Sci* **21** (4), 273.

Hamilton, A., Voinnet, O., Chappell, L. and Baulcombe, D. (2002) Two classes of short interfering RNA in RNA silencing. *EMBO J* **21** (17), 4671.

Hillenkamp, F. and Karas, M. (1990) Mass spectrometry of peptides and proteins by matrix-assisted ultraviolet laser desorption/ionization. *Methods Enzymol* **193**, 280.

Hirai, M.Y., Klein, M., Fujikawa, Y., Yano, M., Goodenowe, D.B., Yamazaki, Y., *et al.* (2005) Elucidation of gene-to-gene and metabolite-to-gene networks in Arabidopsis by integration of metabolomics and transcriptomics. *J Biol Chem* **280** (27), 25590.

Hummel, J., Niemann, M., Wienkoop, S., Schulze, W., Steinhauser, D., Selbig, J., *et al.* (2007) ProMEX: a mass spectral reference database for proteins and protein phosphorylation sites. *BMC Bioinformatics* **8**, 216.

Ideker, T. (2004) Systems biology 101 – what you need to know. *Nat Biotechnol* **22** (4), 473.

Ideker, T., Galitski, T. and Hood, L. (2001a) A new approach to decoding life: systems biology. *Annu Rev Genomics Hum Genet* **2**, 343.

Ideker, T., Thorsson, V., Ranish, J.A., Christmas, R., Buhler, J., Eng, J.K., *et al.* (2001b) Integrated genomic and proteomic analyses of a systematically perturbed metabolic network. *Science* **292** (5518), 929.

Ihmels, J., Levy, R. and Barkai, N. (2004) Principles of transcriptional control in the metabolic network of *Saccharomyces cerevisiae*. *Nat Biotechnol* **22** (1), 86.

Jang, J.C. and Sheen, J. (1997) Sugar sensing in higher plants. *Trends Plant Sci* **2** (6), 208.

Jonak, C., Okresz, L., Bogre, L. and Hirt, H. (2002) Complexity, cross talk and integration of plant MAP kinase signalling. *Curr Opin Plant Biol* **5** (5), 415.

Jonsson, P., Gullberg, J., Nordstrom, A., Kusano, M., Kowalczyk, M., Sjostrom, M., *et al.* (2004) A strategy for identifying differences in large series of metabolomic samples analyzed by GC/MS. *Anal Chem* **76** (6), 1738.

Karas, M. and Hillenkamp, F. (1988) Laser desorption ionization of proteins with molecular masses exceeding 10,000 daltons. *Anal Chem* **60** (20), 2299.

Kell, D.B. (2004) Metabolomics and systems biology: making sense of the soup. *Curr Opin Microbiol* **7** (3), 296.

King, R.D., Whelan, K.E., Jones, F.M., Reiser, P.G., Bryant, C.H., Muggleton, S.H., *et al.* (2004) Functional genomic hypothesis generation and experimentation by a robot scientist. *Nature* **427** (6971), 247.

Kitano, H. (2000) Perspectives on systems biology. *New Gener Comput* **18** (3), 199.

Koch, K.E. (1996) Carbohydrate-modulated gene expression in plants. *Annu Rev Plant Physiol Plant Mol Biol* **47**, 509.

Koller, A., Washburn, M.P., Lange, B.M., Andon, N.L., Deciu, C., Haynes, P.A., *et al.* (2002) Proteomic survey of metabolic pathways in rice. *Proc Natl Acad Sci USA* **99** (18), 11969.

Kose, F., Weckwerth, W., Linke, T. and Fiehn, O. (2001) Visualizing plant metabolomic correlation networks using clique-metabolite matrices. *Bioinformatics* **17** (12), 1198.

Kubota, K., Kosaka, T. and Ichikawa, K. (2005) Combination of two-dimensional electrophoresis and shotgun peptide sequencing in comparative proteomics. *J Chromatogr B Analyt Technol Biomed Life Sci* **815** (1–2), 3.

Larrainzar, E., Wienkoop, S., Weckwerth, W., Ladrera, R., Arrese-Igor, C. and Gonzalez, E.M. (2007) *Medicago truncatula* root nodule proteome analysis reveals differential plant and bacteroid responses to drought stress. *Plant Physiol* **144** (3), 1495.

Levine, E. and Hwa, T. (2007) Stochastic fluctuations in metabolic pathways. *Proc Natl Acad Sci USA* **104** (22), 9224.

Link, A.J., Eng, J., Schieltz, D.M., Carmack, E., Mize, G.J., Morris, D.R., *et al.* (1999) Direct analysis of protein complexes using mass spectrometry. *Nat Biotechnol* **17** (7), 676.

Liu, H., Sadygov, R.G. and Yates, J.R., III (2004) A model for random sampling and estimation of relative protein abundance in shotgun proteomics. *Anal Chem* **76** (14), 4193.

Lorence, A., Chevone, B.I., Mendes, P. and Nessler, C.L. (2004) myo-inositol oxygenase offers a possible entry point into plant ascorbate biosynthesis. *Plant Physiol* **134** (3), 1200.

MacCoss, M.J., Wu, C.C., Liu, H., Sadygov, R. and Yates, J.R., III (2003) A correlation algorithm for the automated quantitative analysis of shotgun proteomics data. *Anal Chem* **75** (24), 6912.

MacKintosh, C. (1998) Regulation of cytosolic enzymes in primary metabolism by reversible protein phosphorylation. *Curr Opin Plant Biol* **1** (3), 224.

May P., Wienkoop, S. Kempa, S. Usadel, B. Christian, N., Rupprecht, I, Weiss, J., Recuenco-Munez, L., Ebenhoh, O. Weckwerth, W., and Wolther, D. (2008) Metabolomics and Proteomics-Assisted Genome Annotation and Analysis of the Draft Metabolic Network of Chlamydomonas reinhardtii. *Genetics* **179**, 157.

Morgenthal, K., Weckwerth, W. and Steuer, R. (2006) Metabolomic networks in plants: transitions from pattern recognition to biological interpretation. Biosystems **83** (2–3), 108.

Morgenthal, K., Wienkoop, S., Scholz, M., Selbig, J. and Weckwerth, W. (2005) Correlative GC-TOF-MS based metabolite profiling and LC-MS based protein profiling reveal time-related systemic regulation of metabolite-protein networks and improve pattern recognition for multiple biomarker selection. *Metabolomics* **1** (2), 109.

Mueller-Linow, M., Weckwerth, W. and Huett, M.T. (2007) Consistency analysis of metabolic correlation networks. *BMC Syst Biol* **1** (1), 44.

Nicholson, J.K., Connelly, J., Lindon, J.C. and Holmes, E. (2002) Metabonomics: a platform for studying drug toxicity and gene function. *Nat Rev Drug Discov* **1** (2), 153.

Nicholson, J.K., Lindon, J.C. and Holmes, E. (1999) 'Metabonomics': understanding the metabolic responses of living systems to pathophysiological stimuli via multivariate statistical analysis of biological NMR spectroscopic data. *Xenobiotica* **29** (11), 1181.

Niittyla, T., Messerli, G., Trevisan, M., Chen, J., Smith, A.M. and Zeeman, S.C. (2004) A previously unknown maltose transporter essential for starch degradation in leaves. *Science* **303** (5654), 87.

Oda, Y., Huang, K., Cross, F.R., Cowburn, D. and Chait, B.T. (1999) Accurate quantitation of protein expression and site-specific phosphorylation. *Proc Natl Acad Sci USA* **96** (12), 6591.

Old, W.M., Meyer-Arendt, K., Aveline-Wolf, L., Pierce, K.G., Mendoza, A., Sevinsky, J.R., *et al.* (2005) Comparison of label-free methods for quantifying human proteins by shotgun proteomics. *Mol Cell Proteomics* **4** (10), 1487.

Oliver, S.G., Winson, M.K., Kell, D.B. and Baganz, F. (1998) Systematic functional analysis of the yeast genome. *Trends Biotechnol* **16** (9), 373.

Ong, S.E., Blagoev, B., Kratchmarova, I., Kristensen, D.B., Steen, H., Pandey, A., *et al.* (2002) Stable isotope labeling by amino acids in cell culture, SILAC, as a simple and accurate approach to expression proteomics. *Mol Cell Proteomics* **1** (5), 376.

Perkins, D.N., Pappin, D.J., Creasy, D.M. and Cottrell, J.S. (1999) Probability-based protein identification by searching sequence databases using mass spectrometry data. *Electrophoresis* **20** (18), 3551.

Rao, C.V., Wolf, D.M. and Arkin, A.P. (2002) Control, exploitation and tolerance of intracellular noise. *Nature* **420** (6912), 231.

Rascher, U., Hutt, M.T., Siebke, K., Osmond, B., Beck, F. and Luttge, U. (2001) Spatiotemporal variation of metabolism in a plant circadian rhythm: The biological clock as an assembly of coupled individual oscillators. *Proc Natl Acad Sci USA* **98** (20), 11801.

Roberts, J.K.M. (2002) Proteomics and a future generation of plant molecular biologists. *Plant Mol Biol* **48** (1–2), 143.

Roessner, U., Luedemann, A., Brust, D., Fiehn, O., Linke, T., Willmitzer, L., *et al.* (2001) Metabolic profiling allows comprehensive phenotyping of genetically or environmentally modified plant systems. *Plant Cell* **13** (1), 11.

Roessner, U., Wagner, C., Kopka, J., Trethewey, R.N. and Willmitzer, L. (2000) Simultaneous analysis of metabolites in potato tuber by gas chromatography-mass spectrometry. *Plant J* **23** (1), 131.

Rolland, F., Moore, B. and Sheen, J. (2002) Sugar sensing and signaling in plants. *Plant Cell* **14**, S185.

Romeis, T. (2001) Protein kinases in the plant defence response. *Curr Opin Plant Biol* **4** (5), 407.

Rontein, D., Nishida, I., Tashiro, G., Yoshioka, K., Wu, W.I., Voelker, D.R., *et al.* (2001) Plants synthesize ethanolamine by direct decarboxylation of serine using a pyridoxal phosphate enzyme. *J Biol Chem* **276** (38), 35523.

Rontein, D., Rhodes, D. and Hanson, A.D. (2003) Evidence from engineering that decarboxylation of free serine is the major source of ethanolamine moieties in plants. *Plant Cell Physiol* **44** (11), 1185.

Ross, P.L., Huang, Y.N., Marchese, J.N., Williamson, B., Parker, K., Hattan, S., *et al.* (2004) Multiplexed protein quantitation in *Saccharomyces cerevisiae* using amine-reactive isobaric tagging reagents. *Mol Cell Proteomics* **3** (12), 1154.

Samoilov, M., Arkin, A. and Ross, J. (2001) On the deduction of chemical reaction pathways from measurements of time series of concentrations. *Chaos* **11** (1), 108.

Sauter, H., Lauer, M. and Fritsch, H. (1988) Metabolic profiling of plants – a new diagnostic technique. *Abstr Pap Am Chem Soc* **195**, 129.

Schena, M., Shalon, D., Davis, R.W. and Brown, P.O. (1995) Quantitative monitoring of gene expression patterns with a complementary DNA microarray. *Science* **270** (5235), 467.

Schmidt, F., Donahoe, S., Hagens, K., Mattow, J., Schaible, U.E., Kaufmann, S.H., *et al.* (2004) Complementary analysis of the Mycobacterium tuberculosis proteome by two-dimensional electrophoresis and isotope-coded affinity tag technology. *Mol Cell Proteomics* **3** (1), 24.

Scholz, M., Gatzek, S., Sterling, A., Fiehn, O. and Selbig, J. (2004) Metabolite fingerprinting: detecting biological features by independent component analysis. *Bioinformatics* **20** (15), 2447.

Scholz, M., Kaplan, F., Guy, C.L., Kopka, J. and Selbig, J. (2005) Non-linear PCA: a missing data approach. *Bioinformatics* **21** (20), 3887.

Shellie, R.A., Welthagen, W., Zrostlikova, J., Spranger, J., Ristow, M., Fiehn, O., *et al.* (2005) Statistical methods for comparing comprehensive two-dimensional gas chromatography-time-of-flight mass spectrometry results: metabolomic analysis of mouse tissue extracts. *J Chromatogr A* **1086** (1–2), 83.

Smolka, M.B., Zhou, H.L., Purkayastha, S. and Aebersold, R. (2001) Optimization of the isotope-coded affinity tag-labeling procedure for quantitative proteome analysis. *Anal Biochem* **297** (1), 25.

Somerville, C. and Dangl, L. (2000) Genomics – plant biology in 2010. *Science* **290** (5499), 2077.

Steuer, R., Kurths, J., Fiehn, O. and Weckwerth, W. (2003a) Interpreting correlations in metabolomic networks. *Biochem Soc Trans* **31**, 1476.

Steuer, R., Kurths, J., Fiehn, O. and Weckwerth, W. (2003b) Observing and interpreting correlations in metabolomic networks. *Bioinformatics* **19** (8), 1019.

Stitt, M., Wilke, I., Feil, R. and Heldt, H.W. (1988) Coarse control of sucrose-phosphate synthase in leaves – alterations of the kinetic-properties in response to the rate of photosynthesis and the accumulation of sucrose. *Planta* **174** (2), 217.

Tanaka, K., Waki, H., Ido, Y., Akita, S., Yoshida, Y., Yoshida, T., *et al.* (1988) Protein and polymer analyses up to m/z 100 000 by laser ionization time-of-flight mass spectrometry. *Rapid Commun Mass Spectrom* **2** (8), 151.

ter Kuile, B.H. and Westerhoff, H.V. (2001) Transcriptome meets metabolome: hierarchical and metabolic regulation of the glycolytic pathway. *Febs Lett* **500** (3), 169.

Thimm, O., Blasing, O., Gibon, Y., Nagel, A., Meyer, S., Kruger, P., *et al.* (2004) MAPMAN: a user-driven tool to display genomics data sets onto diagrams of metabolic pathways and other biological processes. *Plant J* **37** (6), 914.

Timm, A., Nunes-Nesi, T., Parnik, T., Morgenthal, K., Wienkoop, S., Keerberg, O., *et al.* (2008) A cytosolic pathway for the conversion of hydroxypyruvate to glycerate during photorespiration in Arabidopsis. *Plant Cell*, in press.

Trethewey, R.N. (2001) Gene discovery via metabolic profiling. *Curr Opin Biotechnol* **12** (2), 135.

Trethewey, R.N., Krotzky, A.J. and Willmitzer, L. (1999) Metabolic profiling: a Rosetta Stone for genomics? *Curr Opin Plant Biol* **2** (2), 83.

Tweeddale, H., Notley-McRobb, L. and Ferenci, T. (1998) Effect of slow growth on metabolism of Escherichia coli, as revealed by global metabolite pool ("metabolome") analysis. *J Bacteriol* **180** (19), 5109.

Underwood, B.R., Broadhurst, D., Dunn, W.B., Ellis, D.I., Michell, A.W., Vacher, C., *et al.* (2006) Huntington disease patients and transgenic mice have similar procatabolic serum metabolite profiles. *Brain* **129**, 877.

Urbanczyk-Wochniak, E., Luedemann, A., Kopka, J., Selbig, J., Roessner-Tunali, U., Willmitzer, L., *et al.* (2003) Parallel analysis of transcript and metabolic profiles: a new approach in systems biology. *EMBO Rep* **4** (10), 989.

Vance, W., Arkin, A. and Ross, J. (2002) Determination of causal connectivities of species in reaction networks. *Proc Natl Acad Sci USA* **99** (9), 5816.

vanKampen (1992) *Stochastic Processes in Physics and Chemistry* (Amsterdam, Netherlands: Elsevier).

Vantongeren, O.F.R, Vanliere, L., Gulati, R.D., Postema, G. and Boesewinkeldebruyn, P.J. (1992) Multivariate-analysis of the Plankton Communities in the Loosdrecht Lakes – relationship with the chemical and physical-environment. *Hydrobiologia* **233** (1–3), 105.

Veriotti, T. and Sacks, R. (2001) High-speed GC and GC/time-of-flight MS of lemon and lime oil samples. *Anal Chem* **73** (18), 4395.

Wagner, C., Sefkow, M. and Kopka, J. (2003) Construction and application of a mass spectral and retention time index database generated from plant GC/EI-TOF-MS metabolite profiles. *Phytochemistry* **62** (6), 887.

Washburn, M.P., Wolters, D. and Yates, J.R., III (2001) Large-scale analysis of the yeast proteome by multidimensional protein identification technology. *Nat Biotechnol* **19** (3), 242.

Watson, J.T., Schultz, G.A., Tecklenburg, R.E. and Allison, J. (1990) Renaissance of gas-chromatography time-of-flight mass-spectrometry – meeting the challenge of capillary columns with a beam deflection instrument and time array detection. *J Chromatogr* **518** (2), 283.

Webb, J.W., Gates, S.C., Comiskey, J.P. and Weber, D.F. (1986) Metabolic profiling of corn plants using Hplc and Gc/Ms. *Abstr Pap Am Chem Soc* **191**, 70.

Weckwerth, W. (2003) Metabolomics in systems biology. *Annu Rev Plant Biol* **54**, 669.

Weckwerth, W. (2006). In *Methods in Molecular Biology* (Totowa: Humana Press).

Weckwerth, W. (2008) Integration of metabolomics and proteomics in molecular plant physiology – coping with the complexity by data-dimensionality reduction. *Physiol Plant* **132** (2), 176.

Weckwerth, W., Baginsky, S., van Wijk, K., Heazlewood, J.L. and Millar, H. (2008) The multinational Arabidopsis steering subcommittee for proteomics assembles the largest proteome database resource for plant systems biology. *J Proteome Res* **7** (10), 4209–4210.

Weckwerth, W., Loureiro, M.E., Wenzel, K. and Fiehn, O. (2004a) Differential metabolic networks unravel the effects of silent plant phenotypes. *Proc Natl Acad Sci USA* **101** (20), 7809.

Weckwerth, W. and Morgenthal, K. (2005) Metabolomics: from pattern recognition to biological interpretation. *Drug Discov Today* **10** (22), 1551.

Weckwerth, W. and Steuer, R. (2005). In *Metabolomics*, S. Vaidyanathan, G.G. Harrigan R. Goodacre, eds. (Springer).

Weckwerth, W., Tolstikov, V. and Fiehn, O. (2001) Metabolomic characterization of transgenic potato plants using GC/TOF and LC/MS analysis reveals silent metabolic phenotypes. In *Proceedings of the 49th ASMS Conference on Mass spectrometry and Allied Topics, American Society of Mass Spectrometry*, Chicago, USA, p. 1.

Weckwerth, W., Wenzel, K. and Fiehn, O. (2004b) Process for the integrated extraction identification, and quantification of metabolites, proteins and RNA to reveal their co-regulation in biochemical networks. *Proteomics* **4** (1), 78.

Whitehouse, C.M., Dreyer, R.N., Yamashita, M. and Fenn, J.B. (1985) Electrospray interface for liquid chromatographs and mass spectrometers. *Anal Chem* **57** (3), 675.

Whitelegge, J.P. (2004) Mass spectrometry for high throughput quantitative proteomics in plant research: lessons from thylakoid membranes. *Plant Physiol Biochem* **42** (12), 919.

Whitelegge, J.P., Katz, J.E., Pihakari, K.A., Hale, R., Aguilera, R., Gomez, S.M., *et al.* (2004) Subtle modification of isotope ratio proteomics; an integrated strategy for expression proteomics. *Phytochemistry* **65** (11), 1507.

Wienkoop, S., Glinski, M., Tanaka, N., Tolstikov, V., Fiehn, O. and Weckwerth, W. (2004a) Linking protein fractionation with multidimensional monolithic RP peptide chromatography/mass spectrometry enhances protein identification from complex mixtures even in the presence of abundant proteins. *Rapid Commun Mass Spectrom* **18**, 643.

Wienkoop, S. and Weckwerth, W. (2006) Relative and absolute quantitative shotgun proteomics: targeting low-abundance proteins in *Arabidopsis thaliana*. *J Exp Bot* **57** (7), 1529.

Wienkoop, S., Larrainzar, E., Niemann, M., Gonzalez, E.M., Lehmann, U. and Weckwerth, W. (2006a) Stable isotope-free quantitative shotgun proteomics combined with sample pattern recognition for rapid diagnostics. *J Sep Sci* **29** (18), 2793.

Wienkoop, S., Morgenthal, K., Wolschin, F., Scholz, M., Selbig, J. and W. Weckwerth (2008) Integration of metabolomic and proteomic phenotypes: analysis of data covariance dissects starch and RFO metabolism from low and high temperature compensation response in *Arabidopsis thaliana*. *Mol Cell Proteomics* **7** (9), 1725.

Wienkoop, S., Zoeller, D., Ebert, B., Simon-Rosin, U., Fisahn, J., Glinski, M., *et al.* (2004b) Cell-specific protein profiling in *Arabidopsis thaliana* trichomes: identification of trichome-located proteins involved in sulfur metabolism and detoxification. *Phytochemistry* **65** (11), 1641.

Wolters, D.A., Washburn, M.P. and Yates, J.R., III (2001) An automated multidimensional protein identification technology for shotgun proteomics. *Anal Chem* **73** (23), 5683.

Yao, X., Freas, A., Ramirez, J., Demirev, P.A. and Fenselau, C. (2001) Proteolytic 18O labeling for comparative proteomics: model studies with two serotypes of adenovirus. *Anal Chem* **73** (13), 2836.

Yates, J.R., III (2004) Mass spectral analysis in proteomics. *Annu Rev Biophys Biomol Struct* **33**, 297.

Zivy, M. and de Vienne, D. (2000) Proteomics: a link between genomics, genetics and physiology. *Plant Mol Biol* **44** (5), 575.

Annual Plant Reviews (2009) **35**, 290–303
doi: 10.1111/b.9781405175326.2009.000010.x

www.interscience.wiley.com

Annual
Plant
Reviews

Chapter 10

FROM THE IONOME TO THE GENOME: IDENTIFYING THE GENE NETWORKS THAT CONTROL THE MINERAL CONTENT OF PLANTS

Mary Lou Guerinot,[1] Ivan Baxter[3] and David E. Salt[2,3]

[1] *Department of Biological Sciences, Dartmouth College, Hanover, NH, USA*
[2] *Center for Plant Environmental Stress Physiology, Purdue University, West Lafayette, IN, USA*
[3] *Bindley Bioscience Center, Purdue University, West Lafayette, IN, USA*

Abstract: Here we describe ionomics, the quantitative and simultaneous measurement of the elemental composition of living organisms, and its application to the study of plant mineral nutrition. We detail the analytical and bioinformatic approaches that allow for high-throughput screening of the ionome and how they have been applied to the model plant *Arabidopsis thaliana* to elucidate genetic and biochemical pathways necessary for proper ion homeostasis. Ionomics should help us understand which genes are responsible for the ability of plants to adapt to environments that vary widely in mineral content and how these genes ultimately control the ionome of the plant.

Keywords: Arabidopsis; ion homeostasis; ionome; ionomics

10.1 Introduction

Ionomics, the quantitative and simultaneous measurement of the elemental composition of living organisms (Salt *et al.*, 2008), is a new approach for identifying the gene networks that control the mineral content of plants. By systematically measuring the ionome of plants grown over a broad range of growth conditions we are uncovering genes required for proper ion homeostasis. These genes are then, in turn, a powerful resource for the direct manipulation of ionomic traits in crop plants. Mineral deficiencies remain a worldwide health concern despite decades of attention by international health organizations. Whereas post-harvest fortification has proved to be very efficient for some nutrients such as iodine, fortification is particularly difficult for other elements such as iron owing to its rapid oxidation. Biofortification, the process of enriching the nutrient content of crops as they grow, offers a sustainable, low-cost solution to increasing mineral nutrient content (White and Broadley, 2005; Zhu *et al.*, 2007). Indeed, enhanced nutritional value of crops was ranked in the top 10 biotechnologies for improving human health in developing countries (Daar *et al.*, 2002). This chapter provides an overview of the application of ionomics as a high-throughput phenotyping platform. Because the ionome of a plant is the summation of many biological processes, a high-throughput ionomics platform offers a viable system for probing the multiple physiological and biochemical activities that can affect the ionome. With the exception of carbon and oxygen, plants primarily acquire all the elements required for growth from the soil. To achieve this, plants have evolved a complex system of specialized tissues, cellular structures and transport molecules. This system directs the acquisition of ions from the soil solution, and in certain cases, such as iron, this also involves solubilization from the soil matrix. Soluble ions are transported across the surface of the root, in either a passive process with ions moving through the porous cell wall of the root cells, or an active process with ions moving symplastically through the cells of the root. This latter process requires that ions traverse the plasma membrane, a selectively permeable barrier that surrounds cells. Once ions have entered the root, they can either be stored or exported to the shoot. Ion transport to the shoot usually takes place in the xylem, and plants have tight control over ions entering the shoot via the xylem through solute release and absorption by xylem parenchyma cells. Once in the leaf tissue, ions are unloaded from the xylem and distributed throughout the leaf tissue. Ions also move via the phloem. In the case of seeds and grains, phloem sap loading, translocation and unloading are critical processes as mineral nutrients are delivered to seed via the phloem. All these processes, although understood at the physiological level, are not well characterized at the molecular level. Ionomics, in combination with other phenotyping platforms such as transcript profiling and proteomics, offers the potential to close the growing gap between our knowledge of a plant's genotype and the phenotypes it controls.

10.2 Analytical platforms for ionomics

There are a number of options for measuring ion content and distribution in plants. When selecting an analytic technique, one needs to consider which elements one wants to measure, the dynamic range and sensitivity required, sample throughput and whether one wants or needs spatial resolution. Furthermore, the cost per sample can often be the deciding factor. It is also worth noting that because most ionomic analyses are generally comparative, what is important analytically is precision and not accuracy. Precision is a measure of how consistently a result is determined by repeated determinations. Precision is critical if you want to establish that an observed alteration in the ionome is due to the perturbation the experimenter applied to the system, rather than uncontrolled analytical or environmental error. On the other hand, accuracy is a measure of how close your measurement is to the true absolute value. High accuracy in ionomics is only required if the experimenter wants to make conclusive statements about the absolute concentration of particular elements in the ionome.

Methods for elemental analysis either utilize the electronic properties of atoms (emission, absorption and fluorescence spectroscopy) or the nuclear properties of atoms (radioactivity or atomic number). For analysis, elements can either be detected directly (e.g. x-ray fluorescence) or first 'activated' in some way before detection (e.g. neutron activation or plasma ionization). Below is a brief review of some of the most common methods for the simultaneous quantification of multiple elements.

10.2.1 Inductively coupled plasma (ICP)

The goal of ICP is to ionize atoms for their detection by either optical emission spectroscopy (ICP-OES) (also known as atomic emission spectroscopy or ICP-AES) or mass spectroscopy (ICP-MS). ICP generates a plasma or gas in which atoms are present in the ionized state. The sample to be analyzed is pumped into a nebulizer where it is converted into a fine aerosol and then carried into the plasma by an argon gas stream (carrier gas). The plasma, at up to 8000 K, is insulated both electrically and thermally from the instrument, and maintained by a flow of cooling argon gas (coolant gas). Once ionized, the analyte atoms are detected using either an optical emission spectrometer or a mass spectrometer.

10.2.2 Optical emission spectroscopy (ICP-OES)

When ionized atoms in the ICP plasma fall back to ground state, they emit photons at a wavelength characteristic of a given element. For quantification, emitted light from all the atoms introduced into the plasma from the sample can be focused and passed into a spectrometer where an optical filter is used to separate the collected photons by wavelength. A charge injection device detector simultaneously measures the intensities of photons at multiple wavelengths. By comparing these energy intensities to reference standards, a quantitative measure of each element in the sample can be obtained.

10.2.3 Inductively coupled plasma mass spectrometry (ICP-MS)

The ability of ICP to ionize atoms makes it an effective ionization source for detection by mass spectroscopy. Ions from the ICP are focused by a series of ion lenses into the quadrupole mass analyzer. Ions are transmitted through the quadrupole on the basis of their mass to charge ratio and detected by an electron multiplier. Quantification is achieved by comparison to reference standards for each element of interest.

One advantage of ICP-MS over ICP-OES is its greater sensitivity (usually 3–4 orders of magnitude), which, in turn, allows the use of smaller sample sizes. Another advantage of ICP-MS over ICP-OES is that individual isotopes may be measured. One of the main drawbacks of ICP-MS is that the formation of polyatomic ionic species in the plasma can interfere with the measurement of particular elements, although the use of a collision/reaction cell can remove polyatomic species from the plasma before they enter the mass analyzer. Another disadvantage of ICP-MS is that it cannot handle dissolved solids above 0.1%, whereas ICP-OES can handle up to about 3% dissolved solids. Both ICP-OES and ICP-MS have been used successfully for large-scale ionomics projects in both yeast (Eide *et al.*, 2005) and Arabidopsis (Lahner *et al.*, 2003; Rus *et al.*, 2006; Baxter *et al.*, 2008).

10.2.4 X-ray fluorescence (XRF)

XRF is the emission of secondary or fluorescent x-rays from an atom that has been excited by bombardment with high-energy x-rays or gamma-rays. The emitted fluorescence x-rays have energies characteristic of the atom from which they were emitted, and therefore can be used to detect and quantify specific elements in a complex mixture. For the primary excitation of the sample, radiation of sufficient energies is required to allow for the removal of tightly held electrons in the inner shells of atoms. High-energy electrons from outer shells replace these lost electrons, and in the process release fluorescence x-rays, which can be detected for elemental quantification. This high-energy primary excitation radiation can be supplied from a conventional high-voltage x-ray tube to produce a range of x-rays energies allowing excitation and quantification of a broad range of atoms. However, gamma-ray sources can also be used to produce an overlapping range of energies, allowing multi-element detection. Such sealed gamma-ray sources are inherently radioactive, and therefore do not require large power supplies, and can therefore be used in small portable XRF instruments, useful for rapid ionomic analyses in the greenhouse or the field. For example, plant tissue and soil could be analyzed directly in the field, and decisions about collection of fresh tissue for genotyping of segregating mapping populations could be made. Such experiments could be very valuable for the identification of quantitative trait locus (QTL) involved in ionomic adaptation to particular soil conditions. Given that XRF is generally a non-destructive procedure, ionomic analysis by XRF can be performed on living plant specimens without compromising their viability.

A third source of excitation energy for XRF is synchrotron radiation where x-rays are focused into a very intense beam with a small cross-sectional area, allowing analysis of both the distribution and concentration of multiple elements with trace-level detection limits, and with micrometer to submicrometer spatial resolution. Such microXRF techniques can be used to analyze a single region of a sample, or be used for scanning samples in two dimensions, without any sample preparation. Furthermore, XRF microtomography and confocal XRF imaging can be used to map the elemental content of a sample in three dimensions. However, given the intensity of the excitation energies used with these synchrotron-based x-ray beams, these procedures can damage the sample, and will quite possibly be detrimental to the viability of living organisms. A related elemental analysis method, based on the same principles as XRF, is particle (or proton) induced x-ray emission. In this method a beam of protons is generally used to excite atoms in the sample that are detected by their emission of fluorescent x-rays as in XRF.

10.3 Bioinformatics platforms for ionomics

In any large-scale ionomics project, where thousands of samples are to be analyzed over an extended period of time, an information management system is needed to control all aspects of the process. An essential component of this process or any high-throughput phenotyping technique, is the ability to collect the metadata, that is, the data about how the plant was grown, how the sample was harvested and how the sample was analyzed, and associate it with the data coming off the instrument. Although researchers can and should attempt to minimize variation in the environment while performing large screens, some variation is inevitable, and field experiments will of course have large environmental changes. It is therefore incumbent on the researchers to collect as much environmental data as possible and to associate that data with the profiling data so that it can be used in analysis. Without this data, comparisons of data collected across months and years become difficult. Collecting this data in a machine readable fashion requires the use of controlled vocabularies and can be quite time consuming. To make this process workable for labs with limited budgets (i.e. plant labs) data management systems which facilitate easy collection of data and sample tracking are required. The Purdue Ionomics Information Management System (PiiMS), 'an integrated functional genomics platform', is an example of such a system (Baxter et al., 2007), where the work flow has been broken down into four stages; planting, harvesting, drying and analysis. PiiMS is web based so that it can be accessed from every part of the lab, and utilizes saved line catalogues, preset tray maps and drop down menus to minimize the amount of data entry required. PiiMS also provides tools that allow for the retrieval, display and download of the ionomics data, and associated metadata, which it stores.

To further enhance and enrich the ability to extract knowledge from large ionomic data sets, it is also critical to be able to incorporate information from other existing databases. National and international efforts have funded the development of a wide array of information and computational resources. While these resources provide a rich set of genotypic, phenotypic and analytical resources, it requires considerable expertise to find and use appropriate resources, and integration across data sets is very difficult. Web services (Curcin et al., 2005; Neerincx and Leunissen, 2005) provides a means to knit together disparate resources without requiring the very complex task of integrating the information into the core ionomics database. Currently, PiiMS (Baxter et al., 2007) is compatible with the web services branch of BioMOBY (Wilkinson et al., 2005), and developing PiiMS as a client for interaction with the MOBY-S framework for exchange of data and analytical services is currently under development.

10.4 Arabidopsis as a model system for ionomics

10.4.1 Ionomic mutant screens

We originally chose Arabidopsis as the model system for a comprehensive investigation into the genetic basis that underlies mineral nutrition and ion homeostasis in plants. Our proof of concept screen of a fast-neutron mutagenized population of Arabidopsis using ICP-MS identified over 50 ionomic mutants (Lahner et al., 2003). However, with the efficient identification of ionomics mutants, mapping the causal locus became the limiting factor in gene identification. The advent of high-throughput genotyping methodologies, and the availability of a genome-wide knockout collection, has started to relieve some of these difficulties. For example, DNA microarray-based technologies are being used to perform bulk segregant analysis for the rapid localization of causal loci to within a few centiMorgans (Borevitz and Nordborg, 2003; Hazen et al., 2005). Furthermore, new massively parallel sequencing technologies are set to revolutionize the identification of causal mutations, by allowing the rapid sequencing of large regions of DNA to identify polymorphisms between mutant and wild-type plants. Once identified, candidate genes can be rapidly tested by screening for the ionomic phenotype of interest in various types of sequence-indexed insertion lines carrying a mutant allele of the gene of interest. Where a sequence-indexed insertion does not exist, TILLING (targeting induced local lesions in genomes) (Henikoff et al., 2004) is an alternative strategy for the identification of mutant alleles of candidate genes.

XRF has also been used to screen over 100 000 ethyl methanesulfonate mutagenized Arabidopsis seedlings for the identification of mutants with altered ionomes (Delhaize et al., 1993). This screen identified three mutants, pho2 that accumulates threefold higher P in leaves (Delhaize and Randall, 1995), pho1-2 that has decreased levels of P in leaves (Delhaize and Randall, 1995); allelic to pho1-1 identified by Poirier et al. (1991) and man1 (now known as frd3) that accumulates a range of metals in its leaves, including manganese, for which it was originally named (Delhaize, 1996). The genes identified by these mutations have now been cloned and characterized. *FRD3* encodes a citrate effluxer that belongs to the multidrug and toxic efflux family and functions in iron homeostasis by loading citrate into the vasculature (Rogers and Guerinot, 2002; Durrett *et al.*, 2007). *PHO2* encodes an E2 ligase that is the target of a microRNA involved in phosphate homeostasis (Aung *et al.*, 2006; Bari *et al.*, 2006) and *PHO1* encodes a transporter involved in loading phosphate into the xylem (Hamburger *et al.*, 2002). More recently, Young *et al.* (2006) reported on using XRF to screen intact Arabidopsis seed and suggested that this technique could be used as the basis for a high-throughput screen. However, limited beam time probably makes this impractical as a tool for doing primary mutant screens.

10.4.2 Natural variation

Genetic variation occurring among and within natural populations can also be used as a tool for gene discovery. Not surprisingly, Arabidopsis shows abundant natural variation in its ionome. The concentration of all the significant elements in the ionome of healthy Arabidopsis shoot tissue varies over four orders of magnitude, depending on the element and its biological function. Macro nutrients such as Mg, P, K and Ca accumulate to $\geq 10\,000$ μg g^{-1} of the shoot dry weight, whereas micro nutrients such as Fe, Zn, B, Cu, Mn, Mo, Co and Ni range in concentration from 1 to 100 μg g^{-1}. Nonessential, potentially toxic trace elements such Se, As and Cd can accumulate to between 1 and 10 μg g^{-1} without any visible symptoms of toxicity. The concentration of elements measured in Arabidopsis seed varies over five orders of magnitude, a similar scale to that observed in shoots. However, there are several significant differences between the shoot and seed ionome, with certain elements being enriched or reduced relative to other elements from seeds to shoots. For example, on a ug g^{-1} dry weight basis, P does not change concentrations from seed to shoot, but K is ~fourfold lower in the seeds.

Because outcrossing in the wild is rare, the hundreds of Arabidopsis accessions that have been collected are inbred and therefore mostly homozygous, easing analysis and facilitating genetic mapping. Furthermore, unlike laboratory-generated mutations, many generations have passed between the time most naturally occurring mutations arise and the time that they are studied, allowing for recombination to have generated, and natural selection to have identified, genetic combinations suited for particular environments. In addition, the availability of recombinant inbred lines, that is, immortal mapping populations derived from a number of natural accessions, is also a great resource for characterizing genes important for the ionome. For example, the Ler X cvi RIL population has been analyzed for seed cation content (Bentsink *et al.*, 2003; Vreugdenhil *et al.*, 2004). Although none of the QTLs associated with cation levels were very finely mapped, several possible candidate genes were suggested via comparison of the QTL map positions with those of genes known to be involved in cation transport. Because cationic minerals are often complexed with phytate (myo-inositol-1,2,3,4,5,6-hexakisphosphate) in seed, a major QTL that affects phytate levels has also been described. This QTL was narrowed down to 13 genes and the most likely candidate, a gene encoding a vacuolar membrane ATPase subunit, is being pursued. QTLs have also been identified for shoot Cs accumulation (Payne *et al.*, 2004), shoot selenate accumulation (Zhang *et al.*, 2006), seed K, Na, Ca, Mg, Fe, Mn, Zn and P accumulation (Vreugdenhil *et al.*, 2004), and sulfate accumulation (Loudet *et al.*, 2007) in Arabidopsis. Once ionomics QTLs have been identified, genomic tools available for Arabidopsis can be used to locate the genes that underlie these QTLs and thus describe the traits at a molecular level. Such an approach was recently taken to identify the genes responsible for QTL controlling Na in rice and Arabidopsis (Ren *et al.*, 2005; Rus *et al.*, 2006), which interestingly was found to be an HKT (high-affinity K+ transporter) Na-transporter in both species. The gene controlling a major QTL for sulfate accumulation in Arabidopsis was also recently identified to encode adenosine 5′-phosphosulfate reductase, a central enzyme in sulfate assimilation (Loudet *et al.*, 2007).

In addition to using immortal mapping populations of Arabidopsis, large-scale association studies are now possible. Clark *et al.* (2007) used tilling arrays that cover the entire sequenced portion of the Arabidopsis genome with four 25 mers per nucleotide on both strands (aggregate of one billion distinct oligonucleotides) to interrogate 20 accessions of Arabidopsis. These are a subset of the 96 strains analyzed by Nordborg *et al.* (2005). This created a haplotype data set that will not only facilitate large scale association studies, but it will also allow inferences about the microevolution of the Arabidopsis genome (Kim *et al.*, 2007).

We have recently shown that natural variation in shoot Mo content across 92 *Arabidopsis thaliana* accessions is controlled by variation in a mitochondrially localized transporter (*Molybdenum Transporter 1 – MOT1*) (Baxter *et al.*, 2008). A deletion in the *MOT1* promoter is strongly associated with low shoot Mo, occurring in seven of the accessions with the lowest shoot content of Mo. Tomatsu *et al.* (2007) also identified the same gene by mapping the locus responsible for the low Mo levels seen in the Landsberg accession.

10.4.3 Ion distribution in Arabidopsis

Synchrotron-based microXRF has been used successfully for ionomic analysis of both small samples and for two-dimensional imaging of the ionome in several different biological samples, including mycorrhizal plant roots (Yun *et al.*, 1998), intact zooplankton (Ezoe *et al.*, 2002) and intact Se-tolerant and sensitive diamondback moths (Freeman *et al.*, 2006). Recently, XRF microtomography and confocal imaging have also been used for quantitative imaging of the three-dimensional distribution of multiple elements in various plant samples (Isaure *et al.*, 2006; Kim *et al.*, 2006; Nakano and Tsuji, 2006). The recent insightful combination of microXRF imaging with nanogold immunolocalization offers the unique opportunity for the simultaneous colocalization of known cellular structures, such as the mitochondria and chloroplast, with particular features revealed by microXRF elemental mapping (McRae *et al.*, 2006). Such co-localization information could be very important for the biological interpretation of XRF imaging.

Although these XRF imaging techniques do not have the throughput required for ionomics screens, they are proving to be powerful tools for understanding the fundamental biological processes that underlie the ionome. For example, Kim *et al.* (2006) recently used XRF microtomography to quantitatively map the three-dimensional distribution of various elements in intact Arabidopsis seeds. Using this technique, they were able to determine that the vacuolar iron uptake transporter VIT1 is directly involved in storage of iron in the provascular strands of the Arabidopsis embryo, and without such localized stores of iron, *vit1-1* seedlings grow poorly when iron is limiting. XRF has also been recently used to localize arsenic in polished and unpolished rice (Meharg *et al.*, 2008).

10.5 Evolutionary context for ionomics

Over time, plants have adapted to a variety of habitats, many of which have unique mineral signatures. For example, serpentine soils have very low concentrations of Ca ions, very high concentrations of Mg ions and low water carrying capacity. These soils also frequently have elevated levels of heavy metals such as Ni and Co. As such, serpentine soils offer some striking examples of adaptive evolution in plants (as reviewed in Brady *et al.*, 2005). Reciprocal transplantation experiments between populations of plants from serpentine soils and adjacent populations growing on normal soils have convincingly demonstrated the genetic basis of such adaptation (e.g. Nyberg Berglund *et al.*, 2003; Rajakaruna *et al.*, 2003). Interestingly, a large-scale screen for Arabidopsis mutants able to survive in a medium designed to mimic serpentine soils (low Ca^{2+}:Mg^{2+} ratio) identified *cax1* loss of function mutants (Bradshaw, 2005). CAX1 is a vacuolar transporter that sequesters Ca in the vacuole (Hirschi *et al.*, 1996). In addition to being able to survive in a solution with a low Ca^{2+}:Mg^{2+} ratio, *cax1* mutants also have most of the other phenotypes associated with tolerance to serpentine soils, including a higher Mg^{2+} requirement for optimal growth and reduced levels of Mg^{2+} in their leaves. Presumably, mutant plants lacking *cax1* are able to survive on serpentine soils because they can maintain cytoplasmic Ca levels within normal limits. This study is exciting because it suggests that single genes can have very large effects on adaptation of plants to unique environments. Clearly, the next step is to see whether *cax1*-like mutations play a role in the adaptation of natural populations to serpentine soils.

Another group of plants that provide striking examples of microevolutionary adaptation to different mineral environments are the hyperaccumulators, a small but diverse group of plants that are capable of sequestering metals (or metalloids) in their shoot tissues at levels that would be toxic to most organisms. At present, there are only five metals (Cd, Co, Mn, Ni and Zn) and two metalloids (As, Se) for which hyperaccumulation has been reproducibly documented. Approximately 400 plant taxa are known to hyperaccumulate metals, including members of the Brassicaceae, Euphorbiaceae and Asteraceae. In the field, individual plants of a hyperaccumulating species exhibit a wide variation in phenotype, even within a single population. It is important to know what factors lead to this variation, with the two most obvious choices being the concentration of metals in the soil and individual genotypes. The studies conducted to date do not support the idea that the ability to hyperaccumulate a particular metal is a direct function of soil metal content (as reviewed in Pollard *et al.*, 2002). For example, an examination of the genetic variation in the ability to accumulate Zn among and within populations of the hyperaccumulator *Arabidopsis halleri* (a close relative of *A. thaliana*) showed that the Zn concentration of field-collected plants is partly affected by plant genotype but not by total soil Zn (Macnair, 2002).

Studies using controlled crosses, inter-specific hybrids, and molecular markers are beginning to shed light on the genetic control of within population variation in metal tolerance. In general, the inheritance of adaptive, high-level metal tolerance appears to be governed by a small number of genes. A comparison of gene expression in segregating families from a cross between *A. halleri* and its sister non-accumulating species *Arabidopsis petrea*, identified eight genes that were more highly expressed in the accumulator plants, including two metal transporter genes, *NRAMP3* and *ZIP6* (Filatov *et al.*, 2006). Another recent study has also used crosses between *A. halleri* and *A. petrea* to map three QTLs responsible for Zn tolerance (Willems *et al.*, 2007). Interestingly, the three regions co-localize with genes that have been documented to be involved in metal transport: *MTP1-A*, *MTP1-B* and *HMA4*. It therefore now appears likely that Zn hyperaccumulation is achieved by stacking both high expression of the Zn pump, HMA4, in the root pericycle and xylem parenchyma to achieve enhanced xylem loading of Zn (Hanikenne *et al.*, 2008), along with increasing the Zn sink strength of the shoot through high expression of the vacuolar localized Zn pump, MTP1 (Gustin *et al.*, 2009). Comparative transcriptome analyses have also been carried out to identify genes that are more highly expressed in the hyperacummulators *A. halleri* and *Thlaspi caerulescens* compared to their non-accumulating relatives (Becher *et al.*, 2004; Weber *et al.*, 2004; Hammond *et al.*, 2006; Talke *et al.*, 2006; van de Mortel *et al.*, 2006; Weber *et al.*, 2006; Broadley *et al.*, 2007).

It is clear from the studies to date on metal tolerance and hyperaccumulation that there is a strong genetic component to the variation observed in these traits. However, most studies to date have concentrated on the accumulation of one or two elements, and ignored the interaction of the other elements in the ionome. The value of such interactions was recently demonstrated by

Baxter *et al.*, (2008) in a study in which they documented multivariable leaf ionomic signatures for both Fe and P deficiency responses. Ionomics should help us understand which genes are responsible for the ability of plants to adapt to environments that vary widely in mineral content and how these genes ultimately control the ionome of the plant.

Acknowledgements

We would like to gratefully acknowledge helpful discussions with members of our laboratories and with David Eide, Jeff Harper, Julian Schroeder and John Ward who have been our co-PIs on two NSF-funded ionomics projects (DBI-0077378; IOS-0419695). We also acknowledge support from the Office of Basic Energy Sciences at the Department of Energy (DE-FG-2–06ER15809), the Indiana 21st Century Research and Technology Fund (912010479) and the National Institutes of Health (5 R33 DK070290–03).

References

Aung, K., Lin, S.I., Wu, C.C., Huang, Y.T., Su, C.I. and Chiou, T.J. (2006) *pho2*, a phosphate overaccumulator, is caused by a nonsense mutation in a microRNA399 target gene. *Plant Physiol* **141**, 1000–1011.

Bari, R., Datt Pant, B., Sitt, M. and Scheible, W.R. (2006) PHO2, microRNA399, and PHR1 define a phosphate-signaling pathway in plants. *Plant Physiol* **141**, 988–999.

Baxter, I., Lee, J., Guerinot, M.L. and Salt, D.E. (2008) Variation in Molybdenum content across broadly distributed populations of Arabidopsis is controlled by a mitochondrial Molybdenum transporter (MOT1). *PLoS Genet* **4**, e1000004, doi:10.1371/journal.pgen.1000004.

Baxter, I., Ouzzani, M., Orcun, S., Kennedy, B., Jandhyala, S.S. and Salt, D.E. (2007) Purdue ionomics information management system. An integrated functional genomics platform. *Plant Physiol* **143**, 600–611.

Baxter I., Vitek, O., Lahner, B., Muthukumar, B., Borghi, M., Morrissey, J., Guerinot, M.L., and Salts D.E. (2008) The leaf ionome as a multivariable system to detect a plant's physiological status. *Proc Natl Acad Sci USA* **105**, 12081–12086.

Becher, M., Talke, I.N., Krall, L. and Krämer, U. (2004) Cross-species microarray transcript profiling reveals high constitutive expression of metal homeostasis genes in shoots of the zinc hyperaccumulator *Arabidopsis halleri*. *Plant J* **37**, 251–268.

Bentsink, L., Yuan, K., Koorneef, M. and Vreugdenhil, D. (2003) The genetics of phytate and phosphate accumuulation in seeds and leaves of *Arabidopsis thaliana*, using natural variation. *Theor Appl Genet* **106**, 1234–1243.

Borevitz, J.O. and Nordborg, M. (2003) The impact of genomics on the study of natural variation in Arabidopsis. *Plant Physiol* **132**, 718–725.

Bradshaw, H.D. (2005) Mutations in *CAX1* produce phenotypes characteristic of plants tolerant to serpentine soils. *New Phytol* **167**, 81–88.

Brady, K.U., Kruckeberg, A.R. and Bradshaw, H.D. (2005) Evolutionary ecology of plant adaptation to serpentine soils. *Annu Rev Ecol Evol Syst* **36**, 243–266.

Broadley, M.R., White, P.J., Hammond, J.P., Zelko, I. and Lux, A. (2007) Zinc in plants. *New Phytol* **173**, 677–702.

Clark, R.M., Schweikert, G., Toomajian, C., Ossowski, S., Zeller, G., Shinn, P., *et al.* (2007) Common sequence polymorphisms shaping genetic diversity in *Arabidopsis thaliana. Science* **317**, 338–342.

Curcin, V., Ghanem, M. and Guo, Y. (2005) Web services in the life sciences. *Drug Discov Today* **10**, 865–871.

Daar, A.S., Thorsteinsdóttir, H., Martin, D.K., Smith, A.C., Nast, S. and Singer, P.A. (2002) Top ten biotechnologies for improving health in developing countries. *Nature Genetics* **32**, 229–232.

Delhaize, E. (1996) A metal-accumulator mutant of *Arabidopsis thaliana. Plant Physiol* **111**, 849–855.

Delhaize, E. and Randall, P.J. (1995) Characterization of a phosphate-accumulator mutant of *Arabidopsis thaliana. Plant Physiol* **107**, 207–213.

Delhaize, E., Randall, P.J., Wallace, P.A. and Pinkerton, A. (1993) Screening *Arabidopsis* for mutants in mineral nutrition. *Plant Soil* **155/156**, 131–134.

Durrett, T.P., Gassmann, W. and Rogers, E.E. (2007) The FRD3-mediated efflux of citrate into the root vasculature is necessary for efficient iron translocation. *Plant Physiol* **144**, 197–205.

Eide, D.J., Clark, S., Nair, T.M., Gehl, M., Gribskov, M., Guerinot, M.L., *et al.* (2005) Characterization of the yeast ionome: a genome wide analysis of nutrient mineral and trace element homeostasis. *Genome Biol* **6**, R77, doi:10.1186/gb-2005-6-9-r77.

Ezoe, M., Sasaki, M., Hokura, A., Nakai, I., Terada, Y., Yoshinaga, T., *et al.* (2002) Two-dimensional micro-beam imaging of trace elements in a single plankton measured by a synchrotron radiation X-ray fluorescence analysis. *Bunseki Kagaku* **51**, 883–890.

Filatov, V., Dowdle, J., Smirnoff, N., Ford-LLoyd, B., Newbury, H.J. and Macnair, M.R. (2006) Comparison of gene expression in segregating families identifies genes and genomic regions involved in a novel adaptation, zinc hyperaccumulation. *Mol Ecol* **15**, 3045–3059.

Freeman, J.L., Quinn, C.F., Marcus, M.A., Fakra, S. and Pilon-Smits, E.A. (2006) Selenium-tolerant diamondback moth disarms hyperaccumulator plant defense. *Curr Biol* **21**, 2181–2192.

Gustin, J.L., Loureiro, M.E., Kim, D., Na, G., Tikhonova, M., and Salt, D.E. (2009). MTP1-dependent Zn sequestration into shoot vacuoles suggests dual roles in Zn tolerance and accumulation in Zn hyperaccumulating plants. *Plant Journal* (in press).

Hamburger, D., Rezzonico, E., Petétot, J.M.-C., Somerville, C. and Poirier, Y. (2002) Identification and characterization of the Arabidopsis *PHO1* gene involved in phosphate loading to the xylem. *Plant Cell* **14**, 889–902.

Hammond, J.P., Bowen, H.C., White, P.J., Mills, V., Pyke, K.A., Baker, A.J.M., *et al.* (2006) A comparion of the *Thlaspi caerulescens* and *T. avense* shoot transcriptomes. *New Phytol* **170**, 239–260.

Hanikenne, M., Talke, I.N., Haydon, M.J., Lanz, C., Nolte, A., Motte, P., Kroymann, J., Weigel, D., and Krämer, U. (2008) Evolution of metal hyperaccumulation required cis-regulatory changes and triplication of HMA4. *Nature* **453**, 391–395.

Hazen, S.P., Borevitz, J.O., Harmon, F.G., Pruneda-Paz, J.L., Schultz, T.F., Yanovsky, M.J., *et al.* (2005) Rapid array mapping of circadian clock and developmental mutations in Arabidopsis. *Plant Physiol* **138**, 990–997.

Henikoff, S., Till, B.J. and Comai, L. (2004) TILLING. Traditional mutagenesis meets functional genomics. *Plant Physiol* **135**, 630–636.

Hirschi, K.D., Zhen, R.-G., Cunningham, K.W., Rea, P.A. and Fink, G.R. (1996) CAX1, an H^+/Ca^{2+} antiporter from *Arabidopsis. Proc Natl Acad Sci USA* **93**, 8782–8786.

Isaure, M.P., Fraysse, A., Devès, G., Le Lay, P., Fayard, B., Susini, J., *et al.* (2006) Micro-chemical imaging of cesium distribution in *Arabidopsis thaliana* plants and its interaction with potassium and essential trace elements. *Biochmie* **88**, 1582–1590.

Kim, S., Plagnol, V., Hu, T.T., Toomajian, C., Clark, R.M., Ossowski, S., *et al.* (2007) Recombination and linkage disequilibrium in *Arabidopsis thaliana*. *Nat Genet* **39**, 1151–1155.

Kim, S.A., Punshon, T., Lanzirotti, A., Li, L., Alonzo, J.M., Ecker, J.R., *et al.* (2006) Localization of iron in Arabidopsis seed requires the vacuolar membrane transporter VIT1. *Science* **314**, 1295–1298.

Lahner, B., Gong, J., Mahmoudian, M., Smith, E.L., Abid, K.B., Rogers, E.E., *et al.* (2003) Genomic scale profiling of nutrient and trace elements in *Arabidopsis thaliana*. *Nat Biotechnol* **21**, 1215–1221.

Loudet, O., Saliba-Columbani, V., Camilleri, C., Calenge, F., Gaudon, V., Kopivova, A., *et al.* (2007) Natural variation for sulfate content in *Arabidopsis thaliana* is highly controlled by APR2. *Nat Genet* **39**, 896–900.

Macnair, M.R. (2002) Within and between population genetic variation for zinc accumulation in *Arabdiopsis halleri*. *New Phytol* **155**, 59–66.

McRae, R., Lai, B., Vogt, S. and Fahrni, C.J. (2006) Correlative microXRF and optical immunofluorescence microscopy of adherent cells labeled with ultrasmall gold particles. *J Struct Biol* **155**, 22–29.

Meharg, A.A., Lombi, E., Williams, P.N., Scheckel, K.G., Feldmann, J., Raab, A., *et al.* (2008) Speciation and localization of arsenic in white and brown rice. *Environ Sci Technol* **42**, 1051–1057.

Nakano, K. and Tsuji, K. (2006) Development of confocal 3D micro XRF spectrometer and its application to rice grain. *Bunseki Kagaku* **55**, 427–432.

Neerincx, P.B. and Leunissen, J.A. (2005) Evolution of web services in bioinformatics. *Brief Bioinform* **6**, 178–188.

Nordborg, M., Hu, T.T., Ishino, Y., Jhaveri, J., Toomajian, C., Zheng, H., *et al.* (2005) The pattern of polymorphism in *Arabidopsis thaliana*. *PLoS Biol* **3** (7), e196.

Nyberg Berglund, A.B., Dahlgren, S. and Westerbergh, A. (2003) Evidence for parallel evolution and site-specific selection of serpentine tolerance in Cerastium alpinum during the colonization of Scandinavia. *New Phytol* **161**, 199–209.

Payne, K.A., Bowen, H.C., Hammond, J.P., Hampton, C.R., Lynn, J.R., Mead, A., *et al.* (2004) Natural genetic variation in caesium (C) accumulation by *Arabidopsis thaliana*. *New Phytol* **162**, 535–548.

Poirier, Y., Thoma, S., Somerville, C. and Schiefelbein, J. (1991) A mutant of Arabidopsis deficient in xylem loading of phosphate. *Plant Physiol* **97**, 1087–1093.

Pollard, A.J., Powell, K.D., Harper, F.A. and Smith, J.A.C. (2002) The genetic basis of metal hyperaccumulation in plants. *Crit Rev Plant Sci* **21**, 539–566.

Rajakaruna, N., Siddiqi, M.Y., Whitton, J., Bohm, B.A. and Glass, A.D.M. (2003) Differential responses to Na^+/K^+ and Ca^{2+}/Mg^{2+} in two edaphic races of the *Lasthenias california* (Asteraceae) complex: a case for parallel evolution of physiological traits. *New Phytol* **157**, 93–103.

Ren, Z.H., Gao, J.P., Li, L.G., Cai, X.L., Huang, W., Chao, D.Y., *et al.* (2005) A rice quantitative trait locus for salt tolerance encodes a sodium transporter. *Nat Genet* **37**, 1141–1146.

Rogers, E.E. and Guerinot, M.L. (2002) FRD3, a member of the multidrug and toxin efflux family, controls iron deficiency responses in Arabidopsis. *Plant Cell* **14**, 1787–1799.

Rus, A., Baxter, I., Muthukumar, B., Gustin, J., Lahner, B., Yakubova, E., *et al.* (2006) Natural variants of At*HKT1* enhance Na(+) accumulation in two wild populations of Arabidopsis. *PLoS Genet* **2**, e210.

Salt, D.E., Baxter, I. and Lahner, B. (2008) Ionomics and the study of the plant ionome. *Annu Rev Plant Biol* **59**, 709–733.

Talke, I.N., Hanikenne, M. and Kramer, U. (2006) Zinc-dependent global transcriptional control, transcriptional deregulation and higher gene copy number for genes in metal homeostasis of the hyperaccumulator *Arabidopsis halleri*. *Plant Physiol* **142**, 148–167.

Tomatsu, H., Takano, J., Takahashi, H., Watanabe-Takahashi, A., Shibagaki, N. and Fujiwara, T. (2007) An *Arabidopsis thaliana* high affinity molybdate transporter required for efficient uptake of molybdate from the soil. *Proc Natl Acad Sci USA* **104**, 18807–18812.

van de Mortel, J.E., Villanueva, L.A., Schat, H., Kewekkeboom, J., Coughlin, S., Moerland, P.D., *et al.* (2006) Large expression differences in genes for iron and zinc homeostasis, stress response and lignin biosynthesis distinguish roots of *Arabidopsis thaliana* and the related metal hyperaccumulator *Thlaspi caerulescens*. *Plant Physiol* **142**, 1127–1147.

Vreugdenhil, D., Aarts, M.G.M., Koornneef, M., Nelissen, H. and Ernst, W.H.O. (2004) Natural variation and QTL analysis for cationic mineral content in seeds of *Arabidopsis thaliana*. *Plant Cell Environ* **27**, 828–839.

Weber, M., Harada, E., Vess, C., Roepenack-Lahaye, E.V. and Clemens, S. (2004) Comparative microarray analysis of *Arabidopsis halleri* roots identifies nicotianamine synthase, a ZIP transporter and other genes as potential metal hyperaccumulation factors. *Plant J* **37**, 269–281.

Weber, M., Trampczynska, A. and Clemens, S. (2006) Comparative transcriptome analysis of toxic metal responses in *Arabidopsis thaliana* and the Cd2+-hypertolerant facultative metallophyte *Arabidopsis halleri*. *Plant Cell Environ* **29**, 950–963.

White, P.J. and Broadley, M.R. (2005) Biofortifying crops with essential mineral elements. *Trends Plant Sci* **10**, 586–593.

Wilkinson, M., Schoof, H., Ernst, R. and Haase, D. (2005) BioMOBY successfully integrates distributed heterogeneous bioinformatics web services. The PlaNet exemplar case. *Plant Physiol* **138**, 5–17.

Willems, G., Drager, D.B., Courbot, M., Gode, C., Verbruggen, N. and Saumitou-Laprade, P. (2007) The genetic basis of zinc tolerance in the metalophyte *Arabidopsis halleri* ssp. *halleri* (*Brassicaceae*): an analysis of quantitative trait loci. *Genetics* **176**, 659–674.

Young, L.W., Westcott, N.D., Attenkofer, K. and Reaney, M.J.T. (2006) A high throughput determination of metal concentrations in whole, intact *Arabidopsis thaliana* seeds using synchrotron-based x-ray fluorescence spectroscopy. *J Synchrotron Radiat* **13**, 304–313.

Yun, W., Pratt, S.T., Miller, R.M., Cai, Z., Hunter, D.B., Jarstfer, A.G., *et al.* (1998) X-ray imaging and microspectroscopy of plants and fungi. *J. Synchrotron Radiat* **5**, 1390–1395.

Zhang, L., Byrne, P.F. and Pilon-Smits, E. (2006) Mapping quantitative trait loci associated with selenate tolerance in *Arabidopsis thaliana*. *New Phytol* **170**, 33–42.

Zhu, C., Naqvi, S., Gomez-Galera, S., Pelacho, A.M., Capell, T. and Christou, P. (2007) Transgenic strategies for the nutritional enhancement of plants. *Trends Plant Sci* **12**, 548–555.

Annual Plant Reviews (2009) **35**, 304–330
doi: 10.1111/b.9781405175326.2009.000011.x

Chapter 11
DEVELOPMENT AND SYSTEMS BIOLOGY: RIDING THE GENOMICS WAVE TOWARDS A SYSTEMS UNDERSTANDING OF ROOT DEVELOPMENT

Siobhan M. Brady and Philip N. Benfey

Department of Biology and the Institute for Genome Science and Policy, Duke University, Durham, NC, USA

Abstract: Systems biology approaches have recently garnered much interest in plant biology, primarily due to the massive increase of data obtained using genomic approaches. One advantage of using these approaches is the elucidation of emergent behaviour, or a complex pattern, which results from the interaction of simple components. The use of systems biology approaches to understand root development at various scales, including transcriptional networks that regulate cell identity, dynamic cellular behaviour that contributes to root growth and physical root architecture will be explored.

Keywords: systems biology; roots; development; transcriptional network; physical network; modelling.

11.1 Roots and systems biology

Systems theory is the study of the organization of components, and the principles common to the complex organization of these components. The concept of emergence, whereby a complex pattern or behaviour forms from the interaction of simple components, has provoked interest in the use of a systems approach to understand biological networks (Ideker *et al.*, 2001). Biological information is organized in multiple hierarchical levels, from molecular constituents, to cell types and tissues and ultimately into complex networks. These networks are typically robust. Key nodes in the network are often essential and if perturbed, have profound effects on the remainder of the network (Ideker *et al.*, 2001). A systems biology approach outlined by Ideker and colleagues entails the following methods: (1) identification and modelling of all components in the system, (2) system perturbation and monitoring, (3) refinement of the model in response to experimentally observed responses, (4) testing of the model to distinguish between competing hypotheses (Ideker *et al.*, 2001). This chapter will describe research that aims to understand root development using this systems approach. We include descriptions of the molecular, cellular and physical components of the root, their organization and models that describe these aspects of root development in an aim to identify emergent behaviour of these networks.

11.2 Why study roots?

The plant root serves multiple purposes. The root anchors the plant, and through its physical, branched network, explores and exploits the soil. Additionally, its cells absorb and transport nutrients and solutes to the shoot. This exploration is dynamic, plastic and highly responsive to the external environment. The root has many features that make it amenable to study using a systems approach. At the molecular level, the development of novel genomic methods coupled with the simple radial arrangement of root cell types in the model plant, *Arabidopsis thaliana*, has set the platform for elucidating root transcriptional networks. At the cellular scale, the processes of cell division and elongation are coordinately regulated in complex spatiotemporal patterns across the root. Data obtained from increasingly sophisticated image analysis methods combined with kinematic modelling provide information regarding mechanisms that underlie root growth. At the whole organ level, systems theory can also be used to describe aspects of the root's physical network, such as root branching, dynamic growth and global topology.

11.3 Root development in the model plant, *Arabidopsis thaliana*

The Arabidopsis root, in particular, has become a model for studying root development. The root's cylindrical shape, outer tissue rotational symmetry and regular patterning reduce the complexity of its development (Benfey and Scheres, 2000). This simplicity in patterning and a number of molecular techniques that make Arabidopsis a model organism for plants have resulted in the dissection of the developmental genetic pathways that give rise to the final root pattern. Arabidopsis root development is described in Fig. 11.1.

The non-dividing stem cell niche in the Arabidopsis root is named the quiescent centre (QC). Surrounding this niche is four populations of initial cells with stem cell properties who divide and whose daughters differentiate into

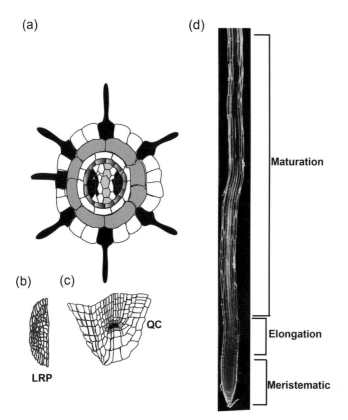

Figure 11.1 Arabidopsis root development. (a) Cross section through an Arabidopsis primary root. The outer tissues are arranged in concentric circles around the vascular cylinder. The epidermis is the outermost tissue and contains hair cells (black) and non-hair cells (white). Interior to the epidermis is the cortex (grey) and then the endodermis (white). The pericycle is the last rotationally symmetric tissue layer and is indicated in shaded grey. In the vascular cylinder, the xylem tissue is arranged in a line through the cylinder with phloem on either side in black. The intervening cells are procambium. (b) Lateral root primordia emerge from pericycle cells located at the xylem pole. (c) A cutaway of the root tip indicating the non-dividing stem cells of the quiescent centre in black, and the initial cells surrounding this non-dividing stem cell niche. (d) The meristematic zone is characterized by rapid cell division, the elongation zone by cell expansion along the root longitudinal axis and the maturation zone by terminal cell differentiation.

specific cell types. Cell types are constrained within cell files and with each division a cell is successively displaced further along the root's longitudinal axis. Distal to the QC, the tiered columella and surrounding lateral root cap protect the QC and initial cells. Proximal to the QC, the epidermis which contains hair and non-hair cells, cortex, endodermis and pericycle cell types are distributed in a rotationally symmetrical fashion around the root's radial axis and are found in concentric cylinders moving from the outside of the root to the inside. This arrangement reduces the dimensionality of the root from three dimensions to two dimensions for these outer cell types. The vascular cylinder is bilaterally symmetrical and is composed of three tissue types: xylem, phloem and procambium. Vascular tissue is found in a diarch arrangement with xylem cells in a line through the vascular tissue and the phloem cells found at the opposite poles to the xylem. Procambial tissue is found in the intervening space between the xylem and phloem. The procambium has meristematic potential and procambial divisions later in the root life cycle give rise to secondary root growth. Arabidopsis is a dicotyledon, and as such its root system consists of a primary root with lateral roots that initiate from pericycle cells at the xylem pole. The root can be further divided into three developmental zones along the root longitudinal axis. Moving from the QC, cells enter the meristematic zone, which is characterized by cuboidal cells and rapid cell division, then the elongation zone where they undergo elongation parallel to the root longitudinal axis, and finally the maturation zone where most cells undergo terminal differentiation. The plant phytohormones auxin and cytokinin play an important role in patterning this root meristematic area and in regulating cell division and elongation of different root cell types (Przemeck *et al.*, 1996; Sabatini *et al.*, 1999; Pitts *et al.*, 1998; Mahonen *et al.*, 2000; Ljung *et al.*, 2005; ; Swarup *et al.*, 2005; De Smet *et al.*, 2007; Ioio *et al.*, 2007).

Sequencing of the Arabidopsis genome and a number of molecular technologies developed for Arabidopsis have rapidly advanced our knowledge of the mechanisms regulating radial pattern organization in the plant root. Elucidating these mechanisms provides an entry point for understanding

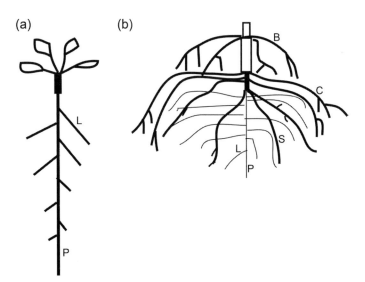

(a)   (b)

Figure 11.2 Root structure of dicots (a) and monocots (b). (a) Dicots have a dominant primary root (P) with lateral roots (L) branching off the primary root. (b) Monocots have a dominant shoot-borne root system. These shoot-borne roots include brace roots (B) and seminal roots (S). The embryonic primary root and lateral roots (L) form the remainder of the root system.

the conservation of developmental mechanisms across plant species, particularly in agronomically important species like cereals. In Arabidopsis and dicots, primary and lateral roots form the majority of the root physical network, while in monocotyledonous plants, like maize and rice, an extensive shoot-borne root system which includes crown and brace roots, predominates (Hocholdinger *et al.*, 2004) (Fig. 11.2). The rotational symmetry of the outer root tissues in other plant species is similar to that of the Arabidopsis root. In maize and rice, for example, differences include an increased number of cell layers for the cortex. Also, the QC consists of 800–1200 cells and is further surrounded by proximal and distal meristems which consist of several hundred cells.

11.4   Systems biology at the molecular level: modelling a root transcriptional network

A number of developmental genetic studies in the Arabidopsis root have identified transcription factors (TFs) as key players in root pattern formation. TFs like *SHORT-ROOT, SCARECROW, WEREWOLF* and *GLABRA2* are all responsible for determining cell-type fate within the root (Cristina *et al.*, 1996; Di Laurenzio *et al.*, 1996; Masucci *et al.*, 1996; Lee and Schiefelbein, 1999; Helariutta *et al.*, 2000; Sabatini *et al.*, 2003). These studies, however, elucidate gene function by analyzing mutants, and thus determined gene function, one to several genes at a time. Five hundred and seventy-seven TFs are expressed in the Arabidopsis root (Birnbaum *et al.*, 2003) and previously mentioned developmental genetic studies have identified only a small subset of these TFs. New techniques developed for plant roots enable genome-scale

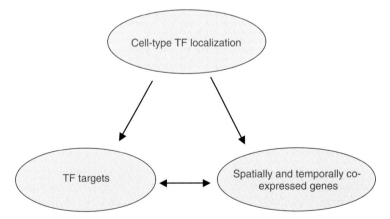

Figure 11.3   Three types of data are needed to model a transcriptional network: (1) cell-type TF localization, (2) groups of co-expressed genes, (3) targets of transcription factors. Information from each data set can be used to infer information from another data set.

detection of cell-type specific transcription profiles, thus allowing researchers to move from single-gene studies to understanding the entire transcriptional network that regulates root development with near cell-specific resolution. This research will be described in the context of a systems biology approach.

In order to elucidate the transcriptional network that regulates root development, all root transcriptional network components need to be identified. In multi-cellular organisms, these component lists should include groups of spatially co-expressed genes, the cell-type localization of all TFs, their respective TF target sequences and should ultimately aim to identify combinatorial TF control of target gene expression in a cell-type specific context (Fig. 11.3). Large scale expression analyses can identify genes that are tightly co-regulated at the level of transcription and that are presumably regulated by a common set of TFs. Bioinformatic and experimental approaches can then identify binding targets of TFs. These data sets must then be integrated to model the root transcriptional network. Resulting models can then be tested by perturbing different components of the network, and comparing network predictions to in vivo transcriptional effects on downstream genes.

In this section, genome-scale methods that identify cell-type specific transcripts will be described. Existing bioinformatic and experimental approaches that identify TF targets will also be discussed. Unfortunately, high-resolution TF target data are still being assembled. Nonetheless, specific examples that utilize these methods and that have described modules of the root transcriptional network will be presented. The elaboration of these examples using genome-scale technologies should ultimately elucidate the transcriptional networks regulating root development.

11.5 Identification of components

11.5.1 Co-expressed cell-type specific transcriptional profiles

11.5.1.1 Expression analysis
Sequencing of plant genomes (*A. thaliana, Oryza sativa ssp. indica, Oryza sativa ssp. japonica*) (The Arabidopsis Genome Initiative, 2000; Goff *et al.*, 2002; Yu *et al.*, 2002) and the availability of platforms to examine genome-scale expression in multiple species sets the stage for understanding the underlying changes in expression that occur during root development. A number of techniques can be used to monitor expression of genes on a whole genome level. These techniques include sequence tag-based methods like Serial Analysis of Gene Expression or Massively Parallel Signature Sequencing (MPSS), or microarray-based methods. In the tag-based methods, sequence tags from a given RNA sample are generated and the tags are sequenced to give quantitative measures of transcript abundance. In microarray-based methods, oligomer probes are either spotted or synthesized on a solid substrate

and queried with fluorescently labelled cRNA generated from a sample of interest. Transcript expression profiles are then measured by fluorescence intensity of the hybridized cRNA.

First-generation expression profiling of plant mRNA populations measured using microarrays were typically generated across entire plants or whole organs. An excellent example of more thorough profiling is the At-GenExpress data set which describes the expression of approximately 23 000 genes across 79 organs and a variety of developmental stages (Schmid et al., 2005; Kilian et al., 2007). This data set also profiles the response of seedlings to a variety of external perturbations (Schmid et al., 2005; Kilian et al., 2007). As described for roots, plant organs are heterogeneous in terms of their tissue composition and profiling mRNA extracted from these tissue mixtures can result in a loss of information about tissue and cell-type specific genomic processes. Two microgenomic techniques, whereby RNA is extracted from specific tissues or cells, and other creative methods of expression profiling have revolutionized the nature of expression profiling in Arabidopsis, maize and rice roots (Birnbaum et al., 2003; Woll et al., 2005; Jiang et al., 2006; Brady et al., 2007). Profiles obtained by these techniques resulted in the first high-resolution map of gene expression within an organ.

11.5.1.2 Fluorescent cell sorting

Genome-wide expression profiling using fluorescent cell sorting of the Arabidopsis root was established in 2003 (Birnbaum et al., 2003, 2005). Green fluorescent protein (GFP) tagged plant lines that mark specific root tissues or cell types were subjected to protoplasting treatment. This treatment enzymatically digests cell walls, and the resulting protoplasts are then sorted based on their fluorescence using a Fluorescence Activated Cell Sorter and RNA is then isolated from the fluorescently marked cells. This RNA is amplified, labelled and hybridized to Affymetrix™ Arabidopsis ATH1 microarray chips, which measure the expression of approximately 23 000 genes. This technique was first applied to five tissues of the Arabidopsis root: the endodermis, the ground tissue (endodermis and cortex), the stele (pericycle, xylem and phloem), non-hair epidermal cells, and the lateral root cap (Birnbaum et al., 2003). This technique has been further extended to profile all cell types except one in the Arabidopsis root (Lee et al., 2006; Levesque et al., 2006; Nawy et al., 2005; Brady et al., 2007). Information gained from these profiles is quite rich and reveals highly tissue- and cell-type specific expression within the root. An additional 392 transcripts were detected in the first cell-sorting experiments in comparison to whole roots profiled in the AtGenExpress data set (Birnbaum et al., 2003). In addition to this increase in detectable transcript number, 51 distinct dominant expression patterns were identified across these 14 profiled cell types that would have never been characterized using the whole root alone (Brady et al., 2007). A profile of 13 developmental time points in a single root has further resulted in 40 distinct dominant expression patterns in developmental time (Brady et al., 2007). These cell-type specific and developmental stage patterns have been coupled together to identify tightly co-expressed genes in both space and time within the Arabidopsis root (Brady *et al.*, 2007). These expression patterns, and their intersections can also be used to identify transcriptional regulatory modules (Brady *et al.*, 2007).

11.5.1.3 Laser capture microdissection

In laser capture microdissection (LCM), plant tissue is fixed and the region of interest is dissected with a laser. RNA is isolated from this fixed tissue, amplified if insufficient RNA is obtained, and hybridized to microarray chips or other expression platforms. This technique has been used to profile the pericycle of wild-type maize roots and the pericycle of the *rum1* maize mutant which is deficient in the initiation of seminal roots and lateral roots. These experiments identified an array of genes that are expressed in the pericycle before lateral root initiation, which are dependent on the action of *RUM1* (Woll *et al.*, 2005). This technique has also been successfully used to profile expression in the maize root cap and in different tissues of the Arabidopsis embryo, including the basal regions of the embryo that give rise to root tissue (Jiang *et al.*, 2006; Spencer *et al.*, 2006). Efforts are also underway to profile expression in rice cell types (http://bioinformatics.med.yale.edu/rc/overview.jspx).

11.5.1.4 Artificial induction systems

Application of the fluorescent cell-sorting technique resulted in one of the first high-resolution expression maps of an organ. Additional creative methods that use mutants combined with LCM, like the study using *rum1* and LCM as previously described, or ones that use mutants or artificial induction systems also provide a means of identifying specific transcriptional programs. Establishment of a lateral root induction system where Arabidopsis roots are grown on a polar auxin transport inhibitor and then transferred to auxin-containing media synchronizes lateral root initiation. RNA isolated from these synchronized roots at various time points was compared to RNA isolated from the dominant auxin signalling mutant, *solitary root1* (*slr1*), to identify genes that are involved in lateral root initiation in an auxin-dependent manner. Analysis of these expression profiles also identified putative negative and positive feedback mechanisms that regulate auxin homeostasis and signalling in the pericycle (Vanneste *et al.*, 2005).

These studies and many others in Arabidopsis provide a wealth of data regarding co-expressed genes that act in highly specific spatial developmental programs. The tight co-expression of these genes suggests one or several common regulating TFs. Identifying these TFs will be instrumental in elucidating the complete transcriptional networks that regulate root development.

11.5.2 Cell-type localization of TFs

11.5.2.1 Post-transcriptional processing

In addition to transcription, a number of events take place that further modulate which transcripts are translated into proteins via post-transcriptional processing and translational regulation. These events must be incorporated into lists that describe the cellular localization of TFs and into resulting models to generate accurate transcriptional networks of root development.

11.5.2.2 Transcriptional and post-transcriptional regulation of TF expression in Arabidopsis roots

Often, the mRNA expression pattern of a gene is used to predict the spatial location of its resulting protein. In plants, however, the degree to which an upstream non-coding sequence and coding sequence contribute to the final mRNA and protein expression patterns was previously unknown. To assess the contribution of these sequences, an unbiased investigation of these processes in the transcription and translation of 61 TFs was performed (Lee *et al.*, 2006). In cases where these sequences alone are insufficient to drive 'correct' transcription or translation, contribution of post-transcriptional regulation can be inferred. For this experiment, 61 TFs were selected based on their enrichment in any one of five tissues (Birnbaum *et al.*, 2003). Their mRNA expression patterns identified in an updated gene expression map of nine radial tissues using the fluorescent cell-sorting technique were compared to the expression pattern conferred by the upstream non-coding region (up to 3 kb) of each TF fused to GFP alone (transcriptional fusion) or fused to the TF coding sequence (translational fusion). For 80% of the transcriptional fusions, regulatory regions in the 3 kb of the 5′ upstream non-coding region were sufficient to recapitulate the mRNA expression pattern. For the remaining 20% of cases, additional regulatory regions or mechanisms are thought to be needed. Confirming this idea, the translational fusion of one of these TFs, *PHABULOSA*, a known microRNA target, recapitulated the expression pattern, whereas the transcriptional fusion did not.

When comparing transcriptional to translational fusions, in cases where GFP was found consistently, different translational and transcriptional expression patterns were found for 29% of the TFs. Two of these cases were due to known regulation by microRNA, and in another six cases, these TFs display expanded expression patterns that are indicative of intercellular protein trafficking. This study demonstrates that expression analysis alone is not sufficient for accurate cell-type localization of TFs. Post-transcriptional regulation in plants does play an important role in the tissue-specific expression and localization of TFs and needs to be integrated into transcriptional network models.

11.5.2.3 MicroRNAs

A large proportion of plant microRNAs target transcripts encoding TFs required for normal growth, development and responsiveness (Jones-Rhoades *et al.*, 2006). MicroRNAs are small endogenous RNAs that regulate gene expression in plants and animals. In plants, microRNAs are approximately 21 nucleotides long and regulate gene expression by binding complementary RNAs and directing cleavage of these by an RNA-induced silencing complex. A large number of microRNA loci exist that have been identified by sequencing, genetics and bioinformatics in many plant species including Arabidopsis, rice and poplar (Jones-Rhoades and Bartel, 2004; Jones-Rhoades *et al.*, 2006; Nakano *et al.*, 2006; Fahlgren *et al.*, 2007). However, no systematic study of microRNA presence in root tissues has been performed. This type of data is currently needed to correctly model this aspect of gene regulation in root transcriptional networks.

11.5.2.4 Additional nuclear and cytoplasmic events that regulate gene expression

Additional nuclear events that regulate gene expression in eukaryotes include splicing, 5′ capping, 3′ polyadenylation, mRNA decay and transport of mRNA from the nucleus to the cytoplasm. Cytosolic mechanisms include regulation of mRNA decay and translation, protein targeting, protein movement and protein degradation. RNA-binding proteins control the majority of these events, and RNA-binding proteins have been shown to associate with unique groups of mRNAs in the nucleus and cytoplasm. It has been proposed that the combinatorial control of these processes by RNA-binding proteins constitutes an additional level of regulatory control for gene expression (Keene, 2003). The degree to which these RNA-binding proteins regulate plant TF expression is unknown, although a number of TFs do show many splice variants and the most recent gene ontology annotates approximately 310 proteins as RNA-binding proteins (Itoh *et al.*, 2004). Immunopurification of plant polyribosomes by epitope tagging of ribosomal proteins produces complexes that include cytosolic ribosomal proteins, ribosome-associated proteins and full-length transcripts (Zanetti *et al.*, 2005). Quantification of these polysomal-associated transcripts by probing microarrays with their associated transcripts provides a direct estimate of transcripts that will be translated. Stress and various sequence features can regulate the amount of polyribosomal loading indicating that translation in plants can be regulated by external conditions and by putative transacting factors that bind to RNA (Kawaguchi and Bailey-Serres, 2005; Zanetti *et al.*, 2005). Deducing this RNA-binding code is also necessary to model transcriptional networks. A related method for identifying the targets of RNA-binding proteins is the ribonucleoprotein-immunoprecipitation microarray (RIP-Chip) which allows for detection of genome-wide targets of ribonucleoproteins by immunoprecipitating RNA-binding protein complexes, dissociating and labelling attached RNA, and hybridizing this RNA to microarrays (Keene *et al.*, 2006). Placing epitope-tagged ribonucleoproteins under the control of tissue-specific promoters can allow for the tissue-specific identification of transcripts associated with these proteins.

11.5.3 TF targets

Thus far we have discussed a variety of mechanisms that can regulate TF expression and translation. In a systems biology approach, these mechanisms must be included in order to accurately model the cell-type localization of all root TFs. Next, the interactions between root TFs and their targets need to be identified and eventually modelled. Public domain resources exist that catalogue known *cis*-elements in upstream gene regions, and other tools exist that allow users to detect enrichment of these elements in co-regulated genes (Davuluri *et al.*, 2003; Guo *et al.*, 2005; O'Connor *et al.*, 2005). These catalogues, however, are by no means complete. Two complementary methods exist that allow for the detection of TF targets. The first of these methods employs chromatin immunoprecipitation. In this technique, TFs are first fixed to their associated DNA, then immunoprecipitated with a specific antibody or via an epitope tag such as GFP. The DNA associated with this complex is then dissociated and identified by quantitative PCR or by labelling this DNA and probing microarray chips (ChIP-chip) to determine genome-wide targets of the TF of interest (Harbison *et al.*, 2004). A second method, a modified yeast one hybrid, was recently developed to identify TF–DNA interactions in the *Caenorhabditis elegans* digestive tract (Deplancke *et al.*, 2004, 2006). In this method, copies of whole promoters are fused with a minimal promoter to two different types of reporter genes and integrated in the yeast genome. These strains are then interrogated with a TF library or with a cDNA library to detect protein–DNA interactions. This method has proven useful in elucidating components of a transcriptional network found in *C. elegans* digestive tissue.

11.6 Component modelling

While data has not been gathered at the genome-scale for TF localization and TF targets, results from the methods that we have described have been integrated to successfully elucidate modules of the root transcriptional network that regulate specific aspects of cell-type patterning. These modules are evidence of a first step towards the ultimate goal of modelling the entire root transcriptional network.

11.6.1 SHR transcriptional network

The *SHORT-ROOT* (*SHR*) TF regulates specification of the root stem cell niche, the asymmetric division of the cortex/endodermal initial and endodermis fate specification (Helariutta *et al.*, 2000; Nakajima *et al.*, 2001). *SHR* mRNA is expressed in the stele, and the SHR protein moves from the stele to the endodermis. The *SCARECROW* (*SCR*) TF is expressed in the endodermis, is necessary for endodermis and QC specification and has been demonstrated to be downstream of *SHR* by various genetic analyses (Nakajima *et al.*, 2001; Sabatini *et al.*, 2003).

Genome-wide meta-analysis was used to identify the targets of *SHR* by combining results from an inducible form of *SHR* driven by its endogenous promoter, ectopic expression of *SHR*, followed by cell sorting and comparisons of the ground tissue layer of mutant to wild-type roots (Levesque *et al.*,

2006). This meta-analysis identified with high statistical confidence eight direct targets that are regulated by SHR in the stele and endodermis. Four of these genes encode putative TFs including *SCR*. While this approach identified direct targets of SHR by combining different expression analyses, in vivo evidence of SHR binding directly to these target promoters was still needed. Chromatin immunoprecipitation (ChIP) with a polyclonal antibody to the GFP portion of the SHR:GFP translational fusion protein and subsequent PCR was used to test for enrichment of the four candidate TF promoters. In vivo evidence was identified for all four TF target promoters being bound by the SHR protein, including the *SCR* promoter (Levesque *et al.*, 2006; Cui *et al.*, 2007). *SCR* expression is reduced in an *scr* and *shr* mutant background, suggesting that SCR also binds its own promoter (Heidstra *et al.*, 2004). ChIP-PCR confirmed this and supports regulation of *SCR* by a SHR/SCR-dependent positive feedback loop. This same assay identified the three remaining SHR target TF promoters, *MAGPIE (MGP)*, *SCR-LIKE3 (SCL3)* and *NUTCRACKER (NUC)* as SCR targets and determined that most binding sites for SHR and SCR at these promoters coincide (Cui *et al.*, 2007). Functional interdependence of the SHR and SCR TFs is due at least in part to the physical interaction of these proteins, as determined by two different assays (Cui *et al.*, 2007). Furthermore, this interaction of SCR and SHR proteins was shown to block SHR movement by sequestering it in the nucleus, and by the SHR/SCR-dependent positive feedback loop for *SCR* transcription. This study elucidated a SHORT-ROOT regulated transcriptional module that regulates endodermis specification by integrating data obtained from expression analysis, TF localization and ChIP-PCR (Levesque *et al.*, 2006) (Fig. 11.4).

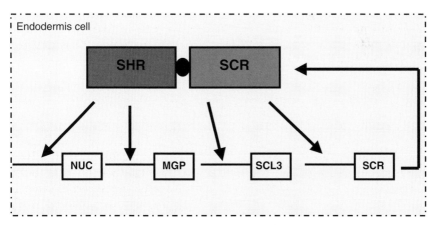

Figure 11.4 A SHR transcriptional module. SHR and SCR transcription factors are functionally interdependent. These TFs physically interact and bind to the promoters of four transcription factors: *NUC, MGP, SCL3* and *SCR*. This positive regulation of *SCR* expression acts as a positive feedback loop, as the SCR protein will go on to interact with SHR, sequester SHR in the nucleus and continue to regulate the transcription of these four transcription factors including itself.

Further studies should delineate the dynamics of this module and the role of these targets in various aspects of SHR function in root development.

11.6.2 Root hair development

Root hair development has primarily been studied in Arabidopsis. In Arabidopsis, root hairs are specified by a positional cue and an associated transcriptional regulatory network. Root hair cells are located over two underlying cortical cells (H-position) while non-hair cells are positioned over a single cortical cell (N-position), and this position-dependent specification results in hair cell files along the longitudinal axis. It has been proposed that epidermal cell files are specified by lateral inhibition which is regulated by differential accumulation of the *WEREWOLF (WER)* TF (Lee and Schiefelbein, 2002). *WER* expression is regulated by positional information and transmission of this information results in slightly higher WER accumulation in the N-position (Lee and Schiefelbein, 1999). It was proposed that WER regulates the expression of the *GL2* and *CAPRICE (CPC)* putative TFs (Lee and Schiefelbein, 2002). GL2 induces the non-hair cell fate in N-position cells, and CPC is proposed to move to the neighbouring H-position cells where it inhibits *GL2* and *CPC* expression and thereby promotes the hair cell fate (Fig. 11.5). Conditional induction of *WER* expression, electromobility shift assays, and a

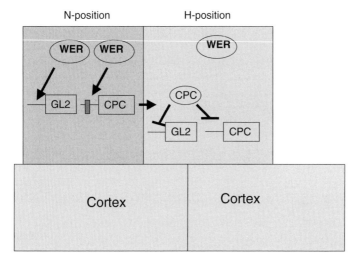

Figure 11.5 WEREWOLF transcriptional regulation of hair and non-hair cell fate. Root hair cells are located over the junction of two underlying cortical cells (H-position) and non-hair cells in the N-position. The WER transcription factor is differentially enriched in the N-position compared to the H-position. WER binds to the *GL2* and to the WER-binding site II in the *CPC* promoter. Expression of *GL2* promotes non-hair cell fate. The CPC protein moves to the H-position and represses expression of *GL2* and *CPC* thus promoting hair cell fate.

yeast one hybrid assay identified the WER-binding site II in the *CPC* promoter as a direct target of the WEREWOLF TF, and provided a direct mechanism for WER in specifying hair and non-hair cell fate via a transcriptional regulatory network (Ryu *et al.*, 2005).

11.7 Systems biology at the cellular level: modelling root growth and the dynamic behaviour of its component cells

11.7.1 Root growth

Growth in plants encompasses two linked cellular processes – cell division and cell expansion. Understanding cell division and expansion and their co-ordination ultimately allows one to understand regulation of root growth, and facilitates a framework to understand its underlying molecular mechanisms. As previously described, roots contain a large number of cells, and groups of these cells divide and elongate in distinct spatial regions along the root's longitudinal axis. Furthermore, cell division and elongation occurs in a temporally dynamic manner. Studies concerning the complex organization of these cell behaviour aspects are well suited to study using a systems biology approach.

Anatomical studies of wild type and mutant roots have characterized cell division and elongation in a static manner. More accurate studies need to capture the dynamics of these processes. This section will describe spatiotemporal cell division and elongation data obtained from image analysis methods, and the integration and modelling of this data using differential equations that describe the kinetics of root growth. These models reveal the cellular mechanisms behind differential cell-type growth, the importance of controlling the number of dividing cells and the conservation of cell elongation rates in roots across species. The plant hormone auxin plays a role in the stimulus-induced elongation of root epidermal cells. Growth predictions made from cell-type specific models of auxin transport that were tested using kinematic models will also be described. Results obtained from this study provide a direct molecular mechanism for gravitropism-induced cell elongation.

11.7.2 Cell expansion and division rates in the Arabidopsis root using kinematic analysis

Beemster and Baskin (1998) employed a kinematic study where graphite particles were placed on Arabidopsis root cells, and the trajectory of these marked cells throughout the root growth zone were quantified to describe root growth and its underlying cellular behaviour. The resulting data was modelled using a series of differential equations (Beemster and Baskin, 1998). This study addressed the relationship between cell production and cell elongation in different cell types, and showed the importance of controlling the number of dividing cells to generate tissues of different cell lengths and to regulate organ enlargement. A subset of the series of equations used to model cell expansion and division rates are given below:

Velocity (μm h^{-1}). The position of individual particles relative to the tip of the root in a pair of images. $X_{i,t}$ is the ith particle at time t. The length of the growth zone was determined as the distance between the QC and the first position where the increase in velocity between successive positions was less than or equal to 0.

$$v(x) = \frac{x_{i,2} - x_{i,1}}{t_2 - t_1}$$

Longitudinal strain rate (% h^{-1}). The derivative of velocity with respect to distance along the root, this derivative essentially measures the rate of elongation:

$$r(x) = 100 \times \frac{\mathrm{d}v}{\mathrm{d}t}$$

Cell length. Determined for cortical and epidermal cell files in the meristematic zone and the elongation zone. The growth zone is a sum of the meristematic and elongation zone intervals.

Cell production rate (μm^{-1} h^{-1}).

$$P(x) = \frac{\mathrm{d}F}{\mathrm{d}x} + \frac{\mathrm{d}\rho(x)}{\mathrm{d}t}; \rho(x) = \text{cell density}; F = \text{cell flux}$$

Cell flux (cells h^{-1}). The rate at which cells flow past a particular position x.

$$F(x) = \frac{v(x)}{l(x)}$$

Cell division rate (cell cell^{-1} h^{-1}). Calculated from $P(x)$ and corrects for cell length.

$$D(x) = P(x) \times l(x)$$

11.7.3 Methods used to smooth and interpolate cell length and velocity data

A number of methods were used to smooth and interpolate cell length and velocity data. The method chosen is quite important as inaccurate smoothing of data could lead to improper conclusions about the relationship between transitions of cell division and cell elongation. In this particular study cubic splines and repeated partial polynomials were used to fit the data. This modelled data was then compared to simulated data which included random noise with the same variance as the original cell length or velocity data to assess its robustness.

These parameters were used to describe root growth at 6 days and 10 days and were able to attribute differences in root growth to changes in cell production, and not to changes in the rate of cell division or elongation (Van Der Weele *et al.*, 2003). These equations demonstrated that cell division rates were fairly constant throughout the meristematic zone, and surprisingly, that cell division continued well beyond the transition to the elongation zone. Cell division and cell elongation rates were also found to be strongly correlated. This study also found differences in cell division parameters between root hair and non-root hair cells. Non-root hair cells have decreased rates of cell production, which was determined not to be caused by slower divisions, but instead to increases in cell length. The size difference between these two types of cells originates in the meristem and is regulated in part through the size of the initial cells.

11.7.4 Using advanced image analysis to refine a kinematic model of root growth

Methods that mark cell velocity manually can be somewhat invasive. Due to the manual nature of marking cells, marks cannot be applied at a high density, and can often be error-prone. Modern methods of image analysis, however, can mark images at high density, compile them in three-dimensional stacks, and in time-lapse imaging, can capture aspects of cell velocity.

A software named RootflowRT (Van Der Weele *et al.*, 2003) works at short time intervals to map two-dimensional velocity within the root meristem and elongation zone. Data obtained from these methods were used to refine the kinematic models used by Beemster and Baskin (1998). This model predicts from somewhat continuous elongation rates over the root growth zone, that cell elongation is a continuous process across the root. Using the higher resolution velocity data in these kinematic models, however, three distinct regions of velocity were found, all separated by abrupt transitions, as opposed to the previously identified sigmoid curves. The first and second regions of velocity correspond to the meristematic and elongation zones respectively, and velocity was linear through each respective region (Fig. 11.6). The data were fit to two different models: the first is a sequence of three linear equations joined at two breakpoints, and the second is by overlapping polynomials. The first fits the data well and when considering the derivative of the velocity plot, or the relative elongation rate, a 'step stool' function is found, indicating a constant rate of elongation for the meristem, a higher constant rate of elongation for the elongation zone and a rapid transition between states. Fitting the data with the overlapping polynomial function, however, reveals a more complex derivative which represents an overall step function with higher spatial frequency fluctuations. When looking at this function across multiple roots, these higher frequency fluctuations appear to be somewhat synchronized amongst roots (Van Der Weele *et al.*, 2003). Mechanistically, these models suggest strict co-regulation of cell cycle machinery and cell

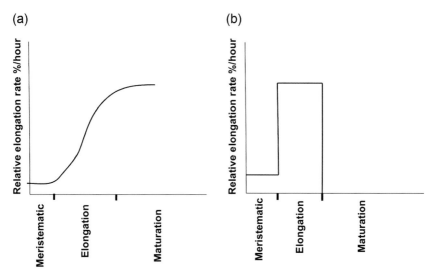

Figure 11.6 (a) Kinematic models constructed with low-resolution cell velocity data identified a sigmoid elongation rate curve that predicted cell elongation as a continuous process across the root. (b) Models using higher resolution velocity data demonstrate three distinct regions of elongation, all separated by abrupt transitions (modified from Beemster and Baskin, 1998 and Van Der Weele *et al.*, 2003, reprinted by permission from American Society of Plant Biologists).

elongation, and the existence of a developmental checkpoint that separates the meristematic zone from the elongation zone, and the elongation zone from the maturation zone. Furthermore, the data fitted with the overlapping polynomial functions suggests some sort of stochastic noise that occurs within each zone across multiple roots.

Velocity information in several unrelated plant species was also collected, and the data fitted using the three linear equation model. The linear regressions in these species correlated quite closely to that of Arabidopsis, and revealed a conservation of elongation rate across species.

In summary, analyzing the dynamic parameters of cellular processes at high resolution, and fitting these data with increasingly sensitive models reveals the cellular mechanisms that underlie organ growth. The striking conservation of elongation rates across plant species further suggests conservation of the molecular mechanisms underlying root growth. Analyzing how these parameters change with perturbation of this system by external stimuli will aid in understanding how the demonstrated phenotypic plasticity of root development is manifested at the level of cellular processes.

11.7.5 Using advanced image analysis to quantify stem cell divisions

While the above-mentioned studies globally modelled the cellular processes of division and elongation, very few studies have addressed the dynamics of stem cell divisions. This is primarily due to the small size and spatial position of these cells, deep within the meristem and surrounding the QC. Stem cell divisions in the root meristem have been described using static anatomical data (Dolan *et al.*, 1993). From this data, it was proposed that the anticlinal division of the columella stem cell and the periclinal division of the epidermal/LRC (lateral root cap) stem cell are synchronized so as to give the same number of cell layers. A novel confocal, time-lapse tracking system (Campilho *et al.*, 2006) that dynamically images the frequencies of these stem cell divisions rejected this proposal. Using this imaging system, weaker coordination than predicted was identified between the two types of divisions (Pearson correlation coefficient = 0.74). The time between each of these divisions was also found to be highly variable, they occurred in an arbitrary order, and showed no evidence of control by a diurnal rhythm. Together, the results obtained from this dynamic imaging reject the possibility of strong synchronization of these two stem cell divisions and demonstrate the need for high quality, high-resolution data when modelling cellular processes.

11.7.6 Combining kinematic and auxin transport models to elucidate tissue-specific regulation of gravitropism

Gravity is perceived in the root via the columella cells (Blancaflor *et al.*, 1998). The root responds to gravity by changing its orientation relative to the gravity vector by differential growth. Auxin was identified as a major contributor to the gravitropic response based on a lateral auxin responsiveness gradient in gravity-sensing tissues, and on the agravitropic phenotypes of many auxin signalling and transport mutants (Bennett *et al.*, 1996; Luschnig *et al.*, 1998; Tian and Reed, 1999; Friml *et al.*, 2002). Auxin transporters that play a role in this response include the auxin efflux transporters, PIN2 and PIN3, and the auxin influx transporter, AUX1 (Bennett *et al.*, 1996; Luschnig *et al.*, 1998; Friml *et al.*, 2002). A systems approach that models spatial auxin flux and cell elongation within the root has provided direct evidence that auxin is indeed the intercellular gravitropic signal (Swarup *et al.*, 2005), and that the LRC and epidermal cells are required for transmission of a lateral auxin gradient that results in differential elongation of the root tissue (Swarup *et al.*, 2005).

Upon gravity stimulus, the PIN3 transporter has been shown to retarget to the basal face of inner columella cells. This retargeting creates an initial lateral auxin gradient (Friml *et al.*, 2002). To determine what role different tissues in the root play in terms of transport of this lateral auxin gradient, the role of each tissue was assessed in terms of auxin transport via AUX1. Expression of AUX1 in LRC and expanding epidermal cells together is required for the gravitropic response, presumably to facilitate delivery of the gravity-induced lateral auxin gradient to elongation-zone tissues (Swarup *et al.*, 2005).

How is this auxin gradient created? Radial diffusion around the circumference of the root, and active auxin transport should play a major role. To address this question, the role of AUX1 and PIN2 in these two processes was

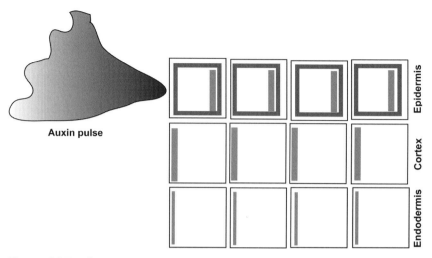

Auxin pulse

Figure 11.7 The auxin transport model used to determine the role of AUX1 and PIN2 in gravity-induced cell elongation. The epidermis, cortex and endodermis are modelled in three rows of cells. AUX1 is located in the epidermis in a square (found on all cell surfaces), and PIN2 and PIN1 are localized on different faces of the cell wall in different cell types (lines).. A gravitopic signal is simulated by application of an auxin pulse which moves as a gradient through these tissues (modified from Swarup *et al.* 2005).

assessed by modelling auxin flux in the root in response to a gravity stimulus (Swarup *et al.*, 2005). This model encompasses approximately 1200 cells, three tissue types (epidermis, cortex and endodermis) and localization of the PIN2, PIN1 and AUX1 transporters (Fig. 11.7). Radial diffusion and facilitated auxin transport were incorporated in the model using diffusive membrane permeability and carrier membrane permeability coefficients, and an IAA diffusion coefficient (Swarup *et al.*, 2005). These tissues were represented in a rectangular array of three rows, with cells indicated in columns. Expression and localization patterns place PIN2 in the epidermis and cortex, weak PIN1 in the epidermis, cortex and endodermis, and AUX1 in the epidermis only (Swarup *et al.*, 2001; Benkova *et al.*, 2003; Blilou *et al.*, 2005). In the model, a gravitropic signal was simulated by supplying an auxin pulse to the apical end of the virtual root and movement of an auxin gradient throughout elongation-zone tissues was monitored. This gradient was simulated in a 'wild-type' virtual root, and simulated in a *pin2* and *aux1* mutant.

In a *pin2* mutant, the model predicts that the mutant would still be able to transport auxin in the elongation zone due to weak epidermal expression of PIN1. Wild-type PIN1 protein is not found in the LRC, and this is sufficient to give rise to the agravitropic phenotype of the *pin2* mutant. The model also predicts correctly that the *aux1* mutant is unable to transport the auxin pulse. The majority of auxin is predicted to remain in the apoplast where it is free to diffuse back to the stele. This diffusion will reduce the scale of the auxin pulse delivered to the central elongation zone of the root, and results in a shallow auxin gradient with significantly decreased levels of auxin in the basal half of the elongation zone that is insufficient to drive a differential growth response. Based on this prediction, root growth was examined in different positions of the elongation zone in the *aux1* mutant using kinematic modelling. In an *aux1* mutant, velocity was decreased relative to wild type only in the basal elongation zone, as predicted by the auxin transport model. Furthermore, measurements of the derivative of velocity indicate that *aux1* cells do not expand to the same amount as in wild type. To determine if AUX1 was sufficient to restore these root growth characteristics, AUX1 was expressed only in the epidermal cells and LRC, and wild-type growth was restored. AUX1 expression in all epidermal cells was also shown to be required for full restoration of this phenotype. While auxin transport in the epidermis via AUX1 was demonstrated to be necessary, the role of auxin signalling in the epidermis was also assessed. Auxin signalling in the LRC and epidermis was selectively disrupted by tissue-specific expression of the mutant *axr3-1* allele, and this disruption completely blocked gravitropism. Further, kinematic analysis of these lines showed that the resulting growth defect was not due to an alteration in basal cell elongation as only a small reduction in basal elongation was observed.

In summary, the simulated model, its predictions and tested biological perturbations have resulted in a mechanistic model of the root gravitropic response. A gravity stimulus at the root tip causes auxin to be asymmetrically distributed and the PIN3 efflux facilitator to retarget to the lower side of the gravity-sensing columella cells (Friml *et al.*, 2002). The resulting lateral auxin gradient is mobilized through the combined action of the auxin influx and efflux facilitators, AUX1 and PIN2 via LRC cells to expanding epidermal cells in the elongation zone. The lateral auxin gradient inhibits the expansion of epidermal cells on the lower side of the root relative to the upper side, causing a differential growth response that ultimately results in root curvature downwards.

Together these four studies demonstrate the need to collect data regarding the dynamics of cell division and cell elongation in order to complement our existing knowledge gained from static anatomical studies of root growth. Further collection of high-resolution measurements, particularly in three dimensions using developing image analysis technology, the subsequent modelling of this data, incorporation of molecular components and conclusions inferred from this large-scale data aid in understanding the spatial and temporal cellular behaviour that guide and power organ growth.

11.8 Systems biology: modelling the root physical network

11.8.1 Root physical network diversity

Root systems are quite diverse among species. The elaboration of root physical networks is influenced by the composition and structure of the external environment, is highly plastic, and the topology of this network is directly related to optimal root physiological function. To understand the organization of this complex topology, attempts have been made to comprehensively describe root taxonomy, to model root growth and ultimately to identify optimal network structure, thus employing many of the tenets of a systems biology approach.

A number of models over the years have been proposed that describe the structure of root systems. Recently, these have moved from describing root structure in two dimensions to three, and they attempt to represent the dynamic response of physical root networks to the external environment. In order to construct these models, an accurate description of the root structure in terms of its component parts and associated parameters is needed. First, an overall description of root taxonomy in multiple species provides a reference point from which information is obtained. Second, a range of quantitative characteristics is annotated to each root axis, like root length, and diameter. Additional terminology is then used to describe the connections between root axes within the network. Finally, these root axis interactions are synthesized in a model that describes global features of the root system and the dynamic responses of these features. Models are then compared to their reference taxonomy descriptions to assess their performance in terms of their ability to accurately capture the complexity of the root physical network. Based on this performance, parameters can be added or changed to fit the model more accurately and these results can aid in understanding the complex interactions between a root and its external environment. This section will describe different generations of models that have been used to describe root architecture.

11.8.2 Describing root taxonomy

A number of attempts have been made to comprehensively describe plant taxonomy on the basis of overall morphology. Creation of a comprehensive atlas of root morphology by species is quite difficult due to the extensive variability and plasticity of the root, and the fine root structure which is easily disturbed and lost upon excavation of the root from the soil. A classification by Cannon (1949) tentatively describes different root networks. Cannon categorized root forms into different types based on the distinction between systems where the primary root emerges from the seed and forms the basis of the root structure, and where extensive adventitious roots form the basis of the root structure from the stem base or other organs. Kutschera (1960) also describes a series of representative root systems from a number of plant species under different soil conditions, providing an information source for phenotypes resulting from the root–soil interaction. Although these taxonomies provide extensive amounts of information, they are not complete. Methods to visually describe the full spectrum of the root system in a non-invasive manner need to be developed to achieve this goal.

11.8.3 Modelling the root physical network

All models that describe the physical root network use a set of parameters that are associated with each root type. The term root type refers to an individual root axis that extends from some point on the root network. The sum of these parameters describes the state of each root type. Root types can display a variety of morphogenetic capacities. Some models allow for a root type to possess only a subset of parameter states dependent on the position of the root within the network hierarchy, while others simplify the model by allowing a root type to be described with the full suite of parameter states.

Single-dimensional root parameters used to describe a root tip include root diameter and length. Two-dimensional characteristics of the physical root network encompass aspects of root branching. A common classification refers to the developmental orders of this branching – primary, secondary, tertiary and so on (Fig. 11.8a). This method is not suitable, however, when roots come from the same junction and when a subjective decision must be made as to which root is of a higher order. A second method of classification uses mathematical trees. These simple trees describe root types as edges and their nodes as the junctions from where the root axis is born. Terminal branches are described as first order, and successive interior links are then characterized

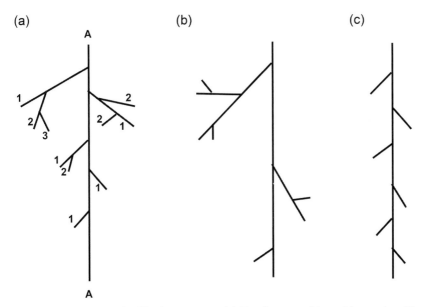

(a)　　　　　(b)　　　　　(c)

Figure 11.8 Root classification systems. (a) Developmental branching orders. The developmental order refers to roots which emerge off the main axis [A] of the root. The primary order is the first to emerge off the main axis, secondary off the primary order axis, tertiary off the secondary order axis and so on. (b) In a dichotomous branching topology, branching generation is equiprobable on all exterior root axis links. (c) In a herringbone branching topology, branching is confined to the main root axis.

by the number of exterior links they serve. Two common topologies in this case include dichotomous (Fig. 11.8b) and herringbone structures (Fig. 11.8c). Parameters associated with these can include inter-node distance, branch angle and time of branch initiation. A second set of parameters describe the dynamic spatial and temporal aspects of these branches, generally in a vector with an associated direction, a scalar value representing elongation and a rate of vector growth. Vector directions can be further influenced by their response to the environment in terms of tropisms, for example gravitropism, or phototropism. Environmental influences on these vectors constrain growth of a root type and in some models are represented as fixed parameters, such as soil or mechanical coefficients and their effect on isotropic or anisotropic growth. Environmental influences can however, be quite variable, and more sophisticated models should represent these influences dynamically. Growth potential is another putative parameter which describes root type growth capacity. The temporal aspect of root growth in models is described within time steps. These two-dimensional characteristics can be further characterized by describing the probabilistic nature of root axis formation.

Fractal analysis has been used to describe three-dimensional root typology (Nielsen *et al.*, 1997; Eshel, 1998; Walk *et al.*, 2004). A fractal is a geometric object that has a fine structure at small scales and that is self-similar. This means that it has some geometric shape that can be subdivided into parts, each of which is approximately a reduced-size copy of the whole. The modular nature of root growth and can create self-similarity at a range of spatial scales. Fractal representations have been therefore used to describe this dimensionality, and have been used to predict the size of the total root system from smaller scales (Nielsen *et al.*, 1997; Eshel, 1998). However, root growth is not always homogenous and growth can decay or branching can change in density depending on root depth. Other models are therefore needed to describe overall root topology. Root Typ is a generic model that analyses global root architecture (Pages *et al.*, 2004). It allows for all root types to take a set number of states, regardless of branch order. This allows for modelling of heterogeneous root systems. Branches can originate from any organ and an alternative form of branching named reiteration is allowed. This allows for axis growth cessation, followed by a number of axes of the same root type to be produced at a significant distance from the original position of axis growth cessation. An estimation of the time of iteration, and the probability of reiteration is considered. Root tips are also allowed to change state, and root tips are allowed to decay or abscise regardless of root position. Together these features capture the heterogeneity of root topology. This model was tested by simulating architecture and comparing to the root taxonomy described by Kutschera (1960) and Pages *et al.* (2004). The model integrates the highest number of different morphogenetic processes, and was able to simulate branch type diversity and mechanical constraints, and growth direction changes in response to depth. However, the model did not perform well in regards to root decay.

In the future, models will need to take into account more parameters, particularly of a physiological nature, to successfully model and describe optimized root architecture. Also, a more comprehensive series of images that describe root architecture in a variety of conditions are needed to test the ability of the model to accurately describe root topology. Higher order geometric models should also prove useful in describing the multidimensionality of root topology and optimal growth path within soil space.

11.9 Future directions

This chapter has described the application of a systems biology approach to describe root transcriptional networks, cellular mechanisms driving root growth and root physical networks. For example, although TFs are undoubtedly essential to determining radial pattern in the Arabidopsis root, many other classes of genes, like signalling molecules, have been demonstrated to play roles in root development. Future additions to the transcriptional network hierarchy entail modelling these additional classes of genes, elucidation of the interactome (a compendium of all protein–protein interactions), and determination of metabolic flux associated with these networks. Recently, rapid methods for metabolic fingerprinting of mutants has been reported (Messerli *et al.*, 2007); however, large scale, high-resolution attempts to determine protein–protein interactions are still lacking. Revealing these additional layers of molecular interactions, and their functional importance in root development will no doubt assist in the manipulation and targeting of favourable characteristics for root and plant growth. Future systems approaches to modelling root development should further describe the molecular mechanisms that underlie the spatial and temporal regulation of cell division and cell elongation, and should aim to simulate optimized root physical networks in higher dimensions, particularly in terms of its dynamic response to external perturbations.

Acknowledgements

The authors would like to thank Jose Dinnney, Anjali Iyer-Pascuzzi, Ji-Young Lee and Terri Long for their critical review of this book chapter. Funding for our research on the systems biology of root development is provided by the NSF, NIH and DARPA. SMB is an NSERC postdoctoral fellow.

References

Beemster, G.T.S. and Baskin, T.I. (1998) Analysis of cell division and elongation underlying the developmental acceleration of root growth in *Arabidopsis thaliana*. *Plant Physiol* **116** (4), 1515–1526.

Benfey, P.N. and Scheres, B. (2000) *Curr Biol* **10** (22), R813.

Benkova, E., Michniewicz, M., Sauer, M., Teichmann, T., Seifertova, D., Jurgens, G., *et al.* (2003) *Cell* **115** (5), 591.

Bennett, M.J., Marchant, A., Green, H.G., May, S.T., Ward, S.P., Millner, P.A., *et al.* (1996) *Science* **273** (5277), 948.

Birnbaum, K., Jung, J.W., Wang, J.Y., Lambert, G.M., Hirst, J.A., Galbraith, D.W., *et al.* (2005) *Nat Methods* **2** (8), 615.

Birnbaum, K., Shasha, D.E., Wang, J.Y., Jung, J.W., Lambert, G.M., Galbraith, D.W., *et al.* (2003) *Science* **302** (5652), 1956.

Blancaflor, E.B., Fasano, J.M. and Gilroy S. (1998) *Plant Physiol* **116** (1), 213.

Blilou, I., Xu, J., Wildwater, M., Willemsen, V., Paponov, I., Friml, J., *et al.* (2005) *Nature* **433** (7021), 39.

Brady, S.M., Orlando, D.A., Lee, J.-Y., Wang, J.Y., Koch, J., Dinneny, J.R., *et al.* (2007) **318** (5851), 801.

Campilho, A., Garcia, B., Toorn, H.V.D., Wijk, H.V., Campilho, A. and Scheres, B. (2006) *Plant J* **48** (4), 619.

Cannon, W.A. (1949) *Ecology* **30**, 542.

Cristina, M., Sessa, G., Dolan, L., Linstead, P., Baima, S., Ruberti, I., *et al.* (1996) *Plant J* **10** (3), 393.

Cui, H., Levesque, M.P., Vernoux, T., Jung, J.W., Paquette, A.J., Gallagher, K.L., *et al.* (2007) *Science* **316** (5823), 421.

Davuluri, R., Sun, H., Palaniswamy, S., Matthews, N., Molina, C., Kurtz, M., *et al.* (2003) *BMC Bioinformatics* **4** (1), 25.

De Smet, I., Tetsumura, T., De Rybel, B., Frey, N.F.D., Laplaze, L., Casimiro, I., *et al.* (2007) *Development* **134** (4), 681.

Deplancke, B., Dupuy, D., Vidal, M. and Walhout, A.J.M. (2004) *Genome Res* **14** (10b), 2093.

Deplancke, B., Mukhopadhyay, A., Ao, W., Elewa, A.M., Grove, C.A., Martinez, N.J., *et al.* (2006) *Cell* **125** (6), 1193.

Di Laurenzio, L., Wysocka-Diller, J., Malamy, J.E., Pysh, L., Helariutta, Y., Freshour, G., *et al.* (1996) *Cell* **86** (3), 423.

Dolan, L., Janmaat, K., Willemsen, V., Linstead, P., Poethig, S., Roberts, K., *et al.* (1993) *Development* **119**, 71.

Eshel, A. (1998) *Plant Cell Environ* **21**, 247.

Fahlgren, N., Howell, M.D., Kasschau, K.D., Chapman, E.J., Sullivan, C.M., Cumbie, J.S., *et al.* (2007) *PLoS ONE* **2** (2), e219, doi.1371/journal.pone.0000219.

Friml, J., Wisniewska, J., Benkova, E., Mendgen, K. and Palme, K. (2002) *Nature* **415** (6873), 806.

Goff, S.A., Ricke, D., Lan, T.-H., Presting, G., Wang, R., Dunn, M., *et al.* (2002) *Science* **296** (5565), 92.

Guo, A., He, K., Liu, D., Bai, S., Gu, X., Wei, L., *et al.* (2005) *Bioinformatics* **21** (10), 2568.

Harbison, C.T., Gordon, D.B., Lee, T.I., Rinaldi, N.J., Macisaac, K.D., Danford, T.W., *et al.* (2004) *Nature* **431** (7004), 99.

Heidstra, R., Welch, D. and Scheres, B. (2004) *Genes Dev* **18** (16), 1964.

Helariutta, Y., Fukaki, H., Wysocka-Diller, J., Nakajima, K., Jung, J., Sena, G., *et al.* (2000) *Cell* **101** (5), 555.

Hocholdinger, F., Park, W.J., Sauer, M. and Woll, K. (2004) *Trends Plant Sci* **9** (1), 42.

Ideker, T., Galitski, T. and Hood, L. (2001) *Annu Rev Genom Human Genet* **2** (1), 343.

Ioio, R.D., Linhares, F.S., Scacchi, E., Casamitjana-Martinez, E., Heidstra, R., Costantino, P., *et al.* (2007) Cytokinins determine Arabidopsis root-meristem size by controlling cell differentiation. Curr Biol **17**, 678–682.

Itoh, H., Washio, T. and Tomita, M. (2004) *RNA* **10** (7), 1005.

Jiang, K., Zhang, S., Lee, S., Tsai, G., Kim, K., Huang, H., *et al.* (2006) *Plant Mol Biol* **60** (3), 343.

Jones-Rhoades, M.W. and Bartel, D.P. (2004) *Mol Cell* **14** (6), 787.

Jones-Rhoades, M.W., Bartel, D.P. and Bartel, B. (2006) *Annu Rev Plant Biol* **57** (1), 19.

Kawaguchi, R. and Bailey-Serres, J. (2005) *Nucleic Acids Res* **33** (3), 955.

Keene, J.D. (2003) *Nat Genet* **33** (2), 111.

Keene, J.D., Komisaraow, J.M. and Friedersdorf, M.B. (2006) *Nat Protocols* **1**, 302.

Kilian, J., Whitehead, D., Horak, J., Wanke, D., Weinl, S., Batistic, O., *et al.* (2007) The AtGenExpress global stress expression data set: protocols, evaluation and model data analysis of UV-B light, drought and cold stress responses. *Plant J* **50**, 347–363.

Kutschera, L. (1960) *Wurzelatlas Mitteleuropaischer Ackerunkrauter und Kulturpflanzen* (Frankfurt am main, Germany: DLG Verlag).

Lee, J.-Y., Colinas, J., Wang, J.Y., Mace, D., Ohler, U. and Benfey, P.N. (2006) *PNAS* **103** (15), 6055.

Lee, M.M. and Schiefelbein, J. (1999) *Cell* **99** (5), 473.

Lee, M.M. and Schiefelbein, J. (2002) *Plant Cell* **14** (3), 611.

Levesque, M.P., Vernoux, T., Busch, W., Cui, H., Wang, J.Y., Blilou, I., *et al.* (2006) *PLoS Biol* **4** (5), e143.

Ljung, K., Hull, A.K., Celenza, J., Yamada, M., Estelle, M., Normanly, J., *et al.* (2005) *Plant Cell* **17** (4), 1090.

Luschnig, C., Gaxiola, R.A., Grisafi, P. and Fink, G.R. (1998) *Genes Dev* **12** (14), 2175.

Mahonen, A.P., Bonke, M., Kauppinen, L., Riikonen, M., Benfey, P.N. and Helariutta, Y. (2000) *Genes Dev* **14** (23), 2938.

Masucci, J.D., Rerie, W.G., Foreman, D.R., Zhang, M., Galway, M.E., Marks, M.D., *et al.* (1996) *Development* **122** (4), 1253.

Messerli, G., Partovi Nia, V., Trevisan, M., Kolbe, A., Schauer, N., Geigenberger, P., *et al.* (2007) *Plant Physiol* **143** (4), 1484.

Nakajima, K., Sena, G., Nawy, T. and Benfey, P.N. (2001) *Nature* **413** (6853), 307.

Nakano, M., Nobuta, K., Vemaraju, K., Tej, S.S., Skogen, J.W. and Meyers, B.C. (2006) *Nucleic Acids Res* **34** (Suppl 1), D731.

Nawy, T., Lee, J.-Y., Colinas, J., Wang, J.Y., Thongrod, S.C., Malamy, J.E., *et al.* (2005) *Plant Cell* **17** (7), 1908.

Nielsen, K.L., Lynch, J.P. and Weiss, H.N. (1997) *Am J Bot* **84** (1), 26.

O'Connor, T.R., Dyreson, C. and Wyrick, J.J. (2005) *Bioinformatics* **21** (24), 4411.

Pages, L., Vercambre, G., Drouet, J.-L., Lecompte, F., Collet, C. and Le Bot, J. (2004) *Plant Soil* **258**, 103.

Pitts, R.J., Cernac, A. and Estelle, M. (1998) *Plant J* **16** (5), 553.

Przemeck, G.K.H., Mattson, J., Hardtke, C.S., Sung, Z.R. and Berleth, T. (1996) *Planta* **200**, 229.

Ryu, K.H., Kang, Y.H., Park, Y.-H., Hwang, I., Schiefelbein, J. and Lee, M.M. (2005) *Development* **132** (21), 4765.

Sabatini, S., Beis, D., Wolkenfelt, H., Murfett, J., Guilfoyle, T., Malamy, J., *et al.* (1999) *Cell* **99** (5), 463.

Sabatini, S., Heidstra, R., Wildwater, M. and Scheres, B. (2003) *Genes Dev* **17** (3), 354.

Schmid, M., Davison, T.S., Henz, S.R., Pape, U.J., Demar, M., Vingron, M., *et al.* (2005) *Nat Genet* **37** (5), 501.

Spencer, M.W.B., Casson, S.A. and Lindsey, K. (2006) *Plant Physiol* **143**, 924.

Swarup, R., Friml, J., Marchant, A., Ljung, K., Sandberg, G., Palme, K., *et al.* (2001) *Genes Dev* **15** (20), 2648.

Swarup, R., Kramer, E.M., Perry, P., Knox, K., Leyser, H.M.O., Haseloff, J., *et al.* (2005) *Nat Cell Biol* **7** (11), 1057.

The Arabidopsis Genome Initiative (2000) Analysis of the genome sequence of the flowering plant Arabidopsis thaliana. *Nature* **408** (6814), 796.

Tian, Q. and Reed, J.W. (1999) *Development* **126** (4), 711.

Van Der Weele, C.M., Jiang, H.S., Palaniappan, K.K., Ivanov, V.B., Palaniappan, K. and Baskin, T.I. (2003) A new algorithm for computational image analysis of deformable motion at high spatial and temporal resolution applied to root growth. Roughly uniform elongation in the meristem and also, after an abrupt acceleration, in the elongation zone. *Plant Physiol* **132**, 1138–1148.

Vanneste, S., De Rybel, B., Beemster, G.T.S., Ljung, K., De Smet, I., Van Isterdael, G., *et al.* (2005) *Plant Cell* **17** (11), 3035.

Walk, T.C., Van Erp, E. and Lynch, J.P. (2004) *Ann Bot* **94** (1), 119.

Woll, K., Borsuk, L.A., Stransky, H., Nettleton, D., Schnable, P.S. and Hochholdinger, F. (2005) *Plant Physiol* **139** (3), 1255.

Yu, J., Hu, S., Wang, J., Wong, G.K.-S., Li, S., Liu, B., *et al.* (2002) *Science* **296** (5565), 79.

Zanetti, M.E., Chang, I.-F., Gong, F., Galbraith, D.W. and Bailey-Serres, J. (2005) *Plant Physiol* **138** (2), 624.

Annual Plant Reviews (2009) **35**, 331–351
doi: 10.1111/b.9781405175326.2009.000012.x

www.interscience.wiley.com

Chapter 12

PERSPECTIVES ON ECOLOGICAL AND EVOLUTIONARY SYSTEMS BIOLOGY

Christina L. Richards,[1] Yoshie Hanzawa,[1] Manpreet S. Katari,[1] Ian M. Ehrenreich,[1,2] Kathleen E. Engelmann[3] and Michael D. Purugganan[1]

[1] *Department of Biology and Center for Genomics and Systems Biology, New York University, New York, NY, USA*
[2] *Department of Genetics, North Carolina State University, Raleigh, NC, USA*
[3] *University of Bridgeport, Bridgeport, CT, USA*

Abstract: Understanding the emergent properties inherent to genome function requires an integrated approach of data from all levels of biology. Molecular biology data alone does not describe the complex interacting functions of organisms, while studies at the level of ecological communities and ecosystems have provided little insight into the molecular underpinnings of adaptation. Merging ecology and evolution into systems biology allows researchers to exploit a wealth of genomic information by incorporating the natural phenotypic, genetic and epigenetic diversity of model systems as well as their diverse ecologies and evolutionary histories. Here, we suggest that systems biology could more fully address the question of how organisms respond to environment if studies incorporated real field settings or experimental manipulation of relevant environmental factors. In addition, although the application of genomic approaches to non-model systems has been slow, we highlight some of the significant progress that has been made. Ecological and evolutionary systems biology will lead to a much more sophisticated understanding of the origins and functions of biological diversity, and serve as a critical component in deciphering how organisms respond to complex environments.

Keywords: *Arabidopsis thaliana*; *Caenorhabditis elegans*; *Drosophila* species; ecological genomics; epigenetics; ecological transcriptome; experimental design; flowering time network; *Fundulus heteroclitus*; natural environment; non-model systems

12.1 Emergent properties of systems biology, ecology and evolution

Understanding biological diversity is a complex question that motivates studies from the molecular level through the level of ecosystems, and at all intervening levels of biological organization. As genomics and systems biology begin to mature, work that synthesizes ecology and evolution with systems biology can begin to explore how these multiple levels of biology contribute to diversity. Systems biology approaches will help to elucidate the complex interplay among different types of biological data (molecular, regulatory, physiological, phenotypic and environmental) and are likely to facilitate unprecedented advances in our understanding of how organisms respond to the environment over both short and long timescales. Ultimately, these approaches can be expanded to more thoroughly address many classic ecological and evolutionary questions about competition, adaptation, invasive species interactions and even global climate change.

It would be difficult to discuss comprehensively in one chapter all the facets that bring ecology, evolution and systems biology together. It is possible, however, to visit the landscape where these disciplines intersect and enrich our understanding of biological processes. In this chapter, we explore some aspects of the possible interfaces between systems biology and ecology, and point out opportunities for further exploration for a new generation of ecological and evolutionary systems biologists. Using examples from several model organisms, particularly *Arabidopsis thaliana*, we illustrate the synergistic interface between systems biology, ecology and evolution.

12.2 Complex environments and ecological systems biology

Organisms in the real world are continuously exposed to multiple environmental signals and must respond appropriately to the dynamic conditions found in nature. Temperature, photoperiod and resource availability are just some key environmental conditions that cue organismal responses. There have been significant advances in dissecting how these and other environmental signals are translated by the organism to appropriate gene expression levels that may ultimately determine phenotypes (Pigliucci, 1996; Schlichting and Smith, 2002); however, with very few exceptions, these studies have been carried out in homogenous environments in controlled laboratory conditions. The natural world, in contrast, is anything but controlled. Dynamic environmental signals can fluctuate temporally during an individual's life cycle and can change spatially according to climate, with varying degrees of predictability. Complex natural environments are the norm, and it is in this context that developmental pathways and physiological states actually function and evolve. It is unclear how these fluctuating signals interact with each other and with an organism's genotype to fine-tune phenotypic response, but clearly the next step in this area of research is to determine how gene expression is modulated in the wild.

Understanding how genes and functional genetic networks are regulated in natural ecological settings in the midst of fluctuating environmental signals and between genetically distinct individuals remain key issues that lie at the intersection of ecology, evolution and systems biology. Although molecular studies have provided data on a wide variety of functionally important genes in various organisms, this data remains largely insufficient to describe the complex interactions that underlie most biological processes, thus highlighting the limitations of reductionist approaches (Mazzochi, 2008). Similarly, studies at the level of populations, ecological communities and ecosystems rarely address the molecular underpinnings of organismal variation and adaptation. Systems biology provides tools and perspectives that can examine the emergent properties inherent in genome function and allow an integrated approach of data from all biological levels (O'Malley and Dupré, 2005; Bonneau et al., 2007; Bruggeman and Westerhoff, 2007).

12.3 Gene networks and the ecological transcriptome

Although the goal of biology is to understand how organismal phenotypes and the genetic architecture of various traits have evolved in nature, gene expression in the wild is poorly understood. This gap in our knowledge is unfortunate, since data on the characteristics of gene expression under natural conditions is a necessary step if we are to determine how genes are regulated in the dynamic, complex environments observed in natural field conditions. It is likely that laboratory expression experiments across several treatments, even with attempts to integrate them, will be unable to mimic how gene expression behaves in natural settings. Microarray technologies allow us to measure the levels and patterns of gene expression in the dynamic ecological settings of field environments, and systems biology will help interpret this massive amount of data so that we can begin to more fully understand the ecological transcriptome.

There are a large number of global gene expression studies in *A. thaliana* (e.g. Harmer et al., 2000; Birnbaum et al., 2003; Schmid et al., 2005) as well as other plant species (e.g. Alba et al., 2005; Li et al., 2006; Starker et al., 2006; Swanson-Wagner et al., 2006). However, most of these studies were undertaken in controlled laboratory conditions, and we could only find five global gene expression studies of wild plant species (as opposed to crop plants) in field conditions (Miyazaki et al., 2004; Taylor et al., 2005; Yang and Loopstra, 2005; Ainsworth et al., 2006; Schmidt and Baldwin, 2006). This limited number of reports clearly demonstrates that there are significant transcriptional differences between controlled and field growth conditions.

One of the earliest field microarray studies was on *Solanum nigrum* using a small gene array. This study showed that methyl jasmonate (MeJa) elicitation in competitive environments resulted in different patterns of transcriptional changes between greenhouse and field-grown plants (Schmidt and Baldwin, 2006). MeJa-treated plants consistently showed differences in transcript levels for genes involved in oxylipin signalling and primary metabolism. However, there were a larger number of downregulated genes in greenhouse-grown compared with field-grown plants, which suggests that the latter are less sensitive to MeJa elicitation. This decreased downregulation of gene transcripts in field plants is thought to be the result of greater exposure to abiotic stresses in the wild (Schmidt and Baldwin, 2006), and clearly demonstrates the importance of measuring response in natural environments.

Another series of studies examined gene transcription responses to increased CO_2 levels in free air CO_2 enrichment (FACE) in *Populus* (Taylor *et al.*, 2005), soybean (Ainsworth *et al.*, 2006) and *A. thaliana* (Miyazaki *et al.*, 2004). Following 6 years of exposure of a *Populus* clone to elevated CO_2 levels, gene expression response depended on the developmental age of the leaves, and ~50 transcripts differed significantly between different CO_2 environments (Taylor *et al.*, 2005). A similar result was also reported in soybean (Ainsworth *et al.*, 2006). The *A. thaliana* study examined both CO_2 and ozone exposure. Microarrays with probes for ~26 000 DNA elements were used to compare transcripts from plants grown in growth chambers to those grown in FACE rings under ambient field conditions. In this study, most changes in gene expression were observed between growth chamber and ambient field conditions rather than between atmospheric treatments. Greater than 1000 transcripts were either up or downregulated between controlled versus field ambient conditions compared with high versus low CO_2 or ozone levels. There was a preponderance of genes associated with general defence reactions, secondary metabolism, redox control, energy provision, protein turnover, signalling and transcription (Miyazaki *et al.*, 2004).

These experiments point out the large-scale differences in expression patterns between growth chamber versus field conditions, but also demonstrate the feasibility of assaying for global transcript abundance levels in field-grown plants using microarray technologies. Still, these studies only deliver a list of possible genes involved in response to these environmental treatments. At best, this approach has identified some novel genes involved in response to different environments (Taylor *et al.*, 2005). Systems biology approaches allow researchers to go beyond the Venn diagram descriptions of differential gene expression (Li *et al.*, 2006; Kammenga *et al.*, 2007; Lee *et al.*, 2008) to formulate dynamic networks that can incorporate changes over time and environments (Bonneau *et al.*, 2007).

One important component of the ecological transcriptome that has largely been overlooked in ecology and evolution is epigenetic effects. Epigenetic effects are the subject of intense study in genomics, and several studies have begun to develop comprehensive maps of epigenetic marks (Vaughn *et al.*, 2007; Zhang *et al.*, 2007; Zilberman *et al.*, 2007; Cokus *et al.*, 2008; Zhang *et al.*, 2008). DNA methylation is the most well-understood epigenetic mark and the few existing studies of natural variation in genome methylation report epigenome polymorphisms in *Gossypium* (Keyte *et al.*, 2006), *Arabidopsis* (Cervera *et al.*, 2002; Riddle and Richards, 2002, 2005), *Oryza* (Ashikawa, 2001; Wang *et al.*, 2004), *Pisum* (Knox and Ellis, 2001) and *Spartina* (Salmon *et al.*, 2005).

There is ever-increasing evidence that heritable variation in ecologically relevant traits can be generated through a suite of epigenetic mechanisms that can alter phenotypes even in the absence of genetic variation (Grant-Downton and Dickinson, 2005, 2006; Jablonka and Lamb, 2005; Rapp and Wendel, 2005; Richards, 2006). Molecular studies using methylation sensitive markers in *Triticum* (Sherman and Talbert, 2002) and *Arabidopsis* (Burn *et al.*, 1993) show that external temperature can change methylation patterns that induce early flowering time. Similarly, one study found a nearly 10% reduction in methylation in induced early flowering lines of *Linim usitatissimum* compared to closely related lines with normal flowering time (Fields *et al.*, 2005). Furthermore, studies in *Arabidopsis, Brassica, Oryza, Spartina* and *Triticum* reveal that methylation patterns can be radically altered by hybridization or polyploidization (Chen and Pikaard, 1997; Comai *et al.*, 2000; Liu *et al.*, 2001; Shaked *et al.*, 2001; Madlung *et al.*, 2002; Salmon *et al.*, 2005).

Epigenetic variation appears to be common in plants and is therefore likely to have effects that are visible to natural selection (reviewed by Rapp and Wendel, 2005; Grant-Downton and Dickinson, 2006). Recent studies also indicate that in some cases, environmentally induced epigenetic changes may be inherited by future generations (Richards, 2006; Whitelaw and Whitelaw, 2006; Bossdorf *et al.*, 2008). In addition, because epigenetic processes are an important component of hybridization and polyploidization events, they may play a key role in speciation and the biology of many invasive species through these processes (Ellstrand and Schierenbeck, 2000; Liu and Wendel, 2003; Rapp and Wendel, 2005; Salmon *et al.*, 2005; Chen and Ni, 2006; Bossdorf *et al.*, 2008). To date, very few ecology and evolution studies have considered the importance of epigenetic effects and most have not gone beyond documenting that they exist (Rapp and Wendel, 2005; Bossdorf *et al.*, 2008). With the appropriate experimental design (Bossdorf *et al.*, 2008), systems biology may be the best context with which to disentangle the contributions of environment, genotype and epigenotype to phenotypic variation.

12.4 Analysis of systems biology data: the role of ecological and evolutionary methods

Ecological and evolutionary information may contribute to systems biology through enhanced functional genome annotation. Placing genes and molecular networks into functional ecological and evolutionary contexts exposes the behaviour of the network in a realistic setting. Expression studies can be rendered even more informative if they incorporate experimental manipulations or treatments of ecologically relevant factors like light, water, nutrients or salt levels or even whole field environments (Fig. 12.1). A large percentage of genes in the genome either encode hypothetical proteins or are of unknown function; ecological and evolutionary studies may be especially useful in proving the basis for expanding the annotation of those genes in the genome. Bioinformatic approaches combine the genome-wide data collected in these types of experiments with previously verified genetic interactions to assemble putative gene interaction networks. Such gene interaction networks may suggest novel associations and functions of genes, and how these relationships change across environments of interest. An important component of the systems biology approach is to use these findings to generate new hypotheses and new experimental designs. Iterating this process can refine the model of these network interactions.

Experimental ecology also provides a long history of developing methods in experimental design and analysis to identify the relevant contributions of environmental factors to organismal response (Bailey, 1981; Sokal and Rohlf, 1995; Scheiner and Gurevitch, 2001). Based on the constraints of the experimental setup and organisms, this may require split plot or repeated-measures designs and include randomization and blocking to control for spatial or systemic contributions to variation, and avoid confounding these elements with variables of interest. Typically, these studies then use analysis of variance (ANOVA) to disentangle the contribution of multiple environmental variables or experimental treatments to variation in ecologically important traits (Sokal and Rohlf, 1995; Scheiner and Gurevitch, 2001). The general model for ANOVA allows for examination of several independent variables by defining the relationship between traits and any number of independent variables, and mixed-model ANOVAs allow for testing of random effects like population and genotype (Littell *et al.*, 2006) within this context.

These approaches have been adapted to the interpretation of gene expression and gene specific modelling of microarray data in particular by fitting a global normalization model incorporating all of the genes, and then running a separate ANOVA for each gene (Wolfinger *et al.*, 2001; Aryoles and Gibson, 2006). In addition, using ANOVA allows for ecological genomic studies to test whether the pattern of gene expression variation is correlated with environmental or ecological variables like temperature, precipitation levels, soil moisture, daylength, photosynthetically active radiation (PAR), age, herbivory and disease status.

12.5 The ecological and evolutionary context of model organisms: the example of *Arabidopsis* and beyond

Much of the advance in molecular and developmental biology in the last few decades owes to the focus of researchers on a handful of model organisms as subjects of experimental study. There has been relatively little attention paid to the natural history of these model species, but that has changed in the last few years as investigators begin to understand the natural contexts in which these organisms live. There has been increasing efforts to use these model species to address fundamental questions in ecology – to exploit the standing natural variation observed in these species and to study their development and behaviour in their natural environments.

The need for a systems biological approach in the study of ecological dynamics can be illustrated by considering recent work with *A. thaliana* (L.) Heynh (Pigliucci, 1998; Koornneef *et al.*, 2004; Shimizu and Purugganan, 2005). This species is a weedy annual plant, occupying disturbed habitats such as the margins of agricultural fields as well as natural ruderal environments. The native range of *A. thaliana* covers Eurasia and Northern Africa, and it is naturalized widely in the world, including in North America and Japan (Hoffmann, 2002). Evolutionary analysis of a set of genome-wide markers suggests that the current species range, which includes most of Eurasia, and parts of North Africa and North America, is the result of the expansion of the species ~17 000 years ago from two glacial refugia in the Iberian Peninsula and Asia (Sharbel *et al.*, 2000). The presence of *A. thaliana* in North America is a recent phenomenon, which is likely due to the migration of Europeans to the continent over the last 300–400 years.

A. thaliana displays a wide range of ecological relationships, including within- and between-species interactions and adaptations to abiotic environments. It responds physiologically and developmentally to a variety of environmental cues, including light, daylength, vernalization, nutrient and water levels (reviewed by Pigliucci, 1998; Koornneef *et al.*, 2004; Shimizu and Purugganan, 2005), can be infected by a wide array of bacterial and fungal pathogens, and is preyed upon by many insect herbivores (Kliebenstein *et al.*, 2002). Despite the role of *A. thaliana* as a model plant system, remarkably little is known about the phenotypic range and performance of this species in the wild. Most of our knowledge of this organism is in the artificial environment of the laboratory, and there is mounting evidence that the behaviour of this organism can differ substantially in the wild. A few field studies of *A. thaliana* have begun to shed light on the ecological genetics of this organism. Early investigations have examined selection in this species at short spatial scales in field conditions (Stratton and Bennington, 1996), and documented selection costs for trichomes as a defence against herbivores (Mauricio and Rausher, 1997). Other studies have looked at the seasonal germination timing in the field (Donohue *et al.*, 2005), fitness costs of *R* disease resistance gene polymorphisms (Tian *et al.*, 2003) and the role of epistasis in fitness-related traits (Malmberg *et al.*, 2005).

Some of the most detailed field studies in *A. thaliana* have focused on the genetic architecture of flowering time, which is arguably one of the most important traits for the ecology and evolution of plant species. A quantitative trait locus (QTL) mapping study for date of bolting (the transition from vegetative to reproductive growth) was undertaken in natural seasonal field environments in Rhode Island and North Carolina (Weinig *et al.*, 2002). This study revealed that photoperiod-specific QTLs found in controlled conditions were undetectable in natural environments, while several QTLs with major effects on flowering time in one or more field environments were undetectable under controlled environment conditions (Weinig *et al.*, 2002). Candidate gene association studies, while not definitive, have suggested that common allelic variation at the flowering time genes *CRY2* (Olsen *et al.*, 2004), *FRI* (Caicedo *et al.*, 2004; Stinchcombe *et al.*, 2004) and *FLC* (Caicedo *et al.*, 2004) is associated with flowering time diversity in natural field conditions. Altogether, these QTL and association mapping studies suggest that the genetic architecture of this life history transition differs significantly between laboratory and natural environments.

While *A. thaliana* has become a robust ecological and evolutionary model system, particularly for field studies, other model genomic organisms offer additional advantages. *Drosophila melanogaster*, a cosmopolitan species, has a long history in the study of evolutionary genetics, and has some of the most advanced tools available for evolutionary studies. In particular, recent advances include the complete genome sequences of 12 closely related species and genome-wide polymorphism data for multiple species (Clark *et al.*, 2007; Stark *et al.*, 2007). Similarly, *Caenorhabditis elegans*, a soil nematode is now emerging as a key model species for quantitative genomics (Li *et al.*, 2006), and systems biology (Gunsalus *et al.*, 2005) (see Chapter 3) could be used to potentially integrate the large amount of data on this organism at the ecological level. There are also a large number of bacterial and fungal systems that can be exploited to study microbial ecology (Whitaker and Banfield, 2006; Wilmes and Bond, 2006; Xu, 2006; Bonneau *et al.*, 2007).

In addition to a wealth of information on model systems, the development of genomic technologies has expanded the range of possible wild species that can be used in ecological and evolutionary systems biology. For example, genome sequencing has been completed in the deciduous forest tree species of *Populus* (Jansson and Douglas, 2007) and the monkey flower *Mimulus guttatus* (Wu *et al.*, 2008), and tools for use in ecological and evolutionary studies in these and other species are becoming available. The completed sequencing of 12 genomes of *Drosophila* species (Clark *et al.*, 2007; Stark *et al.*, 2007) and the near completion of the *A. thaliana* relatives *A. lyrata* and *Capsella rubella* offer opportunities to significantly expand evolutionary systems biology studies beyond model systems as we decipher the commonalities and differences between closely related species. In addition, researchers can increasingly exploit tools and data in species that are closely related to the model systems with genomic resources. For example, microarray chips designed for model taxa can be used for transcriptome studies in non-model species because those probes that do not hybridize can be identified and left out of an analysis (Slotte *et al.*, 2007; Travers *et al.*, 2007). Horvath *et al.* (2003) used this approach with *A. thaliana* microarrays to analyze gene expression in several distant species, including leafy spurge and poplar. With these and other on-going efforts, such as the increasing availability of massively parallel sequencing machines, it is now possible that systems biology can be studied in a large number of relevant taxa (Bonneau *et al.*, 2007).

12.6 Natural variation in genomes and gene networks

The genetic variation extant within species and the phenotypic variation that may result from it are key components that underlie species diversification and adaptation. There have been concerted efforts to understand the extent of molecular variation between individuals and populations, and there has been a recent resurgence in interest in mapping natural genetic variants that may be responsible for natural phenotypic variation. These include genome mapping techniques that scan genomes and map trait loci, such as methods for QTL mapping (Mackay, 2001) and linkage disequilibrium/association mapping (Cardon and Abecasis, 2003).

Coupling systems biology with population variation can provide an evolutionary and ecological context to the large number of global gene expression studies by utilizing multiple sources of information across a diversity of genotypes. One area of interest, for example, is whether key points in a network are responsible for obvious phenotypic differences between genotypes. This can be illustrated by recent studies on variation between different accessions of *A. thaliana* in the important flowering time genes. The network of genes in the flowering time pathway in *A. thaliana* is one of the best-studied regulatory pathways that control a key life history trait in plants (Mouradov *et al.*, 2002; Simpson and Dean, 2002). Two genes in the vernalization pathway in particular, *FRI* and *FLC* are believed to play an essential role in determining response to prolonged cold temperature (Johanson *et al.*, 2000; Weinig *et al.*, 2002; Olsen *et al.*, 2004). *FLC*, which represses flowering, is regulated by *FRI*. An active *FRI-FLC* pathway results in late flowering, whereas an inactive *FRI-FLC* pathway results in early flowering.

Through controlled crosses, researchers have developed nearly isogenic lines (NIL) of *A. thaliana*, which are genetically identical except for one target locus. For example, two of these lines with a functional *FRI* allele have either a non-functional or a functional *FLC* allele (Lee *et al.*, 1993; Michaels and Amasino, 1999). To develop these lines, a functional *FRI* allele from the wild accession Sf2 was introgressed into Col-0, which carries an active *FLC*, resulting in an active *FRI-FLC* pathway (Lee *et al.*, 1993). The inactive *FRI-FLC* pathway was obtained by introducing a loss of function mutation *flc-3* (Michaels and Amasino, 1999).

Figure 12.2 demonstrates how this change in functionality of *FLC* dramatically alters the network of genes involved in the flowering time pathway. These networks are constructed with the multi-network programme encompassed in a software platform for plant systems biology called virtual plant (Gutiérrez *et al.*, 2005) using a publicly available microarray data set

(a)

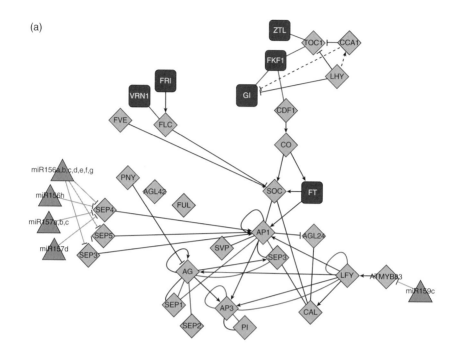

(b)

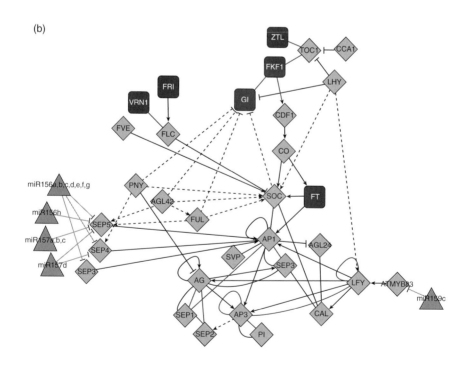

on the early stages of transition to flowering (day 0 to day 7 after induction) (Schmid *et al.*, 2003). The multi-network tool embodied in the virtual plant platform incorporates database information on interactions derived from protein–protein interaction databases (BIND, DIP), protein–DNA interactions (Transfac), miRNA:RNA interactions, coexpression data and literature-based interactions predicted by a text mining tool called GeneWays, as described in Gutierrez *et al.* (2007). The resulting interaction network of flowering time genes shown in Fig. 12.2, shows that losing the functionality of *FLC* results in the addition of 14 edges and loss of 2 edges (denoted by two dashed edges in Fig. 12.2 a and 14 dashed edges in Fig. 12.2b). The loss of functionality of *FLC* also results in the connection of two additional nodes, AGL42 and FUL (compare Figs. 12.2 a and 12.2b). This relatively simple example of using a multi-network approach to depict the changes in interactions among flowering genes allows us to explore the effect of variation in functional genes and their interactions. This approach can then lead to novel biological questions that can be explored experimentally, like addressing the roles of *AGL42* and *FUL* in active *FRI-FLC* plants.

The variability in organismal transcriptomes that arises from natural genetic variants has been documented in a number of different species, including *D. melanogaster* (Jin *et al.*, 2001; Hsieh *et al.*, 2007), yeast (Townsend *et al.*, 2003), *A. thaliana* (Chen *et al.*, 2005; Juenger *et al.*, 2006; Keurentjes *et al.*, 2007; West *et al.*, 2007), *C. elegans* (Li *et al.*, 2006, 2007; Lee *et al.*, 2008) and the fish *Fundulus heteroclitus* (Oleksiak *et al.*, 2002; Whitehead and Crawford, 2006a, b; Burnett *et al.*, 2007). In one of the first studies examining expression variation, microarray analysis in *F. heteroclitus*, showed substantial levels of transcript variation within and between populations (Oleksiak *et al.*, 2002). A study in *D. melanogaster* completed in North Carolina and California populations also showed a bimodal (or at times multimodal) distribution of transcript abundance within populations, indicating variation segregates for gene expression levels (Hsieh *et al.*, 2007).

Several recent studies attempt to map genes that underlie variation in gene expression patterns across the entire genome (as reviewed by Gibson and Weir, 2005; Rockman and Kruglyak, 2007). These expression QTL (eQTL) studies have provided insights into the genetic architecture of natural

---

Figure 12.2 Comparison of the flowering gene networks of plants carrying active and inactive *FRI-FLC* pathway. A strong *FRI* allele from a wild accession Sf2 was introgressed into Col-0 that carries an active *FLC*, resulting in an active *FRI-FLC* pathway. The inactive *FRI-FLC* pathway was obtained by introducing a loss of function mutation *flc-3*. (a) The network with active *FRI-FLC* pathway (*FRI* Sf2, *FLC*). (b) The network with inactive *FRI-FLC* pathway (*FRI* Sf2, *flc-3*). The diamond nodes show transcription factors, the triangle nodes are micro RNAs and the round rectangle nodes are target genes. The black edges with arrowheads suggest transcriptional activation and the black edges with T-ends are transcriptional repression. The grey edges with no end marker are protein–protein interactions. The edges that are present in one network, but missing in the other, are indicated by dashed lines.

expression variation. A recent study in *A. thaliana* using a recombinant inbred line mapping population between the Bay-0 and Sha accessions reveal that approximately 1/3 of the differentially expressed genes were controlled by loci in *cis* (West *et al.*, 2007). Moreover, eQTLs that were regulated in *trans* map to several hotspots that regulated hundreds to thousands of transcript differences, suggesting that the genetic control of transcript level between individuals can be complex. Interestingly, almost all of the >36 000 eQTLs identified were associated with small phenotypic effects. Moreover, by combining eQTL methods with selection of candidate regulatory genes, one can reconstruct regulatory gene networks that are associated with natural variation in expression and other phenotypic traits (Keurentjes *et al.*, 2007).

The interaction of environment with natural genetic variation can provide a glimpse into the basis for expression (and ultimately phenotypic) plasticity in a systems biology context. An experiment with eQTL mapping in *C. elegans* under various growth temperatures demonstrated that nearly 60% of 308 *trans*-acting eQTLs showed a significant eQTL-by-environment interaction, while only 8% of 188 *cis*-acting genes showed an eQTL-by-environment interaction (Li *et al.*, 2007). This suggests that genetic differences in gene expression plasticity are largely regulated in *trans* such that expression variation in groups of genes are driven by individual loci.

Although, these studies were done under controlled laboratory conditions, the transcriptome can differ between genetically distinct *A. thaliana* accessions from different parts of the species range when grown under field conditions. Using microarray technology, the expression levels in the Bay-0 and Sha accessions were assayed at the four leaf seedling stage in a field site in Long Island, New York, in the fall of 2006 (K. Engelmann, D. Nielsen and M. Purugganan, unpublished data). The Bay-0 accession originally came from a fallow field in Bayreuth, Germany (50.0°N, 11.6°E at an altitude of 300–400 m) while Sha is from a mountainous site at Shahdara, Tajikistan (39.3°N and 68.3°E at an altitude of 3300–3400 m).

The correlation in gene expression between the two accessions is high (r = 0.96). But the distribution of *p*-values for differences between the two accessions in the field environment reveals that numerous genes are differentially expressed between the two accessions (Fig. 12.3). From this analysis, 401 genes display significant expression differences between Bay-0 and Sha in the field. Two hundred fifty-five (64%) of these genes are transcription factors or metabolic enzymes, including zinc-finger, homeodomain, bZIP, WRKY-type and myb-like transcription factors, the giberellin response factor RGA1, several auxin-induced proteins and fructose metabolism enzymes. One hundred forty-six genes (36%) encode hypothetical proteins or proteins of unknown function, and studies on their differential expression in ecological field environments may provide clues for further functional annotation of this class of genes. Which of these differences in gene expression, if any, account for the observed accession-specific differences in flowering time under field conditions, remains to be explored. These studies in the wild can also provide an ecological context for understanding the real-world functions of genes and genetic networks, and permit an elucidation of the ecological transcriptome.

12.7 The future of ecological and evolutionary systems biology

At one level, systems biology combines genome-level interaction maps with dynamic modelling at the sub-genome level, where specified inputs and outputs allow the identification of key regulatory components or parameters of the system. Considering the complexities of developing these predictive models, systems biology must develop through close collaboration between experimental, computational and theoretical approaches. Albert and Assmann (see Chapter 1) point out that biochemical reactions within and between cells take place on timescales spanning several orders of magnitude (Papin *et al.*, 2005) and that these timescales are modulated by molecules or complexes and their interactions as well as the environmental conditions (Han *et al.*, 2004; Balázsi *et al.*, 2005). In addition to the dynamic changes in the state of network nodes, the characteristics of biological networks are shaped by dynamic events whose impact occurs on ecological and evolutionary timescales. As Albert and Asmann suggest, we argue that integration of ecological, evolutionary and epigenetic characteristics with transcriptional, metabolic and signal transduction networks is the 'final frontier' of systems biology.

The overarching goal at the intersection of systems biology, ecology and evolutionary biology is to evaluate whether the properties of biological networks as we depict them reflect reality at all levels of biology. Model systems with their full battery of genomic information, combined with the great amount of phenotypic, genetic and epigenetic variation they harbour, can provide a great deal of power to investigate how organisms are able to respond to their ecological milieu. The presence of genotype-by-environment interactions illustrate that the rapidly escalating amount of genomic data and tools applied to model systems in controlled conditions must also reflect natural variation between individuals, populations and species and the importance of understanding how molecular networks behave under real-world ecological conditions.

In addition, recent ecological genomic studies demonstrate that these approaches can be applied not just to well-known model species but to an even broader array of ecologically important organisms. Systems biology should be a critical component of fleshing out how organisms are able to respond to complex environments. This will require effort from all levels of the biological sciences, but should lead to a much more sophisticated understanding of the origins and functions of biological diversity. Moreover, in this world of changing environments, including the global climate, systems biology merged with ecology and evolution may provide predictive insights into adaptive responses of organisms to the future.

Acknowledgements

The authors would like to thank Aviv Madar, Gloria Coruzzi and Alexis Cruikshank for helpful comments and editing. This work was supported by an NSF FIBR grant (EF-0425759), an NSF Plant Genome Research Programme grant (DBI-0319553) and the Guggenheim Foundation.

References

Ainsworth, E.A., Rogers, A., Vodkin, L.O., Walter, A. and Schurr, U. (2006) The effects of elevated CO2 concentration on soybean gene expression. An analysis of growing and mature leaves. *Plant Physiol* **142**, 135–147.

Alba, R., Payton, P., Fei, Z.J., McQuinn, R., Debbie, P., Martin, G.B., *et al.* (2005) Transcriptome and selected metabolite analyses reveal multiple points of ethylene control during tomato fruit development. *Plant Cell* **17**, 2954–2965.

Aryoles, J.F. and Gibson, G. (2006) Analysis of variance of microarray data. *Meth Enzym* **411**, 214–223.

Ashikawa, I. (2001) Surveying CpG methylation at 5′-CCGG in the genomes of rice cultivars. *Plant Mol Biol* **45**, 31–39.

Bailey, R.A. (1981) A unified approach to design of experiments. *J R Stat Soc Ser A* **144** (2), 214–223.

Balázsi, G., Barabási, A.L. and Oltvai, Z.N. (2005) Topological units of environmental signal processing in the transcriptional regulatory network of Escherichia coli. *Proc Natl Acad Sci USA* **102**, 7841–7846.

Birnbaum, K., Shasha, D.E., Wang, J.Y., Jung, J.W., Lambert, G.M., Galbraith, D.W., *et al.* (2003) A gene expression map of the *Arabidopsis* root. *Science* **302**, 1956–1960.

Bonneau, R., Facciotti, M.T., Reiss, D.J., Schmid, A.K., Pan, M., Kaur, A., *et al.* (2007) A predictive model for transcriptional control of physiology in a free living cell. *Cell* **131**, 1354–1365.

Bossdorf, O., Richards, C.L. and Pigliucci, M. (2008) Epigenetics for ecologists. *Ecol Lett* **11**, 106–115.

Bruggeman, F.J. and Westerhoff, H.V. (2007) The nature of systems biology. *Trends Microbiol* **15**, 45–50.

Burn, J.E., Bagnall, D.J., Metzger, J.D., Dennis, E.S. and Peacock, W.J. (1993) DNA methylation, vernalization, and the initiation of flowering. *Proc Natl Acad Sci USA* **90**, 287–291.

Burnett, K.G., Bain, L.J., Baldwin, W.S., Callard, G.V., Cohen, S., Di Giulio, R.T., *et al.* (2007) Fundulus as the premier teleost model in environmental biology: opportunities for new insights using genomics. *Comp Biochem Physiol Part D* **2**, 257–286.

Caicedo, A.L., Stinchcombe, J., Schmitt, J. and Purugganan, M.D. (2004) Epistatic interaction between the *Arabidopsis* FRI and FLC flowering time genes establishes a latitudinal cline in a life history trait. *Proc Natl Acad Sci USA* **101**, 15670–15675.

Cardon, L.R. and Abecasis, G.R. (2003) Using haplotype blocks to map human complex trait loci. *Trends Genet* **19**, 135–140.

Cervera, M.-T., Ruiz-Garcia, L. and Martinez-Zapater, J. (2002) Analysis of DNA methylation in *Arabidopsis thaliana* based on methylation-sensitive AFLP markers. *Mol Genet Genom* **268**, 543–552.

Chen, W.Q.J., Chang, S.H., Hudson, M.E., Kwan, W.K., Li, J.Q., Estes, B., *et al.* (2005) Contribution of transcriptional regulation to natural variations in *Arabidopsis*. *Genome Biol* **6** (4), Art. No. R32.

Chen, Z.J. and Ni, Z. (2006) Mechanisms of genomic rearrangements and gene expression changes in plant polyploids. *Bioessays* **28**, 240–252.

Chen, Z.J. and Pikaard, C.S. (1997) Epigenetic silencing of RNA polymeraseI transcription: a role for DNA methylation and histone modification in nucleolar dominance. *Genes Dev* **11**, 2124–2136.

Clark, A.G., Eisen, M.B., Smith, D.R., Bergman, CM., Oliver, B., Markow, T.A., *et al.* (2007) Evolution of genes and genomes on the Drosophila phylogeny. *Nature* **450**, 203–218.

Cokus, S.J., Feng, S., Zhang, X., Chen, Z., Merriman, B., Haudenschild, C.D., *et al.* (2008) Shotgun bisulphite sequencing of the *Arabidopsis*genome reveals DNA methylation patterning. *Nature* **452**, 215–219.

Comai, L., Tyagi, A.P., Holmes-Davis, K.W.R., Reynolds, S.H., Stevens, Y. and Byers, B. (2000) Phenotypic instability and rapid genome silencing in newly formed *Arabidopsis* allotetraploids. *Plant Cell* **12**, 1551–1567.

Donohue, K., Dorn, L., Griffith, C., Kim, E., Aguilera, A., Polisetty, C.R., *et al.* (2005) Environmental and genetic influences on the germination of *Arabidopsis thaliana* in the field. *Evolution* **59**, 740–757.

Ellstrand, N.C. and Schierenbeck, K.A. (2000) Hybridization as a stimulus for the evolution of invasiveness. *Proc Natl Acad Sci USA* **97**, 7043–7050.

Fields, M.A., Schaeffer, S.M., Krech, M.J. and Brown, J.C.L. (2005) DNA hypomethylation in 5-azacytidine-induced early-flowering lines of flax. *Theor App Genet* **111**, 136–149.

Gibson, G. and Weir, B. (2005) The quantitative genetics of transcription. *Trends Genet* **21**, 616–623.

Grant-Downton, R.T. and Dickinson, H.G. (2005) Epigenetics and its implications for plant biology. 1. The epigenetic network in plants. *Ann Bot* **96**, 1143–1164.

Grant-Downton, R.T. and Dickinson, H.G. (2006) Epigenetics and its implications for plant biology. 2. The 'epigenetic epiphany': epigenetics, evolution and beyond. *Ann Bot* **97**, 11–27.

Gunsalus, K.C., Ge, H., Schetter, A.J., Goldberg, D.S., Han, J.D.J., Hao, T., *et al.* (2005) Predictive models of molecular machines involved in *Caenorhabditis elegans* early embryogenesis. *Nature* **436**, 861–865.

Gutierrez, R.A., Lejay, L.V., Dean, A., Chiaromonte, F., Shasha, D.E. and Coruzzi, G.M. (2007) Qualitative network models and genome-wide expression data define carbon/nitrogen-responsive molecular machines in *Arabidopsis*. *Genome Biol* **8**, R7.

Gutiérrez, R.A., Shasha, D.E. and Coruzzi, G.M. (2005) Systems biology for the virtual plant. *Plant Physiol* **138**, 550–554.

Han, J.D., Bertin, N., Hao, T., Goldberg, D.S., Berriz, G.F., Zhang, L.V., *et al.* (2004) Evidence for dynamically organized modularity in the yeast protein-protein interaction network. *Nature* **430**, 88–93.

Harmer, S.L., Hogenesch, L.B., Straume, M., Chang, H.S., Han, B., Zhu, T., *et al.* (2000) Orchestrated transcription of key pathways in *Arabidopsis* by the circadian clock. *Science* **290**, 2110–2113.

Hoffmann, M.H. (2002) Biogeography of *Arabidopsis thaliana* (L.) Heynh. (Brassicaceae). *J Biogeogr* **29**, 125–134.

Horvath, D.P., Schaffer, R., West, M. and Wisman, E. (2003) *Arabidopsis* microarrays identify conserved and differentially expressed genes involved in shoot growth and development from distantly related plant species. *Plant J* **34**, 125–134.

Hsieh, W.P., Passador-Gurgel, G., Stone, E.A. and Gibson, G. (2007) Mixture modeling of transcript abundance classes in natural populations. *Genome Biol* **8** (6), Art. No. R98.

Jablonka, E. and Lamb, M.J. (2005) *Evolution in Four Dimensions* (Cambridge, MA: MIT Press).

Jansson, S. and Douglas, C.J. (2007) *Populus*: a model system for plant biology. *Ann Rev Plant Biol* **58**, 435–458.

Jin, W., Riley, R.M., Wolfinger, R.D., White, K.P., Passador-Gurgel, G. and Gibson, G. (2001) The contributions of sex, genotype and age to transcriptional variance in *Drosophila melanogaster*. *Nat Genet* **29** (4), 389–395.

Johanson, U., West, J., Lister, C., Michaels, S., Amasino, R. and Dean, C. (2000) Molecular analysis of FRIGIDA, a major determinant of natural variation in *Arabidopsis* flowering time. *Science* **290**, 344–347.

Juenger, T.E., Wayne, T., Boles, S., Symonds, V.V., Mckay, J. and Coughlan, S.J. (2006) Natural genetic variation in whole-genome expression in *Arabidopsis thaliana*: the impact of physiological QTL introgression. *Mol Ecol* **15** (5), 1351–1365.

Kammenga, J.E., Herman, M.A., Ouborg, N.J., Johnson, L. and Breitling, R. (2007) Microarray challenges in ecology. *Trends Ecol Evol* **22**, 273–279.

Keurentjes, J.J.B., Fu, J.Y., Terpstra, I.R., Garcia, J.M., Van Den Ackerveken, G., Snoek, L.B., *et al.* (2007) Regulatory network construction in *Arabidopsis* by using genome-wide gene expression quantitative trait loci. *Proc Natl Acad Sci USA* **104** (5), 1708–1713.

Keyte, A.L., Percifield, R., Liu, B. and Wendel, J.F. (2006) Intraspecific DNA methylation polymorphism in cotton (*Gossypium* hirsutum L.). *J Hered* **97**, 444–450.

Kliebenstein, D., Pedersen, D., Barker, B. and Mitchell-Olds, T. (2002) Comparative analysis of quantitative trait loci controlling glucosinolates, myrosinase and insect resistance in *Arabidopsis thaliana*. *Genetics* **161**, 325–332.

Knox, M.R. and Ellis, T.H.N. (2001) Stability and inheritance of methylation states at PstI sites in Pisum. *Mol Genet Genom* **265**, 497–507.

Koornneef, M., Alonso-Blanco, C. and Vreugdenhil, D. (2004) Naturally occurring genetic variation in *Arabidopsis thaliana*. *Ann Rev Plant Biol* **55**, 141–172.

Lee, I., Bleecker, A. and Amasino, R. (1993) Analysis of naturally occurring late flowering in *Arabidopsis thaliana*. *Mol Gen Genet* **237**, 171–176.

Lee, I., Lehner, B., Crombie, C., Wong, W., Fraser, A.G. and Marcotte, E.M. (2008) A single gene network accurately predicts phenotypic effects of gene perturbation in *Caenorhabditis elegans*. *Nat Genet* **40**, 181–188.

Li, L., Wang, X.F., Stolc, V., Li, X.Y., Zhang, D.F., Su, N., *et al.* (2006) Genome-wide transcription analyses in rice using tiling microarrays. *Nat Genet* **38**, 124–129.

Li, Y., Lvarez, O.A.A., Gutteling, E.W., Tijsterman, M., Fu, J.J., Riksen, J.A.G., *et al.* (2007) Mapping determinants of gene expression plasticity by genetical genomics in *C-elegans*. *PLOS Genet* **2** (12), 2155–2161.

Littell, R.C., Milliken, G.A., Stroup, W.W., Wolfinger, R.D. and Schabenberger, O. (2006) *SAS for Mixed Models* (Cary, NC: SAS Publishing).

Liu, B., Brubaker, C.L., Mergeai, G., Cronn, R.C. and Wendel, J.F. (2001) Polyploid formation in cotton is not accompanied by rapid genomic changes. *Genome* **44**, 321–330.

Liu, B. and Wendel, J.F. (2003) Epigenetic phenomena and the evolution of plant allopolyploids. *Mol Phylogenet Evol* **29**, 365–379.

Mackay, T.F.C. (2001) The genetic architecture of quantitative traits. *Annu Rev Genet* **35**, 303–339.

Madlung, A., Masuelli, R.W., Watson, B., Reynolds, S.H., Davison, J. and Comai, L. (2002) Remodeling of DNA methylation and phenotypic and transcriptional changes in synthetic *Arabidopsis* allotetraploids. *Plant Physiol* **129**, 733–746.

Malmberg, R.L., Held, S., Waits, A. and Mauricio, R. (2005) Epistasis for fitness-related quantitative traits in *Arabidopsis thaliana* grown in the field and in the greenhouse. *Genetics* **171**, 2013–2027.

Mauricio, R. and Rausher, M.D. (1997) Experimental manipulation of putative selective agents provides evidence for the role of natural enemies in the evolution of plant defense. *Evolution* **51**, 1435–1444.

Mazzochi, F. (2008) Complexity in biology. *EMBO Rep* **9**, 10–14.

Michaels, S.D. and Amasino, R.M. (1999) FLOWERING LOCUS C encodes a novel MADS domain protein that acts as a repressor of flowering. *Plant Cell* **11**, 949–956.

Miyazaki, S., Fredricksen, M., Hollis, K.C., Poroyko, V., Shepley, D., Galbraith, D.W., et al. (2004) Transcript expression profiles of *Arabidopsis thaliana* grown under controlled conditions and open-air elevated concentrations of CO_2 and of O_3. *Field Crops Res* **90**, 47–59.

Mouradov, A., Cremer, F. and Coupland, G. (2002) Control of flowering time: interacting pathways as a basis for diversity. *Plant Cell* **14**, S111–S130.

Oleksiak, M.F., Churchill, G.A. and Crawford, D.L. (2002) Variation in gene expression within and among natural populations. *Nat Genet* **32** (2), 261–266.

Olsen, K.M., Haldorsdottir, S., Stinchcombe, J., Weinig, C., Schmitt, J. and Purugganan, M.D. (2004) Linkage disequilibrium mapping of *Arabidopsis* CRY2 flowering time alleles. *Genetics* **157**, 1361–1369.

O'Malley, M.A. and Dupré, J. (2005) Fundamental issues in systems biology. *Bioessays* **27**, 1270–1276.

Papin, J.A., Hunter, T., Palsson, B.O. and Subramaniam, S. (2005) Reconstruction of cellular signalling networks and analysis of their properties. *Nat Rev Mol Cell Biol* **6**, 99–111.

Pigliucci, M. (1996) How organisms respond to environmental changes: from phenotypes to molecules (and vice versa). *Trends Ecol Evol* **11**, 168–173.

Pigliucci, M. (1998) Ecological and evolutionary genetics of *Arabidopsis*. *Trends Plant Sci* **3**, 485–489.

Rapp, R.A. and Wendel, J.F. (2005) Epigenetics and plant evolution. *New Phytol* **168**, 81–91.

Richards, E.J. (2006) Inherited epigenetic variation – revisiting soft inheritance. *Nat Rev Genet* **7**, 395–401.

Riddle, N.C. and Richards, E.J. (2002) The control of natural variation in cytosine methylation in *Arabidopsis*. *Genetics* **162**, 355–363.

Riddle, N.C. and Richards, E.J. (2005) Genetic variation in epigenetic inheritance of ribosomal RNA gene methylation in *Arabidopsis*. *Plant J* **41**, 524–532.

Rockman, M.V. and Kruglyak, L. (2007) Genetics of global gene expression. *Nat Rev Genet* **7**, 862–872.

Salmon, A., Ainouche, M.L. and Wendel, J.F. (2005) Genetic and epigenetic consequences of recent hybridization and polyploidy in Spartina (Poaceae). *Mol Ecol* **14**, 1163–1175.

Scheiner, S.M. and Gurevitch, J. (2001) *Design and Analysis of Ecological Experiments* (New York: Chapman and Hall).

Schlichting, C.D. and Smith, H. (2002) Phenotypic plasticity: linking molecular mechanisms with evolutionary outcomes. *Evol Ecol* **16**, 189–211

Schmid, M., Davison, T.S., Henz, S.R., Pape, U.J., Demar, M., Vingron, M., *et al.* (2005) A gene expression map of *Arabidopsis thaliana* development. *Nat Genet* **37**, 501–506

Schmid, M., Uhlenhaut, N.H., Godard, F., Demar, M., Bressan, R., Weigel, D., *et al.* (2003) Dissection of floral induction pathways using global expression analysis. *Development* **130**, 6001–6012.

Schmidt, D.D. and Baldwin, I.T. (2006) Transcriptional responses of *Solanum nigrum* to methyl jasmonate and competition: a glasshouse and field study. *Func Ecol* **20**, 500–508.

Shaked, H., Kashkush, K., Ozkan, H., Feldman, M. and Levy, A.A. (2001) Sequence elimination and cytosine methylation are rapid and reproducible responses of the genome to wide hybridization and allopolyploidy in wheat. *Plant Cell* **13**, 1749–1759.

Sharbel, T.F., Haubold, B. and Mitchell-Olds, T. (2000) Genetic isolation by distance in *Arabidopsis thaliana*: biogeography and postglacial colonization of Europe. *Mol Ecol* **9**, 2109–2118.

Sherman, J.D. and Talbert, L.E. (2002) Vernalization induced changes of the DNA methylation pattern in winter wheat. *Genome* **45**, 253–260.

Shimizu, K. and Purugganan, M.D. (2005) Evolutionary and ecological genomics of *Arabidopsis thaliana*. *Plant Physiol* **138**, 578–584.

Simpson, G.G. and Dean, C. (2002) *Arabidopsis*, the Rosetta stone of flowering time? *Science* **296**, 285–289.

Slotte, T., Holm, K., McIntyre, L.M., Lagercrantz, U. and Lascoux, M. (2007) Differential expression of genes important for adaptation in Capsella bursa-pastoris (Brassicaceae). *Plant Physiol* **145**, 160–173.

Sokal, R.R. and Rohlf, F.J. (1995) *Biometry* (New York: W.H. Freeman).

Stark, A., Lin, M.F., Kheradpour, P., Pedersen, J.S., Parts, L., Carlson, J.W., *et al.* (2007) Discovery of functional elements in 12 *Drosophila* genomes using evolutionary signatures. *Nature* **450**, 219–232.

Starker, C.G., Parra-Colmenares, A.L., Smith, L., Mitra, R.M. and Long, S.R. (2006) Nitrogen fixation mutants of Medicago truncatula fail to support plant and bacterial symbiotic gene expression. *Plant Phsyiol* **140**, 671–680.

Stinchcombe, J.R., Weinig, C., Ungerer, M., Olsen, K.M., Mays, C., Halldorsdottir, S., *et al.* (2004) A latitudinal cline in flowering time in *Arabidopsis thaliana* modulated by the flowering time gene FRIGIDA. *Proc Natl Acad Sci USA* **101**, 4712–4717.

Stratton, D.A. and Bennington, C.C. (1996) Measuring spatial variation in natural selection using randomly-sown seeds of *Arabidopsis thaliana*. *J Evol Biol* **9**, 215–228.

Swanson-Wagner, R.A., Jia, Y., DeCook, R., Borsuk, L.A., Nettleton, D. and Schnable, P.S. (2006) All possible modes of gene action are observed in a global comparison of gene expression in a maize F-1 hybrid and its inbred parents. *Proc Natl Acad Sci USA* **103**, 6805–6810.

Taylor, G., Street, N.R., Tricker, P.J., Sjodin, A., Graham, L., Skogstrom, O., *et al.* (2005) The transcriptome of *Populus* in elevated CO_2. *New Phytol* **167**, 143–154.

Tian, D., Traw, M.B., Chen, J.Q., Kreitman, M. and Bergelson, J. (2003) Fitness costs of R-gene-mediated resistance in *Arabidopsis thaliana*. *Nature* **423**, 74–77.

Townsend, J.P., Cavalieri, D. and Hartl, D.L. (2003) Population genetic variation in genome-wide gene expression. *Mol Biol Evol* **20** (6), 955–963.

Travers, S.E., Smith, M.D., Bai, J., Hulbert, S.H., Leach, J.E., Schnable, P.S., *et al.* (2007) Ecological genomics: making the leap from model systems in the lab to native populations in the field. *Front Ecol Environ* **5**, 19–24.

Vaughn, M.W., Tanurdzic, M., Lippman, Z., Jiang, H., Carrasquillo, R., Rabinowicz, P.D., *et al.* (2007) Epigenetic natural variation in *Arabidopsis thaliana*. *PLoS Biol* **5**, 1617–1629.

Wang, Y.M., Lin, X.Y., Dong, B., Wang, Y.D. and Liu, B. (2004) DNA methylation polymorphism in a set of elite rice cultivars and its possible contribution to inter-cultivar differential gene expression. *Cell Mol Biol Lett* **9**, 543–556.

Weinig, C., Ungerer, M., Dorn, L.A., Kane, N.C., Halldorsdottir, S., Mackay, T.F.C., *et al.* (2002) Novel loci control variation in reproductive timing in *Arabidopsis thaliana* in natural environments. *Genetics* **162**, 1875–1881.

West, M.A.L., Kim, K., Kliebenstein, D.J., van Leeuwen, H., Michelmore, R.W., Doerge, R.W., *et al.* (2007) Global eQTL mapping reveals the complex genetic architecture of transcript-level variation in *Arabidopsis*. *Genetics* **175** (3), 1441–1450.

Whitaker, R.J. and Banfield, J.F. (2006) Population genomics in natural microbial communities. *Trends Ecol Evol* **21** (9), 508–516.

Whitehead, A. and Crawford, D.L. (2006a) Neutral and adaptive variation in gene expression. *Proc Natl Acad Sci USA* **103**, 5425–5430.

Whitehead, A. and Crawford, D.L. (2006b) Variation within and among species in gene expression: raw material for evolution. *Mol Ecol* **15**, 1197–1211.

Whitelaw, N.C. and Whitelaw, E. (2006) How lifetimes shape epigenotype within and across generations. *Hum Mol Genet* **15**, R131–R137.

Wilmes, P. and Bond, P.L. (2006) Metaproteomics: studying functional gene expression in microbial ecosystems. *Trends Microbiol* **14** (2), 92–97.

Wolfinger, R.D., Gibson, G., Wolfinger, E.D., Bennett, L., Hamadeh, H., Bushel, P.D., *et al.* (2001) Assessing gene significance from cDNA microarray expression data via mixed models. *J Comp Biol* **8**, 625–637.

Wu, C.A., Lowry, D.B., Cooley, A.M., Wright, K.M., Lee, Y.W. and Willis, J.H. (2008) *Mimulus* is an emerging model system for the integration of ecological and genomic studies. *Heredity* **100** (2), 220–230.

Xu, J.P. (2006) Microbial ecology in the age of genomics and metagenomics: concepts, tools, and recent advances. *Mol Ecol* **15** (7), 1713–1731.

Yang, S.H. and Loopstra, C.A. (2005) Seasonal variation in gene expression for loblolly pines (*Pinus taeda*) from different geographical regions. *Tree Physiol* **25**, 1063–1073.

Zhang, X., Clarenz, O., Cokus, S., Bernatavichute, Y.V., Pellegrini, M., Goodrich, J., *et al.* (2007) Whole-genome analysis of histone H3 lysine 27 trimethylation in *Arabidopsis*. *PLoS Biol* **5**, e129.

Zhang, X., Shiu, S., Cal, A. and Borevitz, J.O. (2008) Global analysis of genetic, epigenetic and transcriptional polymorphisms in *Arabidopsis thaliana* using whole genome tiling arrays. *PLoS Genet* **4**, e1000032.

Zilberman, D., Gehring, M., Tran, R.K., Ballinger, T. and Henikoff, S. (2007) Genome-wide analysis of *Arabidopsis thaliana* DNA methylation uncovers an interdependence between methylation and transcription. *Nat Genet* **39**, 61–69.

INDEX

Note: *Italicized page numbers refer to figures and tables*